NUMERICAL SIMULATION AND OPTIMAL CONTROL IN PLASMA PHYSICS

Wiley/Gauthier-Villars Series in
Modern Applied Mathematics

EDITORS

J.-L. Lions (Paris)
J. Dixmier (Paris)

ALREADY PUBLISHED:

AUBIN: *Explicit Methods of Optimization*

BENSOUSSAN and LIONS: *Impulse Control and Quasi-variational Inequalities*

LIONS: *Control of Distributed Singular Systems*

TEMAM: *Mathematical Problems in Plasticity*

BENSOUSSAN: *Perturbation Methods in Optimal Control*

NUMERICAL SIMULATION AND OPTIMAL CONTROL IN PLASMA PHYSICS

With Applications to Tokamaks

Jacques Blum

Joseph-Fourier University, (Grenoble I)

gauthier-villars

JOHN WILEY & SONS

Chichester · New York · Brisbane · Toronto · Singapore

The present English language version was translated by
D. Chillingworth, and produced by
TRANS-INTER-SCIENTIA
P.O. Box 116, Tunbridge Wells, TN2 4XD
Kent, England

Production Management and PAO/CAP
Ref. No. 368-15/PMOC (381)

British Library Cataloguing in Publication Data available

Printed and bound in Great Britain by Butler & Tanner Ltd, Frome and London

Acknowledgements

I wish first of all to express my gratitude to Professor J.-L. LIONS, my mentor for having accepted this book in the series for which he is responsible.

I thank Mr J. TACHON, Head, Department of Physics of Fusion-Plasmas at the C.E.A. at Cadarache, for his encouragement and for the support and interest he gave to this work.

I should also like to thank Mr R. DEI CAS, Mr C. LELOUP and his team and Mr J.-P. MORERA, with whom I have had the great pleasure of working at the Research Department on Controlled Fusion at the C.E.A., at Fontenay-aux-Roses.

My thanks are also due to Mr E.K. MASCHKE for his comments on the drafting of Chapter VI.

Most of the numerical simulations presented in this book have been obtained in collaboration with my friends J. LE FOLL and B. THOORIS, Engineers at CISI-Ingénierie, Saclay, to whom I express my heartfelt thanks.

Some results have been obtained within the framework of contracts concluded with the teams of the JET (Joint European Torus) or NET (Next European Torus) project. I should like to convey my warmest thanks to Messrs T. STRINGER and E. LAZZARO, who have been my contacts in JET, and to Mr O. DE BARBIERI in the NET team.

I am particularly grateful to T. GALLOUËT and J. SIMON who are not only the co-authors of Chapter III but also have assisted by perusing thoroughly my work, informing me of their constructive criticisms.

My thanks are also due to all my former colleagues at the Laboratory for Numerical Analysis at Paris VI.

Finally, I wish to add my appreciation to the translator for the quality of his work, to Mrs A. McMATH and TRANS-INTER-SCIENTIA, who organized the translation project, edited and produced this computerized English version, and to the publisher who made all this possible.

GRENOBLE, May 1988

J. BLUM

Introduction

1. The subject that concerns us in this work is the modelling, numerical simulation and optimal control of equilibrium of the plasma in a Tokamak.

The Tokamak is an experimental device whose purpose is to confine plasma (ionized gas) in a magnetic field in such a way as to control the nuclear fusion of atoms of low mass (hydrogen, deuterium, tritium). The plasma current is obtained by induction from current in primary coils; the Tokamak thus appears as a transformer whose plasma is the secondary, with the primary and the plasma able to be coupled by a magnetic circuit in iron. The currents in the external coils play another role, that of creating and controlling the equilibrium of the plasma. The problem of control of equilibrium has become important from the experimental point of view for the following three reasons :

(i) In the first Tokamaks (Soviet ones in particular) the plasma was surrounded by a thick shell which stabilized its displacements by the eddy currents which were induced in it; this was also the case in TFR 400, the first version of the Tokamak at Fontenay-aux-Roses, which functioned from 1973 to 1976. However, for reasons of convenience of access to the plasma for diagnostic and auxiliary heating purposes, this shell has been dispensed with in the present-day Tokamaks, and the equilibrium of the plasma thus has to be controlled to a large extent by the experimenter. This is the case in TFR 600, the second version of the Tokamak at Fontenay-aux-Roses (without shell) which entered service at the end of 1977 and for which the problem of controlling the equilibrium of the plasma has played an important role.

(ii) In the new generations of Tokamak (JET, TFTR, JT60) which have just entered service and in those which are under construction (TORE Supra) or projected (INTOR, NET) each dimension is multiplied by a factor of sometimes as much as 10 compared with the Tokamaks of the preceding generation, and this requires considerable power. In this case modern control theory becomes indispensible for ensuring control of the displacements, current and shape of the plasma.

(iii) An increasing number of Tokamaks have plasmas with non-circular section. Optimization of the ratio of the kinetic energy of the plasma to the magnetic energy it receives, which represents the yield of the Tokamak, indicates that the meridian cross-section of the plasma ought to be elongated, which is also interesting from the point of view of MHD stability. This is why JET has a D-shape. More diverse configurations such as doublets or divertors have an even more complicated geometry, with for instance magnetic surfaces playing the role of separatrix. In all these cases the static and dynamic control of the form of the plasma has to be handled with the greatest care.

2. The equations which model the behaviour of the plasma are those of MHD (magnetohydrodynamics) (cf. S.I.BRAGINSKII):

$$(0.1)\qquad \left\{\begin{array}{l} \dfrac{\partial n}{\partial t} + \nabla(n\ u) = s \\[2ex] mn\dfrac{du}{dt} + \nabla p = j \times B \\[2ex] \dfrac{3}{2}\dfrac{dp}{dt} + \dfrac{5}{2}\ p\ \nabla u + \nabla Q = s' \\[2ex] \nabla \times E = -\dfrac{\partial B}{\partial t} \\[2ex] E + u \times B = \eta j \\[2ex] \nabla B = 0 \\[2ex] \nabla \times H = j \\[2ex] B = \mu H \end{array}\right.$$

with $\dfrac{d}{dt} = \dfrac{\partial}{\partial t} + u.\nabla$, where n denotes the density of the particles, u their mean velocity, m their mass, p their pressure, j the current density, **B** and **H** the magnetic induction and magnetic field respectively, Q the heat flux, s and s' the source– terms, η the resistivity tensor, E the electric field and μ the magnetic permeability. The viscosity terms are neglected here.

The equilibrium equations for the plasma are thus:

$$(0.2)\qquad \begin{cases} \nabla p = j \times B \\ \nabla B = 0 \\ \nabla \times H = j \\ B = \mu H \;. \end{cases}$$

The first theoretical studies relating to equilibrium of the plasma in a Tokamak are due to H. GRAD – H. RUBIN, C. MERCIER, R. LÜST – A. SCHLÜTER, V.D. SHAFRANOV around 1960.

Here we shall study axisymmetric equilibrium configurations, i.e. those satisfying the following hypothesis : *the configuration is invariant under rotation around the vertical axis of the torus.* We are thus led to a two– dimensional problem in the meridian section of the torus.

3. The first five chapters deal with the stationary problem of axisymmetric equilibrium of the plasma. In particular, we shall study the modelling and numerical simulation of this problem, the mathematical existence of a solution for a simplified model, and the identification and static control of the boundary of the plasma. The last two chapters treat the evolution of equilibrium on the time– scale of thermal diffusion in the plasma, and we shall study the problem of stability and dynamic control of displacements of the plasma.

Chapter I treats the modelling and the numerical simulation of the axisymmetric equilibrium of the plasma. The problem consists of solving equations (0.2) in an axisymmetric configuration. In a meridian section Ω of the torus with coordinates (r,z) we define the flux of the poloidal magnetic field to be $\psi(r,z)$, and the equations for $\psi(r,z)$ are :

$$(0.3)\qquad \begin{cases} \psi = 0 \quad \text{on} \quad \Gamma = \partial\Omega \\ L\psi = \lambda j(r,\psi) 1_{\Omega_p} + j_B \quad \text{in} \quad \Omega \\ I_p = \lambda \int_{\Omega_p} j(r,\psi)\, dS \end{cases}$$

with $\Omega_p = \{M \in \Omega_v | \psi(M) > \sup_D \psi\}$, and $L = -\frac{\partial}{\partial r}(\frac{1}{\mu r}\frac{\partial}{\partial r}) - \frac{\partial}{\partial z}(\frac{1}{\mu r}\frac{\partial}{\partial z})$.

Here Ω_v denotes the vacuum region, Ω_p the plasma (unknown), and j_B the current density in the coils B_i, while $j(r,\psi)$ is the current density function for the plasma, λ is an unknown normalization parameter, I_p is the total plasma current and D is the limiter which is a closed subset of Ω_v (see Fig. 0.1). The boundary Γ_p of the plasma Ω_p is a free boundary, defined as being the flux line in the interior of Ω_v that is tangent to the limiter D. The magnetic permeability μ is a constant μ_0 in the whole of space except in the iron Ω_f (see Fig. 0.1) where μ is a given function $\bar{\mu}(\frac{\nabla^2 \psi}{r^2})$. The operator L is thus a nonlinear elliptic operator.

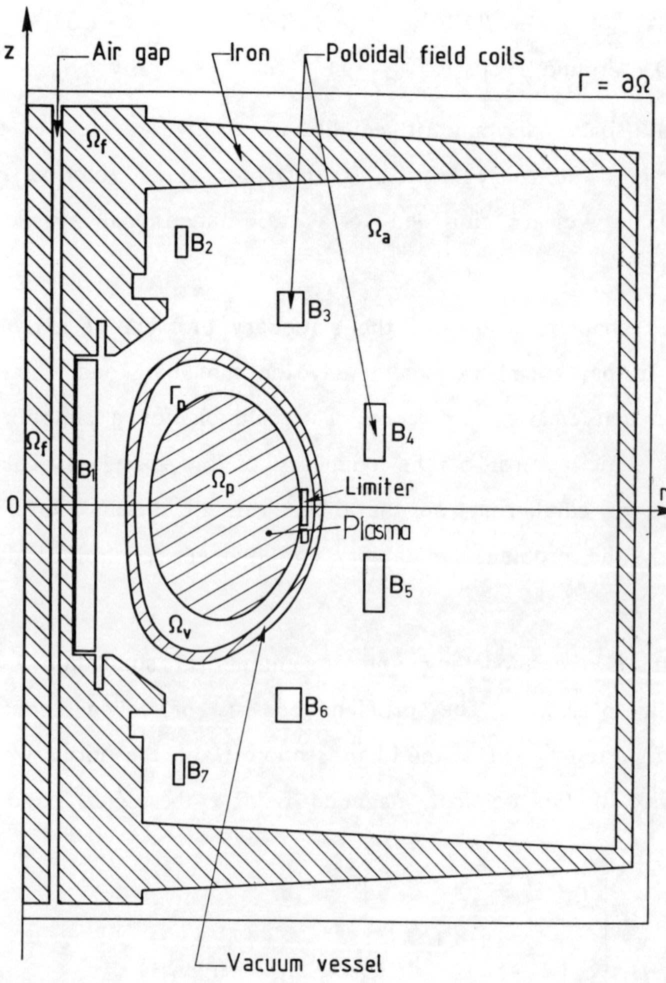

Figure 0.1

The aim of the problem is to calculate the pair (ψ,λ) that is a solution to (0.3), when the current density $j_B(r,z)$ in the coils, the functions $j(r,\psi)$ and $\mu(\frac{-\nabla^2\psi}{r^2})$ and the total plasma current I_p are given. The free boundary Γ_p of the plasma is one of the nonlinearities of the problem.

We give a "weak" formulation of this problem, as well as working out the linearized problem which is useful in the numerical solution (Newton's method), in the control problem and in the study of stability of displacements of the plasma.

Various numerical algorithms (Picard, Marder–Weitzner, Newton) are compared for the treatment of nonlinearities and coupled to the finite element method which is used to solve the linearized elliptic system. These algorithms have allowed simulations to be carried out for a certain number of equilibrium configurations for the following four Tokamaks :

TFR : Tokamak at Fontenay–aux–Roses

JET : Joint European Torus at Culham (G.B.)

TORE Supra : under construction at Cadarache

INTOR : International Tokamak Reactor (projected).

In Chapter II we treat the problem of static control (i.e. at a given instant) of the boundary of the plasma by the currents in the coils. For this problem there are various options : control of the radial position, the elongation or the shape of the plasma. These problems are formulated in the terminology of optimal control of distributed parameter systems (cf. J.L. LIONS [2]). The control of the shape of the plasma can be formulated as follows : define $J(j_B)$ by

$$J(j_B) = \int_{\Gamma_d} [\psi(M)-\psi(F_0)]^2 \, d\sigma \tag{0.4}$$

where ψ is related to j_B by the system (0.3), where Γ_d is the desired boundary of the plasma and F_0 is the point of tangency of Γ_d with the limiter D (see Fig. 0.1). We then search for $j_B^o \in U_{ad}$, where U_{ad} denotes the set of admissible currents, such that

$$J(j_B^o) = \inf_{j_B \in U_{ad}} J(j_B). \tag{0.5}$$

The first order optimality system for these various control problems is set up through the introduction of an appropriate Lagrangian.

This optimal control problem with nonlinear equations of state and non–convex cost function is solved numerically by a sequential quadratic method which consists of solving a sequence of control problems with linear state equations and quadratic cost function.

The solution of the control problem for the shape of the plasma has enabled us to determine, in the case of JET, the currents that give the plasma a D shape; in the case of TORE Supra it has enabled us to evaluate the number of control parameters necessary in order to maintain the plasma circular during a heating phase and, in the case of INTOR, to optimize the currents that realize an asymmetric divertor.

The numerical results presented in these two chapters were obtained with the help of the SCED (Self–Consistent for Equilibrium and Diffusion) code (cf. J. BLUM – J. LE FOLL – B.THOORIS [1] and [2]).

In Chapter III we study the mathematical existence of solutions to the problem (0.3) and the problem (0.5) for a simplified model. This model consists of supposing that we are dealing with a Tokamak without iron (air–transformer Tokamak), so that $\mu = \mu_0$ in the whole of space and consequently the operator L becomes linear. Moreover we suppose that

(0.6) $$j(r,\psi) = \psi - \sup_D \psi .$$

If we put :

$$\psi' = \psi - \sup_D \psi$$

then the system (0.3) becomes :

(0.7) $$\begin{cases} \psi' = \text{constant on } \Gamma \\ L\psi' = \lambda \psi'^{+} 1_{\Omega_V} + j_B \\ I_p = \lambda \int_{\Omega_V} \psi'^{+} dS \end{cases}$$

and

$$\sup_{D} \psi' = 0 \tag{0.8}$$

where Γ is the boundary of the domain of study Ω and where $\psi'^{+} = \sup(\psi',0)$. The system (0.7) with fixed λ has been studied by R. TEMAM [2] and H. BERESTYCKI – H. BREZIS in the case when $\Omega_V = \Omega$ and $j_B = 0$. Those results are generalized here to the study of (0.7) with fixed λ. In particular, using the method of topological degree of J. LERAY – J. SCHAUDER we prove the existence of a connected branch of solutions to (0.7) as λ runs through R^+. From this we deduce sufficient conditions for the existence of a solution to (0.7) – (0.8). In particular if the limiter D is reduced to a point d, we show that the set of points d such that there is a solution to (0.7) – (0.8) with D = {d} is a dense set in Ω_V.

We then give sufficient conditions for the existence of a solution to the control problem (0.5), the functional J now being :

$$J(j_B) = \int_{\Gamma_d} \psi'^2 \, d\sigma \; . \tag{0.9}$$

We establish, under certain hypotheses, the first order optimality system for this control problem. These results were presented in J. BLUM – T. GALLOUET – J. SIMON [1] and [2].

In Chapter IV we study the equilibrium solution branches as a function of the distribution of the currents in the coils. More precisely, we define the parameter Λ as :

$$\Lambda = \frac{I_1}{I_1+I_k} \tag{0.10}$$

where I_1 is the total current in the innermost coil and I_k that in the outermost coil. With the currents I_i in the other coils as well as the sum of the currents being fixed, we vary the parameter Λ and study the function $\psi(\Lambda)$ defined by (0.3) with :

$$j_B = \sum_{i=1}^{k} \frac{I_i}{S_i} 1_{B_i} \tag{0.11}$$

where B_i denotes the meridian section of the ith coil, with 1_{B_i} its characteristic function and S_i the area of B_i.

By a continuation method (cf. H.B. KELLER [1] and [2]) we describe these branches of equilibrium solutions as a function of Λ, and represent them by a diagram relating the radial position Δ_H of the plasma to the parameter Λ. Turning points (limit points) are shown to occur.

Relying on the fact that for given Δ_H there is uniqueness of the equilibrium solution, we introduce a new method for describing solution branches. This method relies on the control of the radial position of the plasma studied in Chapter II.

Finally we show that the limit points correspond to loss of stability of horizontal displacements of the plasma. This allows us to explain displacement instability observed in the Tokamak of Fontenay-aux-Roses.

Chapter V deals with the following inverse problem : the identification of the boundary of the plasma from magnetic measurements. At a certain number of points of the vacuum vessel the experimenter measures the poloidal flux ψ and the poloidal field B_p, which is equal to $1/r\ \partial\psi/\partial n$ where $\partial\psi/\partial n$ denotes the derivative of ψ normal to the vessel. We are then in the position of having a Cauchy problem for an elliptic equation, and thus an "ill-posed" problem in the sense of Hadamard. The aim is to reconstruct the boundary of the plasma from the measurements of ψ and $\partial\psi/\partial n$. Two approaches are used.

The first solves the equation $L\psi = 0$ in the vacuum, i.e. between the vacuum vessel chamber and the boundary Γ_p of the plasma; since Γ_p is unknown, we extend the solution ψ up to a fixed boundary Γ_0' (meaning here $L\psi = 0$) and solve the Cauchy problem on this fixed domain, formulating it as an optimal control problem with regularization. The optimality system having been solved once and for all, this ultra-rapid method (less than 1 ms on CRAY 1) for determining the boundary of the plasma can be applied to the on-line control of the shape of the plasma.

The second method takes account of the equations for ψ in the plasma, i.e. the system (0.3) with $j_B = 0$. The superabundance of boundary conditions on the vacuum vessel chamber allows us to make a parametric identification of the plasma current density, which is a significant problem in itself since at the present time there is no reliable experimental measure for the current density. This method and the code which is derived from it (cf. J. BLUM - J. LE FOLL - B. THOORIS [3]) is at present used in JET to analyze each discharge (cf. M.BRUSATI et al).

Chapter VI treats the evolution of equilibrium of the plasma. On the time-scale of thermal diffusion, the term mn du/dt is negligible compared with ∇p in the system (0.1), so that the equations (0.2) are satisfied at each instant. The plasma thus passes through a succession of equilibrium states which are related by the phenomena of internal diffusion in the plasma governed by equations (0.1). We have to add to these equations the evolution equations of the various circuits and for the currents induced in the vacuum vessel. As the diffusion velocity tangential to the magnetic surfaces is very large relative to the perpendicular diffusion velocity, we may suppose that the densities and temperatures of electrons and ions are constant on each magnetic surface. Then by averaging the equations on each magnetic surface we can rewrite the diffusion equations as spatially one-dimensional equations with respect to a variable which indexes the flux lines (cf. E.K. MASCHKE – J. PANTUSO SUDANO, D.B. NELSON – H. GRAD, F.L. HINTON – R.D. HAZELTINE).

The coupling between the 2-dimensional equilibrium equations and the 1-dimensional diffusion equations has been called the $1\frac{1}{2}$-dimensional transport code by H. GRAD. Its numerical solution is presented here (cf. also J. BLUM – J. LE FOLL) and has allowed simulation of various phases of discharge type in TORE Supra.

In Chapter VII we are interested in the evolution of a plasma of circular cross-section in a Tokamak of large aspect ratio. We present first of all the analytic theory of V.D. SHAFRANOV [3] and [4], which is an expansion to first order in ε of the equilibrium equations, ε being the inverse of the aspect ratio, i.e. $\varepsilon = a/R$ where a and R are the minor radius and the major radius of the plasma, respectively.

We model the map of flux in the machine by algebraic laws deduced from the confrontation between experiment and numerical simulations, starting from a magnetostatic finite element code (cf. J. LE FOLL – B. THOORIS). From these we deduce the system of ordinary differential equations governing the evolution of currents in the various circuits and in the plasma, as well as the evolution of the radial position of the plasma. This system is coupled to the system of 1-dimensional parabolic diffusion equations of the plasma.

This model enables us to study the problem of control of horizontal displacements of the plasma. The method decomposes into a pre-programming or open-loop control, and a feedback or closed-loop control. The pre-programming is calculated in several steps : first, the distribution of the turns in the primary circuit is optimized, then the voltage to apply to the pre-programming circuit is calculated by the methods of optimal control for systems governed by ordinary differential equations, and finally a learning method is applied to optimize the pre-programming from discharge to discharge while taking account of information from the previous discharge.

With the help of a simplified linearized model we study the stability of horizontal displacements of the plasma, and calculate the gain of the feedback that guarantees to the system a certain stability. These methods have enabled us to explain the instability observed in TFR 600 and to optimize the servo-system (cf. J. BLUM – R. DEI CAS).

4. This work has been carried out in collaboration with physicists from the Department of Research on Controlled Fusion of the C.E.A. at Fontenay-aux-Roses (R. DEI CAS, C. LELOUP, J.P. MORERA) and with computation experts from the Service d'Etudes Scientifiques of CISI at Saclay (J. Le FOLL, B. THOORIS).

The practical applications of this study fall into two categories : those relating to the conception phase of a Tokamak, and those dealing with the optimization of discharge behaviour or with the interpretation of results on an existing machine.

In the first category we cite in particular :

- optimization of the position and number of magnetic probes (pick-up coils) enabling the plasma boundary in JET to be identified (cf. Chapter V)
- optimization of the control system for the shape of the plasma in TORE Supra, and in particular the determination of the number of control parameters necessary to maintain the plasma circular during a heating phase (cf. Chapter II and VI).

In the second category we may mention :

- explanation of the instability of horizontal displacement observed in TFR 600 and optimization of the servo-system (cf. Chapters IV and VII)
- the method for identifying the boundary of the plasma, and the internal magnetic surfaces, which is used systematically for analyzing discharges in JET (cf. Chapter V).

5. Bibliographic notes concerning the work of other authors in this field are given in each chapter. Numerical simulation of MHD instabilities is not dealt with in this work; we refer to R. GRUBER – J. RAPPAZ and the references therein for this type of problem.

Table of contents

1. A free boundary problem: the axisymmetric equilibrium of the plasma in a tokamak

I.1 THERMONUCLEAR FUSION AND THE TOKAMAK DEVICE :

The reaction of nuclear fusion is the source of the radiation energy emitted by the stars: these reactions take place under gigantic pressures (10^{11} atm) due to gravitational forces and the masses of the stars.

To control thermonuclear fusion in the laboratory, the simplest reaction envisaged for the first fusion reactors is the Deuterium– Tritium reaction:

$$ {}^{2}_{1}D + {}^{3}_{1}T \rightarrow {}^{4}_{2}He + {}^{1}_{0}n + 17.6 \text{ Mev} . $$

The energy balance for such a "thermonuclear" environment should be positive, that is to say that the energy liberated by the nuclear reaction ought to be greater than the thermal energy losses of the ionized gas called the plasma. This condition, known as Lawson's criterion, can be written:

$$ n\tau > f(T) $$

where n is the charged particle density, τ is the energy confinement time, T is the temperature of the plasma, and where the function f is represented in Fig. I.1 as the limit of the ignition domain. For a temperature of 10 Kev ($\sim$100 million degrees), $n\tau$ has to be at least equal to 10^{14} $cm^{-3} \times s$. Research in thermonuclear fusion is aimed at realizing an experimental device which permits confinement of the plasma while satisfying Lawson's criterion.

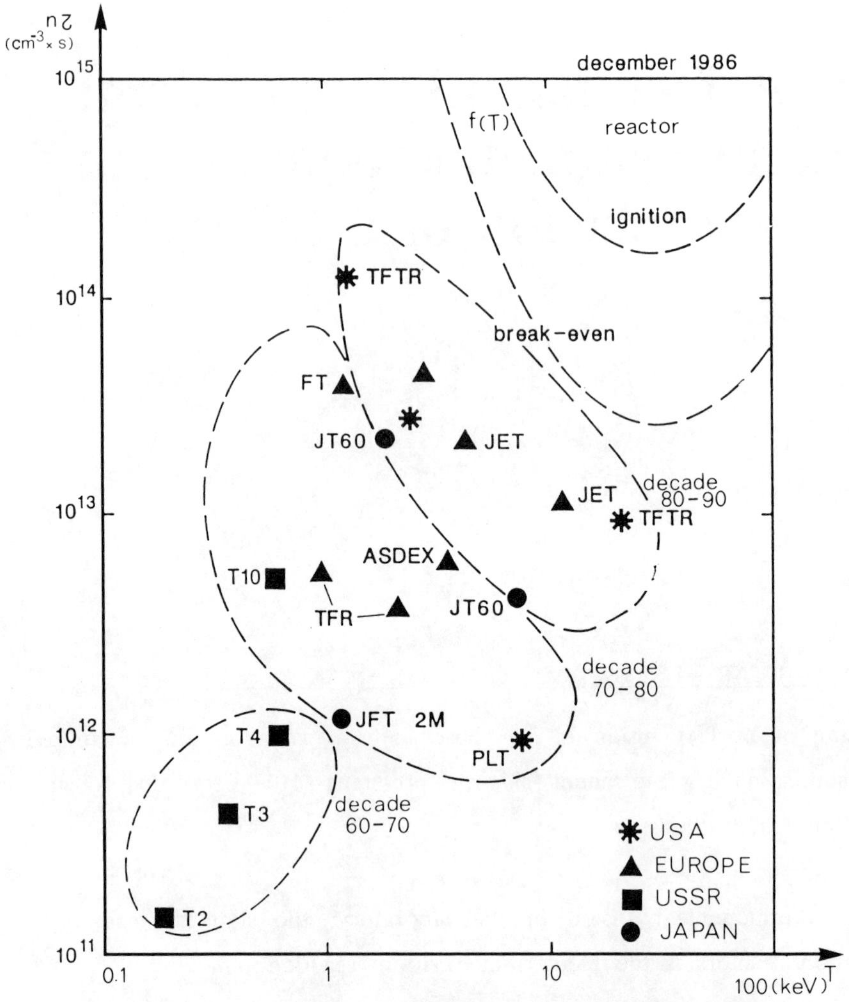

Figure I.1 : *Diagram Representing the Ignition Domain* (Lawson's *Criterion)*

Two approaches are possible. One aims to confine the plasma for a very short time τ but at a very high density n : this is fusion by inertial confinement, where laser beams (or beams of electrons or ions) converge on a target (plasma) in order to bring it to the thermonuclear state.

The second approach is that of magnetic confinement where the ionised particles are confined within a magnetic field. A charged particle essentially describes a helix centred on the field line, and in order to confine the particle it suffices to maintain this field line on a closed surface, which necessarily must have a toroidal configuration. The density is much lower than in inertial confinement and is of the order of 10^{14} cm^{-3}; the confinement time must then be more significant (of the order of a second). The Tokamak device corresponds to this second method (cf. L.A. ARTSIMOVITCH, A. SAMAIN).

a)

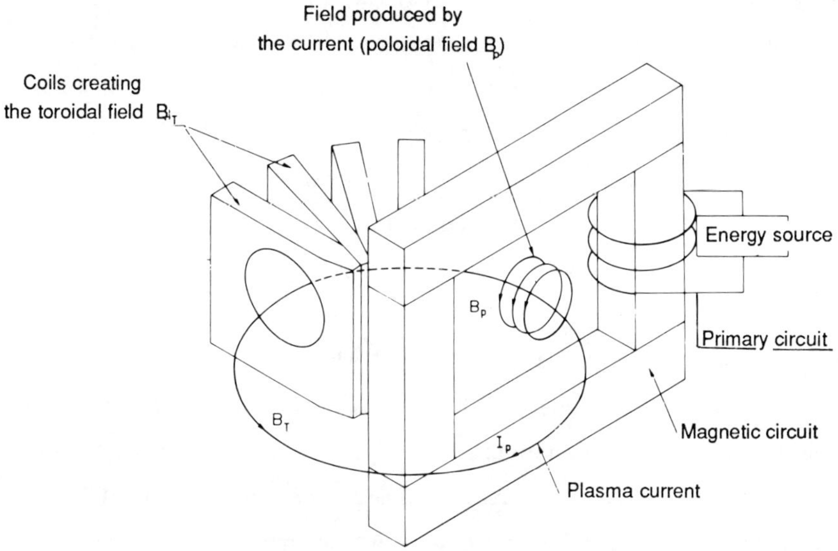

b)

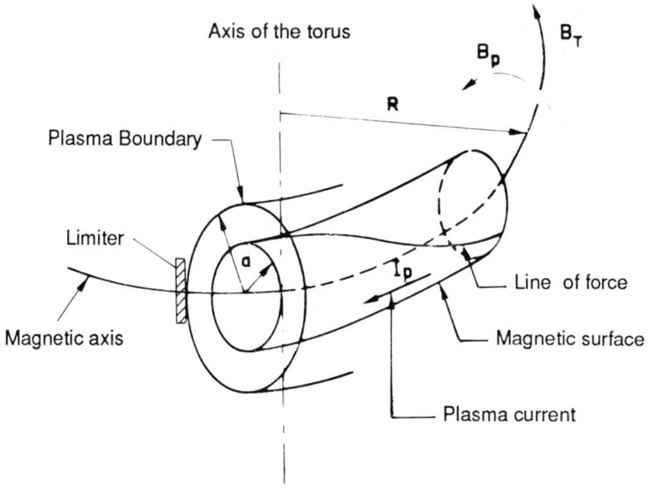

Figure I.2 : Schematic diagram of the principle of a Tokamak.

Figure I.2a represents a schematic diagram of the Tokamak principle. The magnetic field in the plasma region is the resultant of a poloidal field B_P, generated by the plasma current I_P, and a toroidal field B_T produced by coils wound around the torus. The plasma current I_P is obtained by induction from currents in poloidal field coils; the plasma thus appears as the secondary of a transformer whose poloidal field coils constitute the primary, and the magnetic circuit is the main element of coupling between primary and plasma. The toroidal field is needed because of a stability condition : the field lines in fact have to turn at least once around the axis of the torus while they make one revolution in the poloidal sense (cf. C. MERCIER). These field lines generate magnetic surfaces which have the topology of nested tori (cf. Fig. I.2b).

The magnetic axis corresponds to the case where the magnetic surface degenerates into a closed curve. The boundary of the plasma is a particular magnetic surface, defined by its tangency with the limiter. The object of this chapter is to simulate the plasma boundary and the magnetic surfaces, with the currents in the coils, the total plasma current, its current density function and the magnetic permeability of the iron being the data for the problem.

I.2 – MATHEMATICAL MODELLING OF AXISYMMETRIC EQUILIBRIUM FOR THE PLASMA IN A TOKAMAK

I.2.1 Maxwell's Equations and the Equilibrium Equation for the Plasma.

The equations which govern the equilibrium of a plasma in the presence of a magnetic field are on the one hand Maxwell's equations and on the other hand the equilibrium equations for the plasma itself.

Maxwell's equations as follows are satisfied in the whole of space (including the plasma) :

(I.1) $$\nabla B = 0$$

(I.2) $$\nabla \times H = j$$

(I.3) $$B = \mu H$$

where **B** and **H** represent the induction and the magnetic field respectively, μ is the magnetic permeability and **j** is the current density. The relation (I.1) is the equation of conservation of magnetic induction, equation (I.2) is Ampère's Theorem, and the relation (I.3) is linear in air, vacuum and plasma, where $\mu = \mu_0$, and nonlinear in iron where μ is a function of **B**.

The equilibrium equation for the plasma is

$$\nabla p = \mathbf{j} \times \mathbf{B}. \tag{I.4}$$

This equation (I.4) signifies that the plasma is in equilibrium when the force ∇p due to the kinetic pressure p is equal to the force of the magnetic pressure $\mathbf{j} \times \mathbf{B}$. We deduce immediately from (I.4) that

$$\mathbf{B} \cdot \nabla p = 0 \tag{I.5}$$

$$\mathbf{j} \cdot \nabla p = 0\ . \tag{I.6}$$

Thus in a plasma in equilibrium the field lines and the current lines lie on isobaric surfaces (p = const.); these surfaces, generated by the field lines, are called magnetic surfaces. In order that they should remain within a bounded volume of space it is necessary that they have toroidal topology. Each surface p = C (except for a set of values of C of measure zero) is traversed ergodically by the flux lines (cf. M.D. KRUSKAL – R.M. KULSRUD), and we suppose that these surfaces form a family of nested tori. The innermost torus degenerates into a curve which we call the magnetic axis.

I.2.2 The Hypothesis of Axial Symmetry.

In the cylindrical coordinate system (r,z,ω) (see Fig. I.3) the hypothesis of axial symmetry consists in supposing that the magnetic induction **B** is independent of the toroidal angle ω.

Let us rewrite equations (I.1) – (I.4) using this hypothesis of axisymmetry. Equation (I.1) becomes

$$\frac{1}{r}\frac{\partial}{\partial r}(rB_r) + \frac{\partial B_z}{\partial z} = 0\ . \tag{I.7}$$

Let C be the circle centred on the axis of the torus and passing through some point M, and let D be the disc having C as its circumference (see Fig. (I.3)). Define the variable ψ by

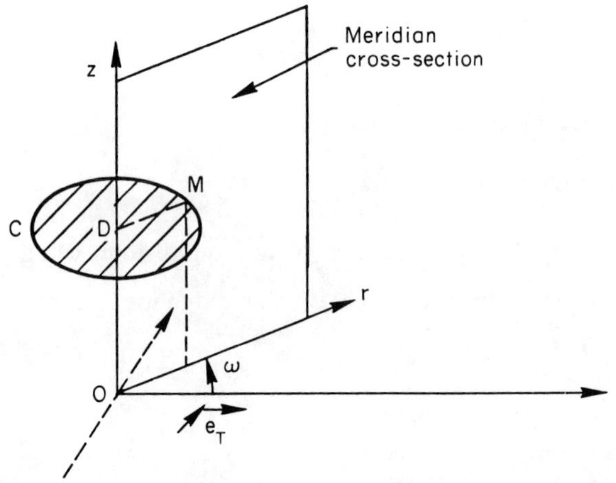

Figure I.3 : Diagram of Cylindrical Coordinates

(I.8) $$\psi(r,z) = \int_0^r r' B_z \, dr' = \frac{1}{2\pi} \int_D B_z \, dS = \frac{1}{2\pi} \int_D \mathbf{B}.d\mathbf{S} \; .$$

Up to a factor of 2π the quantity ψ is thus equal to the poloidal flux (across D) of the magnetic field **B**. By (I.7) and (I.8) we can thus write

(I.9) $$\begin{cases} B_r = -\dfrac{1}{r}\dfrac{\partial\psi}{\partial z} \\ B_z = \dfrac{1}{r}\dfrac{\partial\psi}{\partial r} \; . \end{cases}$$

As far as the toroidal component $\mathbf{B}_T$ of the induction **B** is concerned, we define f by

(I.10) $$\mathbf{B}_T = \frac{f}{r} \, \mathbf{e}_T \; .$$

Then the magnetic induction **B** can be written as :

(I.11) $$\begin{cases} \mathbf{B} = \mathbf{B}_P + \mathbf{B}_T \\ \mathbf{B}_P = \dfrac{1}{r} \, [\nabla\psi \times \mathbf{e}_T] \\ \mathbf{B}_T = \dfrac{f}{r} \, \mathbf{e}_T \end{cases}$$

where e_T is the unit vector in the toroidal sense and where $\mathbf{B_P}$ denotes the poloidal component of **B**. According to (I.11), in an axisymmetric configuration the magnetic surfaces are generated by the rotation of the flux lines ψ = constant around the vertical axis O_z of the torus.

From equations (I.2), (I.3) and (I.11) we obtain the following expression for j :

$$\text{(I.12)} \qquad \begin{cases} j = j_P + j_T \\[1em] j_P = \dfrac{1}{r} \left[\nabla \left[\dfrac{f}{\mu}\right] \times e_T\right] \\[1em] j_T = (L\psi) e_T \end{cases}$$

where j_P and j_T are the poloidal and toroidal components respectively of j, and the operator L is defined by

$$\text{(I.13)} \qquad L\cdot = -\frac{\partial}{\partial r} \left[\frac{1}{\mu r} \frac{\partial \cdot}{\partial r}\right] - \frac{\partial}{\partial z} \left[\frac{1}{\mu r} \frac{\partial \cdot}{\partial z}\right] .$$

The expressions (I.11) and (I.12) for **B** and **j** are valid in the whole of space (air, iron, plasma) since they involve only Maxwell's equations and the hypothesis of axial symmetry.

I.2.3 *The Grad–Shafranov Equation.*

If we consider now the plasma region, the relation (I.5) implies that ∇p is collinear with $\nabla\psi$, and therefore p is constant on each magnetic surface : we can denote this by

$$\text{(I.14)} \qquad p = p(\psi) .$$

Relation (I.6) combined with the expression (I.12) for **j** implies that ∇f is collinear with ∇p, and therefore that f is likewise constant on each magnetic surface :

$$\text{(I.15)} \qquad f = f(\psi) .$$

The equilibrium relation (I.4) combined with the expressions (I.11) and (I.12) for **B** and j implies that :

$$\text{(I.16)} \qquad \nabla p = \frac{L\psi}{r} \nabla\psi - \frac{f}{\mu_0 r^2} \nabla f .$$

If we use the notation

$$\frac{\nabla p}{\nabla\psi} = \frac{\partial p}{\partial\psi} , \quad \frac{\nabla f}{\nabla\psi} = \frac{\partial f}{\partial\psi} ,$$

then (I.16) can be written :

$$(I.17) \qquad L\psi = r\,\frac{\partial p}{\partial \psi} + \frac{1}{2\mu_0 r}\,\frac{\partial f^2}{\partial \psi}\ .$$

This equation (I.17) is called the Grad–Shafranov equilibrium equation (cf. H. GRAD – H. RUBIN, V.D. SHAFRANOV [1], R. LÜST – A. SCHLÜTER, C. MERCIER). The operator L is an elliptic linear operator since μ is equal to μ_0 in the plasma. By (I.12), the right–hand side of (I.17) represents the toroidal component of the plasma current density. It involves the functions p and f which are solutions to the diffusion equations, and we shall see in Chapter VI this coupling between equilibrium and diffusion equations. A simplified method consists of being given a priori the laws $p(\psi)$ and $f(\psi)$; then (I.17) becomes a usual nonlinear elliptic equation. This process in which the functions p and f are deduced from experimental observation is used in Chapters I to V.

I.2.4 The Equations for ψ in the Domain Ω.

Making the assumption of axial symmetry, we work in a meridian section of the Tokamak represented by the half–plane $(r > 0, z)$. We suppose moreover that the machine and the experimental conditions are symmetric with respect to the equatorial plane, which restricts the domain to the quadrant $(r > 0, z > 0)$ represented in Fig.I.4 ; the z–axis is the vertical axis of the torus and the r–axis is the section of the equatorial plane. We limit the study to the interior Ω of the domain OABC, where the points A, B and C are taken sufficiently far away from the magnetic circuit in order that the flux may be considered to be zero on AB and BC.

The domain Ω includes the magnetic circuit Ω_f (iron), the air Ω_a (including the coils B_i), the air–iron interface (Γ_{af}), the plasma Ω_p and its boundary Γ_p :

$$\Omega = \Omega_f \cup \Omega_a \cup \Omega_p \cup \Gamma_{af} \cup \Gamma_p\ .$$

In Fig.I.4 the section Ω_{cv} of the vacuum vessel (included in Ω_a) and the vacuum region Ω_v, which is inside the vacuum vessel and is where the plasma is situated $(\Omega_p \subset \Omega_v)$, are also shown. Finally the limiter which prevents the plasma from touching the vacuum vessel is indicated by D.

Let us now write down the equations for ψ in the various regions.

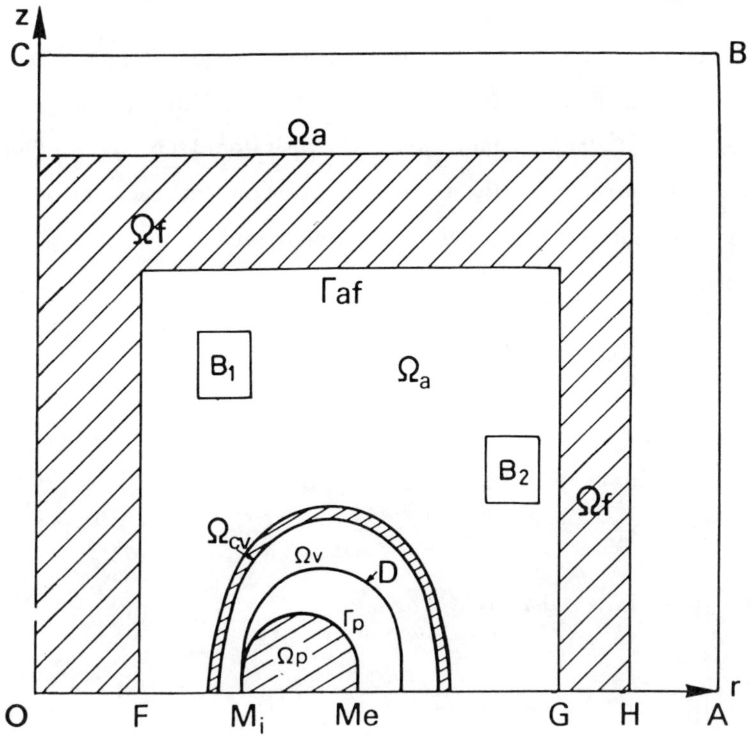

Figure I.4 : *Definition of* Ω *(Meridian Section of the Torus).*

I.2.4.1 *The Magnetic Circuit* Ω_f :

The magnetic permeability μ of the iron is a given function $\bar{\mu}$ of B_P^2. By (I.11) :

$$B_P^2 = \frac{1}{r^2} |\nabla \psi|^2$$

hence

$$\mu = \bar{\mu}(B_P^2) = \bar{\mu}\left(\frac{(\nabla\psi)^2}{r^2}\right) . \tag{I.18}$$

Since there is no current circulating in the iron, the equation for ψ follows from (I.12) :

$$L\psi = 0 \quad \text{in } \Omega_f \tag{I.19}$$

where L is give by (I.13) and μ by (I.18). This is a nonlinear elliptic equation.

I.2.4.2 *The Air* Ω_a :

This region contains the coils B_i, the vacuum vessel Ω_{cv} and the part of the vacuum region Ω_v which is exterior to the plasma. In this domain μ is equal to μ_0 (magnetic permeability of the vacuum) and the operator L is then elliptic and linear. We suppose that the current density j_T is homogeneous in each of the k coils B_i in which there is a total current I_i and whose meridian section has area S_i. By (I.12) the equation for ψ becomes :

(I.20) $$L\psi = j_B \quad \text{in } \Omega_a$$

where

$$j_B = \frac{I_i}{S_i} \quad \text{in } B_i,\ i \in \{1,\dots,k\}$$

$$j_B = j_{cv} \quad \text{in } \Omega_{cv}$$

$$j_B = 0 \quad \text{in } \Omega_a - \left[\bigcup_{i=1}^{k} \bar{B}_i \cup \bar{\Omega}_{cv}\right]$$

and where j_{cv} is the toroidal component of the current density in the vacuum vessel.

I.2.4.3 *The Plasma* Ω_p :

The equation governing the behaviour of ψ in the plasma is the Grad–Shafranov equation (I.17). If we limit ourselves to the stationary problem, we can take the functions $p(\psi)$ and $f(\psi)$ as given, being deduced from experimental observations. First of all we suppose that p and f^2 have the same type of dependence on ψ. The right–hand side of equation (I.17) which represents the toroidal component of the current density of the plasma can then be written :

(I.21) $$j_T(r,\psi) = \lambda h(r) g(\psi_N)$$

with

$$h(r) = \frac{r\beta}{R_0} + \frac{R_0(1-\beta)}{r} \quad \text{and} \quad \psi_N = \frac{\psi-\psi_a}{\psi_p-\psi_a}$$

where R_0 denotes the major radius of the vacuum vessel and β a parameter whose physical significance we shall give later. The scalar ψ_a represents the value of ψ on the magnetic axis, which we suppose to be unique in Ω_p , and ψ_p is the value of ψ on the boundary Γ_p of the plasma. The quantity $\psi_N(M)$ denotes the fraction of the poloidal flux inside the magnetic surface passing through the point M relative to the total poloidal flux $(\psi_p - \psi_a)$ in the plasma; ψ_N is called the "normalized" poloidal flux. In the simple topological configurations that we propose to study here, ψ increases from the edge of the plasma up to the magnetic axis so that ψ_N belongs to the interval [0,1]. The magnetic axis corresponds to $\psi_N = 0$ and the edge of the plasma to $\psi_N = 1$. The parameter λ is a coefficient defined by the data of the total plasma current I_p :

$$I_p = \lambda \int_{\Omega_p} h(r) g(\psi_N) \, dS \, . \qquad \text{(I.22)}$$

The equation for ψ in the plasma then becomes by (I.12) and (I.21) :

$$L\psi = \lambda h(r) g(\psi_N) \quad \text{in} \quad \Omega_p \qquad \text{(I.23)}$$

where λ is defined by (I.22). This equation is the Grad–Shafranov equation (I.17) corresponding to the following functions $p(\psi)$ and $f(\psi)$:

$$\text{(I.24)} \qquad \begin{cases} p(\psi) = \dfrac{\lambda\beta}{R_o} \displaystyle\int_{\psi_p}^{\psi} g(\psi_N) \, d\psi \\[2ex] f(\psi) = \left[2\lambda\mu_o R_o (1-\beta) \displaystyle\int_{\psi_p}^{\psi} g(\psi_N) d\psi + f_o^2\right]^{\frac{1}{2}} \end{cases}$$

where f_o is the value of f taken to be constant outside the plasma and where p vanishes at the edge of the plasma. The function $g(\psi_N)$ is at the disposal of the experimenter for describing the current density profile of the plasma. We shall see an example of this later.

I.2.4.4 Boundary Conditions :

For boundary conditions we take (cf. Fig.I.4) :

$$\text{(I.25)} \qquad \begin{cases} \psi = 0 \quad \text{on} \quad \Gamma_o = AB \cup BC \cup CO \\[1ex] \dfrac{\partial \psi}{\partial z} = 0 \quad \text{on} \quad \Gamma_1 = OA \, . \end{cases}$$

In fact ψ is taken to be zero on OC by definition of the poloidal flux (I.8). We suppose furthermore that few flux lines leave the magnetic circuit Ω_f, so that at a certain distance from the iron we may suppose that the flux is zero; we take A, B and C sufficiently far away from the iron that ψ can be considered to be zero on AB and BC. The condition on OA is a symmetry condition that arises from the fact that we are considering configurations that are symmetric relative to the equatorial plane.

I.2.4.5 *Interface Conditions* :

We suppose there is no surface current on the air–iron interface Γ_{af} ; the tangential component of **H** is therefore continuous, according to Ampère's theorem. The equation of conservation of **B** implies the continuity of the normal component of **B**. On the basis of (I.3) and (I.11) we can then write

$$\text{(I.26)} \quad \begin{cases} \left(\frac{1}{\mu}\frac{\partial\psi}{\partial n}\right)_f = \left(\frac{1}{\mu}\frac{\partial\psi}{\partial n}\right)_a \\ (\psi)_f = (\psi)_a \end{cases} \quad \text{on } \Gamma_{af}$$

where the suffix f denotes the quantity taken in the iron Ω_f, the suffix a denotes the quantity taken in the air Ω_a, and $\partial/\partial n$ denotes the normal derivative on Γ_{af} in the direction pointing outwards from Ω_f. Since the magnetic permeability μ in the iron is very different from μ_0 there is a jump discontinuity in $\partial\psi/\partial n$ at this interface.

We suppose that there is also no surface current density on the boundary Γ_p of the plasma. We then have

$$\text{(I.27)} \quad \begin{cases} \left(\frac{\partial\psi}{\partial n}\right)_e = \left(\frac{\partial\psi}{\partial n}\right)_i \\ (\psi)_e = (\psi)_i \end{cases} \quad \text{on } \Gamma_p$$

where the suffices e and i denote quantities taken in the exterior and interior of the plasma Ω_p, and $\partial/\partial n$ is the normal derivative on Γ_p in the outward direction from Ω_p.

I.2.4.6 *The Free Boundary* Γ_p *of the Plasma* :

The boundary Γ_p of the plasma is defined as being the outermost flux line which is inside the limiter. This limiter D is a closed subset of the vacuum region Ω_v . It is represented in Fig.I.4 as a semicircle; it can also consist of a finite number of rails which will be represented by points. Its rôle is to prevent the plasma from touching the vacuum vessel : it is an obstacle across which the plasma cannot pass.

If D is a continuous curve (see Fig.I.5a), Γ_p will be the equipotential of ψ which is tangent to D from the inside and is contained within Ω_v. If D is made up of a finite number of points ($\{D_1,D_2,D_3\}$ in Fig.I.5b) then Γ_p will be the equipotential of ψ contained in Ω_v which passes through one of these points and is such that all the other points of the limiter are external to the plasma. This condition translates in all these cases to

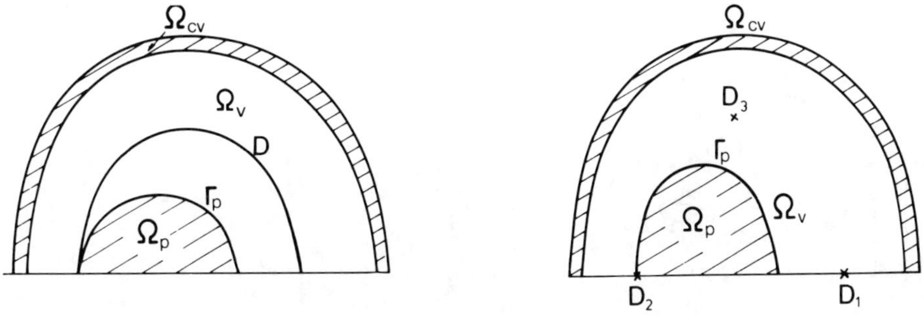

Figure I.5 : Examples of limiters

(I.28) $$\Gamma_p = \{M \in \Omega_v \mid \psi(M) = \sup_D \psi\}$$

using the sign convention $I_p > 0$.

The interface conditions (I.27) and the supplementary condition (I.28) define the free boundary Γ_p. The plasma Ω_p is then defined as follows :

(I.29) $$\Omega_p = \{M \in \Omega_v \mid \psi(M) > \sup_D \psi\} \ .$$

I.2.4.7 The Problem (P_I) :

To summarize, the triple (ψ,λ,Γ_p) satisfies the following equations :

$$
(I.30)\quad \left\{
\begin{array}{l}
L\psi = 0 \quad \text{in} \quad \Omega_f \\
L\psi = j_B \quad \text{in} \quad \Omega_a \\
L\psi = \lambda h(r)\; g(\psi_N) \quad \text{in} \quad \Omega_p \\
I_p = \lambda \displaystyle\int_{\Omega_p} h(r)\; g(\psi_N) dS \\
\psi = 0 \quad \text{on} \quad \Gamma_o \\
\dfrac{\partial\psi}{\partial z} = 0 \quad \text{on} \quad \Gamma_1 \\
\left\{
\begin{array}{l}
\left(\dfrac{1}{\mu}\dfrac{\partial\psi}{\partial n}\right)_f = \left(\dfrac{1}{\mu_o}\dfrac{\partial\psi}{\partial n}\right)_a \\
(\psi)_f = (\psi)_a
\end{array}
\right. \quad \text{on} \quad \Gamma_{af} \\
\left\{
\begin{array}{l}
\left(\dfrac{\partial\psi}{\partial n}\right)_e = \left(\dfrac{\partial\psi}{\partial n}\right)_i \\
(\psi)_e = (\psi)_i
\end{array}
\right. \quad \text{on} \quad \Gamma_p \\
\Gamma_p = \{M \in \Omega_v \mid \psi(M) = \sup\limits_D \psi\} \\
\Omega_p = \{M \in \Omega_v \mid \psi(M) > \sup\limits_D \psi\}
\end{array}
\right.
$$

where the operator L is defined by (I.13), μ by (I.18) and j_B by (I.20).

The problem (P_I) that we wish to solve is the following : given the functions $\bar{\mu}(B_p^2)$, $h(r)$ and $g(\psi_N)$, the currents I_i in the coils B_i, the current density j_{cv} in the vacuum vessel and the total current I_p of the plasma, find the triplet (ψ,λ,Γ_p) satisfying equations (1.30).

<u>*Remark I.1.*</u> :

The system (I.30) can also be written in the following way :

$$
(I.31)\quad \left\{
\begin{array}{l}
L\psi = j_B + \lambda h(r) g(\psi_N) 1_{\Omega_p} \\
I_p = \lambda \displaystyle\int_{\Omega_p} h(r) g(\psi_N) dS \\
\psi = 0 \quad \text{on} \quad \Gamma_o \\
\dfrac{\partial\psi}{\partial z} = 0 \quad \text{on} \quad \Gamma_1 \\
\Omega_p = \{M \in \Omega_v \mid \psi(M) > \sup\limits_D \psi\}
\end{array}
\right.
$$

where the support of j_B is $\bigcup_{i=1}^{k} B_i \cup \Omega_{cv}$ and 1_{Ω_p} is the characteristic function for Ω_p. This formulation will be used in Chapter III in order to study the existence of a solution of this problem for a simplified model. □

Remark I.2: *Physical interpretation of the parameter* β :

Let us calculate the ratio of the kinetic pressure to the magnetic pressure due to the poloidal field, using (I.11), (I.21) and (I.24) :

$$\frac{|\nabla p|}{|j \times B_p|} = \frac{r^2 \beta}{r^2\beta + R_o^2(1-\beta)} .$$

If we position ourselves at a point such that $r = R_o$ where R_o is the major radius of the torus we then have

$$\beta = \left[\frac{|\nabla p|}{|j \times B_p|}\right]_{r=R_o} .$$ □

Remark I.3: *Choice of the function* $g(\psi_N)$:

In Tokamaks whose magnetic surfaces have circular section, experimental observations allow the toroidal current density to be expressed as a function of ρ/a where a is the minor radius of the plasma and ρ is the minor radius of the magnetic surface under consideration :

$$j_T(\rho) = \lambda \left[1 - \frac{\rho^2}{a^2}\right]^{\gamma} .$$

In a neighbourhood of the magnetic axis, ψ_N behaves as ρ^2/a^2 (cf. C. MERCIER). A possible extension of the expression for $j_T(\rho)$ to Tokamaks of arbitrary cross–section consists in taking a function $g(\psi_N)$ of the form

(I.32) $$g(\psi_N) = \delta + \zeta(1-\psi_N^{\alpha})^{\gamma}$$

with $\alpha > 0$, $\gamma > 0$.

Let us consider some important particular cases :

. $\delta = 0$: this is the case of current density which vanishes at the boundary of the plasma.

. $\delta = 0$, $\zeta = 1$, $\alpha = 1$, $\gamma = 1$: $g(\psi_N) = 1-\psi_N$: this is the so-called "linear" case (called thus because g depends linearly on ψ_N) ; it corresponds to functions p and f^2 that are parabolic in ψ.

. $\delta = 1$, $\zeta = 0$: $g(\psi_N) = 1$: this is the case of flat current.

Other types of functions g (polynomial or exponential in ψ_N, for example) can equally well be considered.

Note also that functions $g_1(\psi_N)$, $g_2(\psi_N)$ that are different for p and f can be considered in the expressions (I.24) for $p(\psi)$ and $f(\psi)$. □

Remark I.4. :

After integration of (I.23) over Ω_p, equation (I.22) can also be written in the following way :

$$I_p = -\int_{\Gamma_p} \frac{1}{\mu_0 r} \frac{\partial\psi}{\partial n} d\Gamma \quad \text{(I.33)}$$

The integral on the right hand side can also be taken over any contour of Ω_a included in Ω_v and containing the plasma Ω_p. The relation (I.33) is the integral form of Ampère's theorem.

Likewise on integrating the first equation of (I.31) over Ω we obtain

$$I_p + \sum_{i=1}^{k} I_i + \int_{\Omega_{cv}} j_{cv}\, dS = -\int_{\Gamma} \frac{1}{\mu r} \frac{\partial\psi}{\partial n} d\Gamma \quad \text{(I.34)}$$

where Γ is the boundary of Ω ($\Gamma = \Gamma_0 \cup \Gamma_1$). □

I.2.5 Weak Formulation of the Problem (P_I) .

I.2.5.1 The Weak Problem (P'_I) :

Let $L^{p_o}(\Omega)$ (respectively $L^{p_o}(\Omega;\ r\ dr\ dz)$) be the space of real-valued functions on Ω that are p_oth-power integrable with respect to Lebesgue measure drdz (respectively, the measure rdrdz). Up to a factor of 2π, the measure rdrdz is equal to the volume element dV integrated with respect to ω and comes from the axisymmetric character of the problem.

(I.35) $$V^{p_o}(\Omega) = \{\varphi \in L^{p_o}(\Omega) \text{ such that } \frac{1}{r} \nabla\varphi \in [L^{p_o}(\Omega; rdrdz)]^2 \text{ and } \varphi = 0 \text{ on } \Gamma_o\}$$

with $\nabla\varphi = (\frac{\partial\varphi}{\partial r}, \frac{\partial\varphi}{\partial z})$ and p_o being a scalar between 1 and $+\infty$.

Recall the Sobolev space $W^{1,p_o}(\Omega)$ is defined as follows (cf. J.L. LIONS [1], R.A. ADAMS) :

$$W^{1,p_o}(\Omega) = \{\varphi \in L^{p_o}(\Omega) \text{ such that } \nabla\varphi \in [L^{p_o}(\Omega)]^2]\}.$$

The space $V^{p_o}(\Omega)$ is contained in $W^{1,p_o}(\Omega)$. Let p_o' be the conjugate scalar to p_o, i.e. such that

$$\frac{1}{p_o} + \frac{1}{p_o'} = 1 .$$

For each pair $(\psi,\varphi) \in V^{p_o}(\Omega) \times V^{p_o'}(\Omega)$, define the following form a_μ :

(I.36) $$a_\mu(\psi,\varphi) = \int_\Omega \frac{1}{\mu r} \nabla\psi\cdot\nabla\varphi \, dS$$

with $dS = drdz$, where μ is μ_o in Ω_a and Ω_p and is the function $\bar{\mu}(\frac{\text{grad}^2\psi}{r^2})$ in Ω_f. Since $\mu \geqslant \mu_o$ everywhere we have $\mu^{-1} \in L^\infty(\Omega)$. As $r^{-1} \nabla\psi$ belongs to $[L^{p_o}(\Omega;rdrdz)]^2$ and $r^{-1} \nabla\varphi$ belongs to $[L^{p_o'}(\Omega;rdrdz)]^2$ we deduce that $r^{-1}\nabla\psi\cdot\nabla\varphi$ belongs to $L^1(\Omega)$, which makes the expression (I.36) for $a_\mu(\psi,\varphi)$ meaningful.

The problem (P_I') is defined as follows : given the functions $\bar{\mu}(B_p^2)$, $h(r)$, $g(\psi_N)$, the currents I_i and I_p, and the current density j_{cv}, we seek the pair $(\psi,\lambda) \in V^{p_o}(\Omega) \times R$, with $p_o > 2$, such that

(I.37)

$$\begin{cases} a_\mu(\psi,\varphi) = \sum_{i=1}^{k} \frac{I_i}{S_i} \int_{B_i} \varphi \, dS + \int_{\Omega_{cv}} j_{cv} \, \varphi \, dS + \lambda \int_{\Omega_p} h(r) g(\psi_N) \varphi dS. \\ \qquad \forall \varphi \in V^{p_o'}(\Omega) \\ I_p = \lambda \int_{\Omega_p} h(r) g(\psi_N) dS \end{cases}$$

with $\Omega_p = \{M \in \Omega_v \mid \psi(M) > \sup \psi\}$.
D

Remark I.5. :

The scalar p_0 has been chosen strictly greater than 2 so that if $\psi \in V^{p_0}(\Omega)$ then $\psi \in C^0(\bar{\Omega})$. In fact $V^{p_0}(\Omega) \subset W^{1,p_0}(\Omega)$ and according to the Sobolev embedding theorems (cf. J.L. LIONS [1], R.A. ADAMS) if $p_0 > 2$ then $W^{1,p_0}(\Omega) \subset C^0(\bar{\Omega})$. The continuity of ψ allows meaning to be given to the quantities $\sup_D \psi$ and $\sup_{\Omega p} \psi$ which occur in the definitions of Ω_p and ψ_N. Another possibility is to look for ψ in the space $V^2(\Omega) \cap C^0(\bar{\Omega})$, the test function φ being taken in $V^2(\Omega)$ which is a subspace of $H^1(\Omega)$. □

Remark I.6. "Variational" formulation of the problem for Γ_p *fixed* :

If we write the equation for ψ in the plasma Ω_p in the form (I.17) the first equation of (I.37) is

$$a_\mu(\psi,\varphi) = \sum_{i=1}^{k} \frac{I_i}{S_i} \int_{B_i} \varphi dS + \int_{\Omega_{cv}} j_{cv}\, \varphi\, dS$$

(I.38)

$$+ \int_{\Omega_p} [r \frac{\partial p}{\partial \psi} + \frac{1}{2\mu_0 r} \frac{\partial f^2}{\partial \psi}]\varphi\, dS, \quad \forall \varphi \in V^{p_0'}(\Omega) .$$

If we now suppose that the boundary Γ_p of the plasma is fixed then equation (I.38) with $p_0 = p_0' = 2$ is none other than Euler's equation for the following variational problem :

$$\text{(I.39)} \qquad J(\psi) = \inf_{\psi' \in V^2(\Omega)} J(\psi')$$

with

$$J(\psi') = \frac{1}{2} \int_\Omega \pi(\frac{\nabla^2 \psi'}{r^2})\, r\, dS - \sum_{i=1}^{k} \frac{I_i}{S_i} \int_{B_i} \psi'\, dS$$

$$- \int_{\Omega_{cv}} j_{cv}\, \psi'\, dS - \int_{\Omega_p} p(\psi') r\, dS - \frac{1}{2\mu_0} \int_{\Omega_p} \frac{f(\psi')^2}{r}\, dS$$

where π denotes the primitive of the function $1/\bar{\mu}$ in Ω_f and is equal to $(1/\mu_0)$Id in Ω_a and Ω_p, where Id is the identity operator.

The formulation (I.39) is a variational formulation of the problem when the boundary Γ_p of the plasma is fixed.

If we restrict ourselves to the plasma domain Ω_p with Γ_p fixed, the Grad–Shafranov equation (I.17) is the Euler equation for the following variational problem :

(I.40) $$J'(\psi) = \inf_{\psi' \in V^2(\Omega_p)} J'(\psi')$$

with

$$J'(\psi') = \frac{1}{2\mu_0}\int_{\Omega_p} \frac{\nabla^2 \psi'}{r}\, dS - \int_{\Omega_p} p(\psi')r\, dS - \frac{1}{2\mu_0}\int_{\Omega_p} \frac{f(\psi')^2}{r}\, dS$$

$$= \left[\int \frac{B_P^2}{2\mu_0} dV - \int p\, dV - \int \frac{B_T^2}{2\mu_0} dV\right] \frac{1}{2\pi}$$

where these latter integrals are taken over the torus generated by rotation of Ω_p about the axis Oz. □

I.2.5.2 *"Equivalence" of the problems* (P_I) *and* (P_I') *:*

Let us assume that the boundary Γ of Ω and the air–iron interface Γ_{af} are regular. We then have the following proposition :

Proposition I.1 :

Let $(\psi, \lambda) \in V^{p_0}(\Omega) \times R$ be a solution of the problem (P_I) (i.e. satisfying equations (I.30)) with Γ_p regular and $p_0 > 2$. If, moreover, the restriction of $\mu^{-1}r^{-1}\ \nabla\psi$ to Ω_f (Ω_a,Ω_p respectively) belongs to $[W^{1,p_0}(\Omega_f)]^2$ ($[W^{1,p_0}(\Omega_a)]^2$, $[W^{1,p_0}(\Omega_p)]^2$, respectively) then the pair (ψ, λ) is a solution to (P_I') (i.e. satisfies (I.37)).

Conversely, if $(\psi, \lambda) \in V^{p_0}(\Omega) \times R$ is a solution to (P_I') with Γ_p regular and $p_0 > 2$, and if also $\mu^{-1}r^{-1}\ \nabla\psi$ belongs to $[W^{1,p_0}(\Omega)]^2$, then (ψ,λ,Γ_p) is a solution to (P_I). □

The proof of Proposition I.1 rests on the following lemma (cf. J. NECAS) :

Lemma I.1:

Let Ω_0 be a bounded open set in R^2 with regular boundary Γ_0. Let **u** be a vector in $[W^{1,p_0}(\Omega_0)]^2$ and v a function in $W^{1,p_0'}(\Omega_0)$ where p_0' is the scalar conjugate to p_0. Then we have the following formula for integration by parts :

$$\int_{\Omega_o} v\ \nabla\cdot u\ dS = -\int_{\Omega_0} u\cdot\nabla v\ dS + \int_{\Gamma_0} (u\cdot n)v\ d\Gamma \tag{I.41}$$

where **n** is the unit vector normal to Γ_0 . This lemma is a strict consequence of Green's formula (cf. J. NECAS). □

Proof of Proposition I.1 :

Let $(\psi, \lambda, \Gamma_p)$ be a solution to problem (P_I) with $\psi \in V^{p_0}(\Omega)$ and Γ_p regular. Let φ be a function in $V^{p_0}(\Omega)$.

Multiply equations (I.19), (I.20) and (I.23) by φ and integrate over Ω_f, Ω_a, Ω_p respectively. If the restriction of $1/\mu r\ \nabla\psi$ to Ω_f (Ω_a, Ω_p respectively) belongs to $[W^{1,p_0}(\Omega_f)]^2$ ($[W^{1,p_0}(\Omega_a)]^2$, $[W^{1,p_0}(\Omega_p)]^2$ respectively) we may apply (I.41) with $\mathbf{u} = 1/\mu r\ \nabla\psi$, $v = \varphi$ and $\Omega_0 = \Omega_f$ (Ω_a, Ω_p respectively); we obtain

$$\text{(I.42)}\quad \left\{\begin{aligned} &\int_{\Omega_f} \frac{1}{\mu r}\ \nabla\psi\cdot\nabla\varphi\ dS - \int_{\Gamma_{af}} \frac{1}{\mu r}\frac{\partial\psi}{\partial n}\ \varphi\ d\Gamma \\ &+ \int_{OF} \frac{1}{\mu r}\frac{\partial\psi}{\partial z}\ \varphi\ dr + \int_{GH} \frac{1}{\mu r}\frac{\partial\psi}{\partial z}\ \varphi\ dr = 0 \end{aligned}\right.$$

$$\text{(I.43)}\quad \left\{\begin{aligned} &\int_{\Omega_a} \frac{1}{\mu_0 r}\ \nabla\psi\cdot\nabla\varphi\ dS + \int_{\Gamma_{af}} \frac{1}{\mu_0 r}\frac{\partial\psi}{\partial n}\varphi\ d\Gamma \\ &+ \int_{\Gamma_p} \frac{1}{\mu_0 r}\frac{\partial\psi}{\partial n}\varphi\ d\Gamma + \int_{FM_i} \frac{1}{\mu_0 r}\frac{\partial\psi}{\partial z}\ \varphi\ dr + \int_{M_eG} \frac{1}{\mu_0 r}\frac{\partial\psi}{\partial z}\ \varphi dr \\ &+ \int_{HA} \frac{1}{\mu_0 r}\frac{\partial\psi}{\partial z}\varphi\ dr = \sum_{i=1}^{k} \frac{I_i}{S_i} \int_{B_i} \varphi\ dS + \int_{\Omega_{cv}} j_{cv}\ \varphi\ dS \end{aligned}\right.$$

$$\text{(I.44)}\quad \left\{\begin{aligned} &\int_{\Omega_p} \frac{1}{\mu_0 r}\ \nabla\psi\cdot\nabla\varphi\ dS - \int_{\Gamma_p} \frac{1}{\mu_0 r}\frac{\partial\psi}{\partial n}\ \varphi\ d\Gamma \\ &+ \int_{M_iM_e} \frac{1}{\mu_0 r}\frac{\partial\psi}{\partial z}\ \varphi\ dr = \lambda \int_{\Omega_p} h(r)g(\psi_N)\ \varphi\ dS \end{aligned}\right.$$

where $\partial/\partial n$ denotes the normal derivative in the outward direction from Ω_f and Ω_p, and where the points F, G, H, M_i and M_e on the r-axis are represented in Fig. I.4. By adding (I.42), (I.43) and (I.44) and using the boundary and interface conditions (I.25), (I.26) and (I.27) we obtain the first equation of (I.37). The pair (ψ,λ) is then a solution to (P_I').

Conversely, if (ψ,λ) satisfies (I.37) and if we consider φ belonging to $\mathcal{D}(\Omega_f)$ ($\mathcal{D}(\Omega_a)$, $\mathcal{D}(\Omega_p)$ respectively), by applying Green's formula we obtain (I.19) ((I.20), (I.23) respectively). Since ψ belongs to $V^{p_o}(\Omega)$ with $p_o > 2$ it belongs to $C^0(\bar{\Omega})$ and its continuity is thus assured on Γ_{af} and Γ_p. If we suppose moreover that $1/\mu r\ \nabla\psi$ belongs to $[W^{1,p_o}(\Omega)]^2$ then we deduce the interface conditions (I.26) and (I.27). Now consider a function φ in $V^{p_o}(\Omega)$, multiply equations (I.19), (I.20) and (I.23) by φ and integrate over Ω_f, Ω_a, Ω_p. Thanks to the fact that $1/\mu r\ \nabla\psi$ belongs to $[W^{1,p_o}(\Omega)]^2$, we may use (I.41) and thus obtain equations (I.42) to (I.44). Adding these equations, using the interface conditions (I.26) and (I.27) and subtracting the first equation of (I.37) we obtain the following relation:

$$\text{(I.45)} \qquad \int_{OA} \frac{1}{\mu r}\frac{\partial\psi}{\partial z}\,\varphi\, dr = 0\ , \quad \forall\varphi \in V_{p_o'}(\Omega)\ .$$

From this we deduce the symmetry condition (I.25). The triple $(\psi,\ \lambda,\ \Gamma_p)$ thus satisfies equations (I.30). □

The weak formulation (P_I') of the problem thus includes in a natural way the boundary and interface conditions. This is what gives it interest and justifies its use in the numerical solution of the problem.

Remark I.7. Regularity of ψ :

Because of the discontinuity of $\partial\psi/\partial n$ at the air–iron interface Γ_{af}, the function ψ does not belong to $C^1(\bar{\Omega})$. Let us restrict ourselves to the domain Ω_{ap} lying between the iron core and the return arms, namely the air–plasma region (not including the air which is on the outside of the iron). The operator L is elliptic with coefficients infinitely differentiable in Ω_{ap}; the right–hand–side in the equation for ψ, namely the current density, is in $L^p(\Omega_{ap})$ for all $p \geqslant 1$. The solution ψ is then in $W^{2,p}_{loc}(\Omega_{ap})$ (cf. J.L. LIONS – E. MAGENES); it belongs in particular to $W^{2,p}(\Omega_v)$ and thus to $C^{1,\alpha}(\bar{\Omega}_v)$ for all $\alpha<1$. The function ψ is not in general in $C^2(\bar{\Omega}_v)$; indeed if the plasma current density j_T does not vanish on Γ_p, the second normal derivative $\partial^2\psi/\partial n^2$ is discontinuous on Γ_p. □

Remark I.8. Regularity of Γ_p :

Although the function ψ is not of class C^2 in the neighbourhood of Γ_p, it is clear that Γ_p can be of class C^2 or even C^∞. But certain configurations are such that Γ_p can have multiple points.

First we define the "doublet" and "droplet" configurations:

. The "droplet" is the configuration represented in Fig.I.6a ; the complete plasma (Ω_p and its reflection in the r axis) is not connected.

. The "doublet" is the configuration represented in Fig.I.6b ; one line of flux in the interior of the plasma has a hyperbolic point : it is called the separatrix of the plasma.

The limit configuration between doublet and droplet, represented in Fig.I.6c, is such that Γ_p is the separatrix and therefore presents the point P as a multiple point.

Likewise configurations with divertor, as in I.6d, are configurations with multiple points; the divertor is a magnetic device whose purpose is to extract the particles situated at the boundary of the plasma and thus to reduce the emission of impurities near the plasma centre. The plasma boundary Γ_p includes a hyperbolic point, X– point, namely the point P' in Fig.I.6d. □

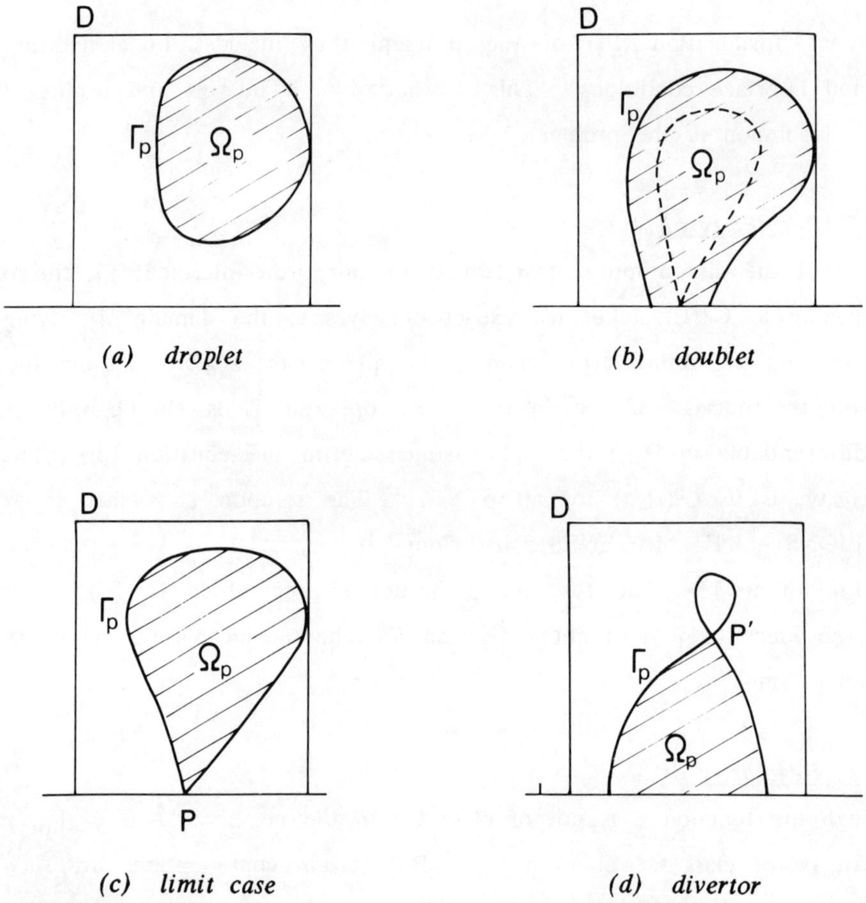

Figure I.6 : Particular equilibrium configurations

I.3 LINEARIZATION OF THE PROBLEM:

The problems (P_I) and (P'_I) contain three nonlinearities :

. the magnetic permeability μ of the iron which is a function of $\frac{\nabla^2\psi}{r^2}$

. the plasma current density function $j_T(r,\psi)$

. the free boundary Γ_p of the plasma which is a nonlinearity in itself.

The linearization of the problem (P'_I) is very useful as much for the numerical solution of the problem as for the study of the linear stability of displacements of the plasma or for the solution of control problems which will be treated in the following chapters.

Let I be the vector in R^k with components I_i representing the currents in the k coils B_i. Let us give to I the increment $\tilde{I}$; we propose to calculate the increments $\tilde{\psi}$ of ψ and $\tilde{\lambda}$ of λ to first order around the equilibrium solution (ψ,λ) corresponding to the currents I_i and satisfying equations (I.37). We define $\tilde{\psi}$ and $\tilde{\lambda}$ as the derivatives in the sense of Gâteaux (cf. J. CEA) of ψ and λ with respect to I, in the direction $\tilde{I}$:

$$\text{(I.46)} \quad \begin{cases} \tilde{\psi} = \lim\limits_{\theta\to 0} \dfrac{\psi(I+\theta\tilde{I})-\psi(I)}{\theta} \\ \\ \tilde{\lambda} = \lim\limits_{\theta\to 0} \dfrac{\lambda(I+\theta\tilde{I})-\lambda(I)}{\theta} \end{cases} .$$

We make the following hypotheses for the pair $u = (\psi,\lambda)$ corresponding to the vector I :

H1) $\sup\limits_D \psi$ is attained at one and only one point M_o of D, which is then the unique point of contact between Γ_p and D. If D is a regular curve in Ω_V, then ψ is assumed to be of class C^2 on D in a neighbourhood of M_o and to have a nondegenerate maximum at M_o.

H2) $\sup\limits_{\Omega_p} \psi$ is attained at one and only one point M_1 which is an interior point of Ω_p and is the magnetic axis. We suppose ψ to be of class C^2 in a neighbourhood of M_1 and the point M_1 to be a nondegenerate elliptic point.

H3) $\nabla\psi$ vanishes nowhere on Γ_p.

Remark I.9 :

The configuration in Figs. I.6c, I.6d do not satisfy the hypothesis H3. We shall see later how the numerical solution of this problem nevertheless allows us to simulate these configurations. □

On Ω_p define the form e_ψ, linear on $V^{p_o}(\Omega)$, by the value of $e_\psi(\psi')$ at each point M of Ω_p, where $\psi' \in V^{p_o}(\Omega)$:

(I.47) $$e_\psi(\psi')(M) = \frac{[\psi(M_o)-\psi(M_1)]\psi'(M)+[\psi(M)-\psi(M_o)]\psi'(M_1)+[\psi(M_1)-\psi(M)]\psi'(M_o)}{[\psi(M_o)-\psi(M_1)]^2}$$

For every pair $u = (\psi, \lambda)$ which is a solution of (P_I') we define the following bilinear forms a_u', b_u and d_u on the space $V^{p_o}(\Omega) \times V^{p_o'}(\Omega)$:

(I.48) $$a_u'(\psi',\varphi) = -2\int_{\Omega_f} \frac{\mu'}{\mu^2 r^3} (\nabla\psi.\nabla\psi')(\nabla\psi.\nabla\varphi)dS$$

(I.49) $$d_u(\psi',\varphi) = \lambda\left[\int_{\Omega_p} h(r)g'(\psi_N)e_\psi(\psi')\varphi\ dS + \int_{\Gamma_p} h(r)g(1)[\psi'(M_o)-\psi'(M)]\left[\frac{\partial\psi}{\partial n}\right]^{-1} \varphi\ d\Gamma\right]$$

(I.50) $$b_u(\psi',\varphi) = a_\mu(\psi',\varphi) + a_u'(\psi',\varphi) - d_u(\psi',\varphi)$$

$$\forall(\psi',\varphi) \in V^{p_o}(\Omega) \times V^{p_o'}(\Omega)$$

where $\mu'(B_p^2)$ is the derivative of $\bar{\mu}(B_p^2)$, $g'(\psi_N)$ is the derivative of $g(\psi_N)$ and where a_μ is defined by (I.36).

Finally we define the form $c_u(\varphi)$ by :

(I.51) $$c_u(\varphi) = \int_{\Omega_p} h(r)g(\psi_N)\varphi\ dS \quad , \quad \forall\varphi \in V^{p_o'}(\Omega) .$$

We can then formally establish the following proposition :

<u>Proposition I.2</u> :

If the hypotheses H1, H2 and H3 are satisfied for a pair (ψ,λ) which is a solution to (P_I'), and if the pair $(\tilde{\psi},\tilde{\lambda}) \in V^{p_0}(\Omega) \times R$ defined by (I.46) exists, then it is a solution to the following system:

$$\text{(I.52)} \quad \begin{cases} b_u(\tilde{\psi},\varphi) - \tilde{\lambda}\, c_u(\varphi) = \sum_{i=1}^{k} \frac{\tilde{I}_i}{S_i} \int_{B_i} \varphi \, dS, \quad \forall \varphi \in V^{p_0'}(\Omega) \\ d_u(\tilde{\psi},1) + \tilde{\lambda} c_u(1) = 0 \end{cases}$$

where the forms b_u, c_u and d_u are defined by (I.49) to (I.51). □

The proof of the proposition rests on the following lemmas :

<u>Lemma I.2</u> :

Let p_0 be a real number greater than 2. If hypothesis H1 is satisfied, the map which associates $\psi_p = \sup_D \psi$ to $\psi \in V^{p_0}(\Omega)$ is Gâteaux differentiable and its derivative is the map which associates $\tilde{\psi}(M_0)$ to $\tilde{\psi} \in V^{p_0}(\Omega)$.

<u>Proof of Lemma I.2</u> :

Let us study successively the case where D consists of a finite number of points, and where D is a regular curve in Ω_V. If D consists of a finite number of points and if $\sup_D \psi$ is attained at just one of the points M_0, then for θ sufficiently small $\sup(\psi + \theta\tilde{\psi})$ is likewise attained at M_0 and

$$\text{(I.53)} \quad \lim_{\theta \to 0} \frac{1}{\theta} \left[\sup_D (\psi+\theta\tilde{\psi}) - \sup_D \psi\right] = \tilde{\psi}(M_0) \; .$$

If D is a regular curve, denote by s the arc–length coordinate of a point M of D, with s_0 the coordinate of the point M_0 where $\sup_D \psi$ is attained. Consider the function $\psi(s)$ on D ; by hypothesis Hl the maximum of ψ on D is nondegenerate, and so

$$\text{(I.54)} \quad \psi'(s_0) = \frac{d\psi}{ds}(s_0) = 0 \; , \; \psi''(s_0) = \frac{d^2\psi}{ds^2}(s_0) < 0 \; .$$

For s close to s_0 we have

$$\text{(I.55)} \quad \psi(s) = \psi(s_0) + \frac{1}{2}\psi''(s_0)(s-s_0)^2 + o(s-s_0)^2 \; .$$

and so for s sufficiently close to s_o :

$$\psi(s) < \psi(s_0) + \frac{1}{4}\psi''(s_0)(s-s_0)^2 \quad . \tag{I.56}$$

Consider a function $\tilde{\psi} \in V^{p_0}(\Omega)$. It belongs to $W^{1,p_0}(\Omega)$ and hence, according to J.L. LIONS [1], to the space of Hölder continuous functions $C^{0,\alpha}(\bar{\Omega})$ with $\alpha = 1 - 2/p_0$. Therefore for s sufficiently close to s_0 there exists a constant C such that

$$\tilde{\psi}(s) < \tilde{\psi}(s_0) + C(s-s_o)^{\alpha} \quad . \tag{I.57}$$

For θ sufficiently small, $\sup_D(\psi+\theta\tilde{\psi})$ is attained at a point M'_0 with arc-length coordinate s'_0 which approaches s_0 as $\theta \to 0$ (by compactness of D). We have :

$$(\psi+\theta\tilde{\psi})(s'_o) > (\psi+\theta\tilde{\psi})(s_o) \quad . \tag{I.58}$$

Using (I.56) and (I.57) with $s = s'_0$ the inquality (I.58) then becomes :

$$\frac{1}{4}\psi''(s_0)(s'_0-s_0)^2 + \theta C(s'_0-s_0)^{\alpha} > 0$$

whence

$$s'_o-s_o < \left[-\frac{4C\theta}{\psi''(s_o)}\right]^{\frac{1}{2-\alpha}} \quad . \tag{I.59}$$

By (I.55) and (I.59) we have

$$|\psi(s'_0) - \psi(s_0)| < \frac{(16C^2)^{\frac{1}{2-\alpha}}\,\theta^{\frac{2}{2-\alpha}}}{|\psi''(s_0)|^{\frac{\alpha}{2-\alpha}}} \quad .$$

Since $\frac{2}{2-\alpha} = \frac{2p_0}{2+p_0} > 1$ for $p_0 > 2$:

$$\lim_{\theta\to 0}\frac{1}{\theta}\left[\sup_D(\psi+\theta\tilde{\psi}) - \sup_D \psi\right] = \tilde{\psi}(s_o) \quad . \tag{I.60}$$ □

Remark I.10 :

Lemma 1.2 is still satisfied if the second sentence in the hypothesis H1 is modified as follows: Let ℓ be the first integer such that $\psi^{(\ell)}(s)$ is non zero at $s = s_0$; suppose the function ψ to be of class C^{ℓ} on D in a neighbourhood of M_0 . Hypothesis H1 corresponding to $\ell = 2$ can thus be extended to all even values of ℓ. □

Lemma I.3 :

If hypothesis H2 is satisfied, the map which associates $\psi_a = \sup_{\Omega p} \psi$ to $\psi \in V^{p_0}(\Omega)$ is Gâteaux differentiable and its derivative is the map which associates $\tilde{\psi}(M_1)$ to $\tilde{\psi} \in V^{p_0}(\Omega)$. □

Proof of Lemma I.3 :

Let H denote the Hessian matrix at M_1, namely

$$H = \begin{bmatrix} \dfrac{\partial^2 \psi}{\partial r^2} & \dfrac{\partial^2 \psi}{\partial r \partial z} \\ \\ \dfrac{\partial^2 \psi}{\partial r \partial z} & \dfrac{\partial^2 \psi}{\partial z^2} \end{bmatrix} .$$

In a neighbourhood of M_1 the function ψ can be written

$$\text{(I.61)} \qquad \psi(M) = \psi(M_1) + \frac{1}{2}\, {}^tX\, H\, X + o(\|MM_1\|^2)$$

where X is the vector MM_1 and ${}^tX$ is its transpose.

As the point M_1 is nondegenerate elliptic the quadratic form associated to H is negative definite. Therefore there exists a constant positive C_1 such that in a neighbourhood of M_1 :

$$\text{(I.62)} \qquad \psi(M) < \psi(M_1) - C_1 \|MM_1\|^2 .$$

Let $\tilde{\psi}$ be a function $V^{p_0}(\Omega)$; it belongs to $C^{0,\alpha}(\Omega)$ with $\alpha = 1 - 2/p_0$ and so there exists a constant C_2 such that in a neighbourhood of M_1 we have :

$$\text{(I.63)} \qquad \tilde{\psi}(M) < \tilde{\psi}(M_1) + C_2 \|MM_1\|^{\alpha} .$$

For θ sufficiently small, $\sup_{\Omega p_-}(\psi + \theta\tilde{\psi})$ is attained at a point M_1' which approaches M_1 as $\theta \to 0$ (by compactness of $\bar{\Omega}_p$). We have :

$$\text{(I.64)} \qquad (\psi+\theta\tilde{\psi})(M_1') > (\psi+\theta\tilde{\psi})(M_1) .$$

Using (I.62) and (I.63) with $M = M_1'$ the inequality (I.64) becomes :

$$\theta C_2 \|M_1 M_1'\|^{\alpha} > C_1 \|M_1 M_1'\|^2$$

or

$$\text{(I.65)} \qquad \|M_1 M_1'\| < \left[\frac{\theta C_2}{C_1}\right]^{\frac{1}{2-\alpha}} .$$

By (I.61) and (I.65) there exists a constant C_3 depending on α such that

$$|\psi(M_1') - \psi(M_1)| < C_3\ \theta^{\frac{2}{2-\alpha}} .$$

Since $\frac{2}{2-\alpha} = \frac{2p_o}{2+p_o} > 1$ for $p_o > 2$:

$$\text{(I.66)} \qquad \lim_{\theta \to 0} \frac{1}{\theta} \left[\underset{\Omega_p}{\sup}\ (\psi+\theta\tilde{\psi}) - \underset{\Omega_p}{\sup}\ \psi\right] = \tilde{\psi}(M_1)$$ □

Lemma I.4 :

If hypotheses H1 and H3 are satisfied the variation $\tau(M)$ of a point M of Γ_p corresponding to a perturbation $\tilde{\psi}$ of ψ is characterized by :

$$\text{(I.67)} \qquad (\tau \cdot n)\ (M) = \frac{\tilde{\psi}(M_o) - \tilde{\psi}(M)}{\frac{\partial \psi}{\partial n}(M)}\ , \quad \forall M \in \Gamma_p$$

where **n** is the unit vector normal to Γ_p at M, and $\partial\psi/\partial n$ is the normal derivative of ψ at that point. □

Formal proof of Lemma I.4 :

The boundary Γ_p of the plasma is characterized by (I.28) which can be written

$$\text{(I.68)} \qquad \psi(M) = \psi(M_o)\ , \quad \forall M \in \Gamma_p .$$

Using Lemma 1.2, the linearization of (I.68) can be written as follows :

$$\text{(I.69)} \qquad \tilde{\psi}(M) + \nabla\psi \cdot \tau = \tilde{\psi}(M_o)$$

where τ represents the variation MM' of the point M of Γ_p, while the variation of ψ is the function $\tilde{\psi} \in V^{\,o}(\Omega)$. If we now use the hypothesis H3 then the equation (I.69) takes the form (I.67). □

This lemma can be justified mathematically (cf. F. MURAT – J. SIMON).

Proof of Proposition I.2 :

The linearization of equations (I.37) relies mainly on the use of a lemma from continuum mechanics justified mathematically in J. SIMON [1] and references therein, and which can here be written in the following way :

$$(I.70)\qquad (D_\psi \int_{\Omega_p} h(r)g(\psi_N)\varphi\, dS, \tilde{\psi}) = \int_{\Omega_p} h(r)(D_\psi g(\psi_N), \tilde{\psi})\varphi\, dS + \int_{\Gamma_p} h(r)g(1)\varphi(\tau\cdot\mathbf{n})d\Gamma ,$$

where $(D_\psi\cdot, \tilde{\psi})$ denotes the Gâteaux derivative in the $\tilde{\psi}$– direction and where τ is the variation of the free boundary Γ_p associated to the variation $\tilde{\psi}$ of ψ.

Under the hypotheses H1 and H2, and using Lemmas I.2 and I.3, the linearization of the normalised flux ψ_N defined in (I.21) is given by the expression I.47 for $e_\psi(\tilde{\psi})$. Thanks to the hypothesis H3 and Lemma I.4 the variation τ of Γ_p is characterised by (I.67). The relations (I.52) are then the linearizations of equations (I.37), through use of the derivation formula (I.70). □

<u>*Remark I.11*</u> *: Linearization of Problem* (P_I) *:*

Denote by δ_{Γ_p} the measure on Γ_p defined by :

$$(\delta_{\Gamma_p}, v) = \int_{\Gamma_p} v\, d\Gamma , \quad \forall v \in C^0(\bar{\Omega})$$

and by L' the operator which is zero on Ω_a and Ω_p and is defined on Ω_f as follows :

$$L'. = \nabla\cdot \left[\frac{2\mu'}{\mu^2 r^3} (\nabla\psi\cdot\nabla\cdot)\nabla\psi\right]$$

the operators ∇ and $\nabla\cdot$ being taken in the (r, z)– plane.

The linearization of (I.31) for the pair (ψ,λ) is

$$(I.71)\qquad \begin{cases} \tilde{\psi} = 0 \quad \text{on} \quad \Gamma_0 \\ \dfrac{\partial\tilde{\psi}}{\partial z} = 0 \quad \text{on} \quad \Gamma_1 \\ (L+L')\tilde{\psi} - [\lambda h(r)g'(\psi_N)e_\psi(\tilde{\psi}) + \tilde{\lambda} h(r)g(\psi_N)]1_{\Omega_p} \\ \qquad - \lambda h(r)g(1)(\dfrac{\partial\psi}{\partial n})^{-1}[\tilde{\psi}(M_0)-\tilde{\psi}(M)]\,\delta_{\Gamma_p} = \displaystyle\sum_{i=1}^{k} \frac{\tilde{I}_i}{S_i} 1_{B_i} \\ d_u(\tilde{\psi},1) + \tilde{\lambda} c_u(1) = 0 . \end{cases}$$

The system (I.52) corresponds to the "weak" formulation of (I.71).

In Chapter III we shall see sufficient conditions for the existence of a solution to the linearized problem for a simplified model. □

I.4 NUMERICAL METHODS OF SOLUTION :

I.4.1 Approximation by finite elements : the problem $(P_I'^{\ell})$:

We introduce a triangulation $\mathcal{C}_\ell$ on $\bar{\Omega}$ using triangles T of diameter less than or equal to ℓ; by definition we have

(I.72) $$\bar{\Omega} = \bigcup_{T \in \mathcal{C}_\ell} T$$

with the condition that if T_1 and T_2 are two distinct triangles we have

. either $T_1 \cap T_2 = \emptyset$

. or T_1 and T_2 have a common vertex

. or T_1 and T_2 have a common edge.

The triangulation of $\bar{\Omega}$ is carried out in such a way that Γ_{af} and the boundaries of the coils B_i coincide with edges of triangles. This is possible for the free boundary Γ_p of the plasma only if a new mesh is constructed at each iteration. Another method has been used here for treating Γ_p: the mesh is fixed and independent of Γ_p; the free boundary will then cross the grid of the triangulation. If the limiter D consists of a finite number of points, these are taken as vertices of the mesh; if D is a curve in Ω_v it is approximated by a polygonal line D_ℓ in Ω_v consisting of a union of edges of triangles of $\mathcal{C}_\ell$.

Associated to the triangulation $\mathcal{C}_\ell$ is the space V_ℓ defined by

(I.73) $$V_\ell = \{\varphi \in C^0(\bar{\Omega}) \mid \forall T \in \mathcal{C}_\ell \ , \ \varphi|_T \in P_1 \text{ and } \varphi = 0 \text{ on } \Gamma_0\}$$

where P_1 denotes the vector space of polynomials of degree 1 with real coefficients in the variables r and z. This space V_ℓ approximates the space $V^{p_0}(\Omega)$ for $p_0 > 2$ (cf. P.G. CIARLET – P.A. RAVIART).

Let γ_ℓ be the set of vertices of $\mathcal{C}_\ell$ belonging to Γ_0 and ω_ℓ its complement in the set of all vertices of $\mathcal{C}_\ell$. Let N_ℓ be the cardinality of ω_ℓ. Each element φ of V_ℓ is determined by the values taken at the N_ℓ vertices M_j of ω_ℓ. For each vertex M_j of ω_ℓ we define $\varphi_i \in V_\ell$ by :

$$\varphi_i(M_j) = \delta_{ij} \ , \ 1 \leqslant i,j \leqslant N_\ell .$$

The set of the N_ℓ functions φ_i constitutes a basis for V_ℓ. We look for $(\psi_\ell, \lambda_\ell) \in \times V_\ell \times R$ satisfying the following discrete system that approximates the equations (I.74) :

(I.74)
$$\begin{cases} a_{\mu_\ell}(\psi_\ell,\varphi) = \displaystyle\sum_{i=1}^{k} \frac{I_i}{S_i}\int_{B_i} \varphi \, dS + \int_{\Omega_{cv}} j_{cv}\varphi \, dS \\ \qquad + \lambda_\ell \displaystyle\int_{\Omega_p^\ell} h(r)g(\psi_N^\ell)\,\varphi \, dS, \qquad \forall\varphi \in V_\ell \\ I_p = \lambda_\ell \displaystyle\int_{\Omega_p^\ell} h(r)g(\psi_N^\ell)\, dS \\ \text{with } \Omega_p^\ell = \{M \in \Omega_V \,|\, \psi_\ell(M) > \sup_{D_\ell} \psi_\ell\} \end{cases}$$

with $\mu_\ell = \bar{\mu}(\frac{1}{r^2}\nabla^2\psi_\ell)$, $\psi_N^\ell = \dfrac{\psi_\ell - \psi_a^\ell}{\psi_p^\ell - \psi_a^\ell}$;

here ψ_p^ℓ is the value of ψ_ℓ on Γ_p^ℓ i.e. the maximum of ψ_ℓ on D_ℓ, which is always attained at a vertex of $\mathcal{C}_\ell$ because of the linear character of the functions of V_ℓ in each triangle; ψ_a^ℓ represents the maximum of ψ_ℓ in Ω_p^ℓ and it is necessarily attained at a vertex of the mesh. Finally, the vector $\nabla\psi_\ell$ is constant on each triangle of $\mathcal{C}_\ell$ by definition of V_ℓ, and μ_ℓ is thus constant on each triangle T. Recall that

(I.75)
$$\Gamma_p^\ell = \{M \in \Omega_V \,|\, \psi_\ell(M) = \sup_{D_\ell} \psi_\ell\} .$$

It is clear that the boundary of the plasma is then a polygonal line in Ω_V.

The numerical method of integration used in solving (I.74) consists of taking for each integral over a triangle T the product of the area of the triangle with the value of the integrand at the centre of gravity of the triangle. This is compatible with the finite elements P_1, linear in each triangle, that are used here, as is shown in P.G. CIARLET.

Problem $(P_I'^\ell)$ consists of finding a pair $(\psi_\ell, \lambda_\ell) \in V_\ell \times R$ which is a solution to (I.74). The difficulty of solving $(P_I'^\ell)$ resides in the nonlinearities which are :

. the magnetic permeability μ_ℓ of the iron, which is a function of $\nabla^2\psi_\ell$

. the current density function of the plasma, which involves $g(\psi_N^\ell)$

. the free boundary Γ_p^ℓ .

Various methods have been tried for treating these nonlinearities that permit reduction to the solution of a sequence of linear problems. We shall study successively the iteration methods of Picard, Marder– Weitzner and Newton– Raphson.

In what follows we solve the problem $(P_I'^\ell)$, but for ease of writing the index ℓ will be omitted from the equations.

Remark I.12 :

The term $a_\mu(\psi,\varphi)$ is meaningless when computed analytically in a neighbourhood of the z– axis for functions ψ, $\varphi \in V_\ell$, because $\nabla\psi$ and $\nabla\varphi$ are constant on each triangle and $1/r$ is not integrable in the neighbourhood of the z– axis. Physically, the function ψ behaves like r^2 as $r \to 0$ and then the term $a_\mu(\psi,\varphi)$ makes sense; if it is computed by the numerical integration formula above, where $1/r$ is evaluated at the centre of gravity of the triangle, then the correct analytical expression is obtained in a neighbourhood of $r = 0$ up to a factor of 9/8 (in dimension 1 the exact expression is recovered). For finite elements P_1 the numerical integration formula at the centre of gravity of the triangle is therefore better than analytic integration. This fact has already been mentioned in ZIENKIEWICZ. □

I.4.2 Algorithms of Picard type :

Suppose that we have to solve the equation

$$\mathcal{L}\psi = F(\psi). \tag{I.76}$$

where $\mathcal{L}$ is a linear operator and F a nonlinear operator. The Picard iterations, also called fixed point iterations, are written

$$\mathcal{L}\psi^{n+1} = F(\psi^n). \tag{I.77}$$

If the derivative operator F' of F exists, a necessary condition for convergence of this algorithm is that the spectrum of the operator $\Phi(\psi) = \mathcal{L}^{-1}F'(\psi)$ be contained within the unit ball.

I.4.2.1 *The algorithm AP1* :

If the operator $\mathcal{L}$ is nonlinear (as is the case of the operator L defined by (I.13)) then the Picard iterations for the model problem (I.76) can be written

$$\mathcal{L}^n \psi^{n+1} = F(\psi^n). \tag{I.78}$$

with $\mathcal{L}^n = \mathcal{L}(\psi^n)$.

We are given $(\psi^0, \lambda^0) \in V_\ell \times R$. From $u^n = (\psi^n, \lambda^n)$ we compute $u^{n+1} = (\psi^{n+1}, \lambda^{n+1}) \in V_\ell \times R$ such that

$$\left\{\begin{array}{l} a_{\mu^n}(\psi^{n+1},\varphi) = \sum_{i=1}^{k} \frac{I_i}{S_i} \int_{B_i} \varphi \, dS + \int_{\Omega_{cv}} j_{cv} \, \varphi \, dS \\ \qquad + \lambda^n \int_{\Omega_p^n} h(r)g(\psi_N^n) \, \varphi \, dS, \quad \forall \varphi \in V_\ell \\ \Omega_p^{n+1} = \{M \in \Omega_v | \psi^{n+1}(M) > \sup_{D_\ell} \psi^{n+1}\} \\ I_p = \lambda^{n+1} \int_{\Omega_p^{n+1}} h(r)g(\psi_N^{n+1}) dS \end{array}\right. \tag{I.79}$$

To specify the algorithm completely it is necessary to indicate the way in which μ^n is computed. If μ^n is calculated from B_p^2 by $\mu^n = \bar{\mu}(1/r^2 \, \nabla^2\psi^n)$, the algorithm turns out to be divergent in the majority of cases. This divergence has already been demonstrated in R. GLOWINSKI – A. MARROCCO, and is uniquely tied to the magnetostatic character of the problem, as B_p^2 and thus μ are varying considerably from one iteration to another.

In order to make the algorithm converge we have to express μ as a function of a magnetostatic invariant of the problem. For this we use Ampère's theorem. The contour $\mathcal{C}_o$ is represented in Fig 1.7 and the quantity $\int_{\mathcal{C}_o} H_p \cdot d\ell$, where H_p is the poloidal component of the magnetic field, is thus equal to the algebraic sum of Ampère turns (i.e. the sum of the currents in the coils, in the vacuum vessel and in the plasma). This is verified at the end of the first iteration by an application of Green's formula. Therefore it is natural to choose H_p as a magnetostatic invariant and to consider the function $\mu = \tilde{\mu}(H_p^2)$ instead of $\mu = \bar{\mu}(B_p^2)$.

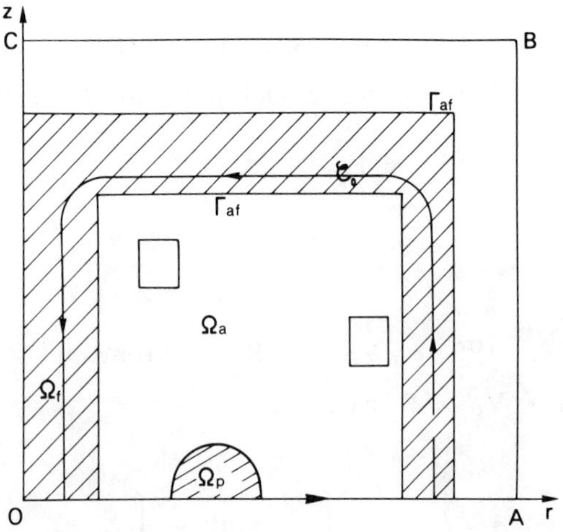

Figure I.7 : The contour $\mathscr{C}_0$ within the iron.

We thus compute μ^n from $(H_p^n)^2$ by

(I.80) $$\mu^n = \tilde{\mu}\left(\frac{|\nabla^2 \psi^n|}{r^2(\mu^{n-1})^2}\right)$$

This expression for μ^n is the one used in (I.79). It should be noted that since μ^{n-1} comes into the calculation of $(H_p^n)^2$, the algorithm is no longer a classical Picard algorithm of the type of (I.78); we call it *AP1*.

The equation for ψ^{n+1} in (I.79) requires the solution of a linear system whose matrix $(a_{\mu^n}(\varphi_i,\varphi_j))_{1 \leqslant i,j \leqslant N_\ell}$ is symmetric, positive definite and sparse. We use a Cholesky method in order to invert this linear system. For calculating the integral

$$\int_{\Omega_p^n} h(r)g(\psi_N^n)\varphi \, dS$$

that appears in the right-hand-side, we take into account the triangles within the plasma and those cut by Γ_p^n ; for the latter we compute the integral over the part of the triangle interior to Ω_p^n.

I.4.2.2 Algorithm *AP2* :

For the model problem (I.76) with $\mathscr{L}$ nonlinear, this algorithm may be written

(I.81) $$\mathscr{L}^{n+1}\psi^{n+1} = F(\psi^n).$$

We are given $(\psi^0, \lambda^0) \in V_\ell \times R$. From $u^n = (\psi^n, \lambda^n)$ we compute

$$
(I.82)\quad \left\{
\begin{aligned}
& a_{\mu^{n+1}}(\psi^{n+1},\varphi) = \sum_{i=1}^{k} \frac{I_i}{S_i}\int_{B_i} \varphi \, dS + \int_{\Omega_{cv}} j_{cv}\,\varphi\, dS \\
& \qquad + \lambda^n \int_{\Omega_p^n} h(r)g(\psi_N^n)\,\varphi\, dS, \quad \forall\varphi \in V_\ell \\
& \Omega_p^{n+1} = \{M \in \Omega_v \,|\, \psi^{n+1}(M) > \sup_{D_\ell} \psi^{n+1}\} \\
& I_p = \lambda^{n+1} \int_{\Omega_p^{n+1}} h(r)g(\psi_N^{n+1})\,dS \quad .
\end{aligned}
\right.
$$

The sole difference from the algorithm (*AP1*) is that the nonlinearity due to $\bar{\mu}(B_p^2)$ is here treated implicitly in order to avoid the divergence of the algorithm related to explicit involvement of the nonlinearity $\bar{\mu}(r^{-2}\nabla^2\psi)$. The fixed-point iterations are thus brought to bear upon the two nonlinearities $g(\psi_N)$ and Ω_p.

Let $u = (\psi, \lambda) \in V_\ell \times R$ be a solution to the problem $(P_I'^{\ell})$. Let M_u be the linear operator which to $(\psi_1, \lambda_1) \in V_\ell \times R$ associates $(\psi_2, \lambda_2) \in V_\ell \times R$ such that

$$
(I.83)\quad \left\{
\begin{aligned}
& a_\mu(\psi_2,\varphi) + a_u'(\psi_2,\varphi) = d_u(\psi_1,\varphi) + \lambda_1 c_u(\varphi) \\
& d_u(\psi_2,1) + \lambda_2 c_u(1) = 0
\end{aligned}
\right.
$$

where the forms a_μ, a_u', d_u and c_u are defined by (I.36), (I.48), (I.49) and (I.51). If M_u is well defined then a necessary condition for the convergence of the algorithm (*AP2*) is that the spectral radius $\rho(M_u)$ of M_u be smaller than 1. This spectral radius is calculated by the power method.

Each iteration of the algorithm (I.81) consists of solving a nonlinear problem. In practice $\mathscr{L}$ can be linearized, and (I.81) replaced by

$$
(I.84)\quad (\mathscr{L}^n + \mathscr{L}'^n)\psi^{n+1} = \mathscr{L}'^n\psi^n + F(\psi^n)
$$

which involves the derived operator $\mathscr{L}'^n$. This is then a mixed Picard–Newton algorithm because $\mathscr{L}$ is treated by linearization (Newton) and F by a fixed-point approach (Picard). In (I.82) $\bar{\mu}(B_p^2)$ is likewise linearized so that the first equation in (I.82) is replaced by

$$a_{\mu^n}(\psi^{n+1},\varphi) + a'_{u^n}(\psi^{n+1},\varphi) = a'_{u^n}(\psi^n,\varphi) + \sum_{i=1}^{k} \frac{I_i}{S_i}\int_{B_i}\varphi dS$$

(I.85)

$$+ \int_{\Omega_{cv}} j_{cv}\,\varphi\, dS + \lambda^n \int_{\Omega_p^n} h(r)g(\psi_N^n)\,\varphi\, dS, \quad \forall\varphi \in V_\ell$$

with $\mu^n = \bar{\mu}(\frac{\nabla^2\psi^n}{r^2})$, $u^n = (\psi^n,\lambda^n)$.

The two other equations in (I.82), which allow Ω_p^{n+1} and λ^{n+1} to be calculated remain unchanged. The algorithm modified in this way is thus a mixed Picard–Newton algorithm since the non-linearity $\bar{\mu}(B_p^2)$ is treated by Newton's method and the two other nonlinearities by Picard's method.

The system (I.85) whose matrix has entries $(a_{\mu^n}(\varphi_i, \varphi_j) + a'_{u^n}(\varphi_i,\varphi_j))_{1\leqslant i,j\leqslant N_\ell}$ is solved by a Crout method (cf. A.S. HOUSEHOLDER).

I.4.3 *Marder–Weitzner Algorithm AMW* :

The algorithms of Picard type converge only when the linearized operator is contracting. In the other cases B. MARDER – H. WEITZNER have applied an algorithm which for the model equation (I.76) with $\mathscr{L}$ linear is written as follows :

$$\text{(I.86)}\quad \begin{cases} \mathscr{L}\psi^{n+1/3} = F(\psi^n) \\ \mathscr{L}\psi^{n+2/3} = F(\psi^{n+1/3}) \\ \psi^{n+1} = (1-\alpha)\psi^n + 2\alpha\psi^{n+1/3} - \alpha\psi^{n+2/3} . \end{cases}$$

If we put $\Phi(\psi) = \mathscr{L}^{-1}F'(\psi)$ it is clear that convergence of (I.86) requires that the spectrum of $(1-\alpha)I + 2\alpha\Phi(\psi) - \alpha\Phi^2(\psi)$ be contained in the unit ball. This is possible if 1 is not an eigenvalue of $\Phi(\psi)$ and if α is taken sufficiently small. M. SERMANGE has shown that if the complexification of $\Phi(\psi)$ is compact with real eigenvalues different from 1, and if α is sufficiently small, then the algorithm (I.86) is convergent.

In order to adapt the Marder–Weitzner algorithm to our problem it is enough to define the following algorithm *AMW* based on *AP2* :

$$u^n \overset{(I.82)}{\rightarrow} u^{n+1/3} \overset{(I.82)}{\rightarrow} u^{n+2/3} \rightarrow u^{n+1} = (1-\alpha)u^n + 2\alpha u^{n+1/3} - \alpha u^{n+2/3} .$$

The main inconvenience with this algorithm is the slowness of its convergence.

I.4.4 Algorithms of Newton type

I.4.4.1 Newton Algorithm *AN* :

Newton–Raphson iterations for the model problem (I.76) have the following form for linear :

$$(\mathscr{L}-F'(\psi^n))\psi^{n+1} = F(\psi^n)-F'(\psi^n)\psi^n \qquad \text{(I.87)}$$

and for $\mathscr{L}$ nonlinear :

$$(\mathscr{L}^n+\mathscr{L}'^n - F'(\psi^n))\psi^{n+1} = F(\psi^n) + \mathscr{L}'^n\psi^n - F'(\psi^n)\psi^n. \qquad \text{(I.88)}$$

Thus we are led to use the linearization of the discrete problem $(P_I'^{\ell})$. To do this we refer to the calculation of the linearization of the problem (P_I') carried out in I.3. In order to apply Newton's algorithm it is necessary that the linearization of (P_I') should exist and be invertible for each of the successive iterates ψ^n. Let us examine first of all what becomes of the hypotheses H1, H2 and H3 in the case of the discrete problem and the finite element formulation P_1 of I.4.1. We use the fact that ψ is linear and $\nabla\psi$ constant in each triangle; moreover D_ℓ either consists of a finite number of vertices of the triangulation $\mathscr{C}_\ell$ or is the union of a finite number of edges of triangles of $\mathscr{C}_\ell$. The discrete versions of hypotheses H1, H2, H3 thus take the form:

H_1') $\sup_{D_\ell}\psi$ is attained at a unique vertex of $\mathscr{C}_\ell$, which is the point of contact between Γ_p and D_ℓ.

H_2') $\sup_{\Omega_p}\psi$ is attained at a unique vertex M_1 of $\mathscr{C}_\ell$, which is in the interior of Ω_p and is the magnetic axis.

H_3') $\nabla\psi$ does not vanish on Γ_p.

A consequence of hypothesis H_3' is that Γ_p is a polygonal line in Ω_v. The hypotheses H_1', H_2', H_3' are satisfied <u>except</u> if there is equality between the values of ψ at certain particular vertices of the grid.

Suppose now that the solution to the discrete problem $(P_I'^{\ell})$ satisfies the hypotheses H_1', H_2', H_3', and that the matrix of the linear system is invertible. In this case, because of the finite element method chosen these hypotheses are still satisfied in a neighbourhood of the solution in the space V_ℓ; the linearized operator exists and is invertible in this neighbourhood, allowing Newton's method to be used. This is not true for the continuous problem for which we have differentiability at the solution to the problem but not in a neighbourhood.

Using the equations (I.52) we can write the Newton algorithm in the following form :

We take $(\psi^0, \lambda^0) \in V_\ell \times R$, and given $u^n = (\psi^n, \lambda^n)$ we compute $u^{n+1} = (\psi^{n+1}, \lambda^{n+1})$ by :

$$\text{(I.89)} \quad \begin{cases} b_{u^n}(\psi^{n+1},\varphi) - \lambda^{n+1} c_{u^n}(\varphi) = \sum_{i=1}^{k} \frac{I_i}{S_i} \int_{B_i} \varphi \, dS \\ \qquad + \int_{\Omega_{cv}} j_{cv} \, \varphi \, dS + a'_{u^n}(\psi^n,\varphi), \ \forall \varphi \in V_\ell \\ d_{u^n}(\psi^{n+1},1) + \lambda^{n+1} c_{u^n}(1) = I_p \end{cases}$$

since $e_{\psi^n}(\psi^n) = 0$ and $d_{u^n}(\psi^n, \varphi) = 0$.

The two equations of (I.89) constitute a linear system whose unknowns are the N_ℓ values of ψ^{n+1} at the vertices of the mesh and the scalar λ^{n+1}. The matrix K^n for this system decomposes into the sum of a symmetric matrix K_0^n and a matrix K_1^n of rank less than or equal to four. The matrix K_1^n contains the two columns corresponding respectively to the vertices M_0^n and M_1^n as well as the last column and the last row of K^n corresponding respectively to the unknown λ^{n+1} and to the last equation of (I.89). To invert the linear system associated to K^n we can factorize K^n in such a way as to lead to the inversion of linear systems with matrix K_0^n that can be solved by a Crout method. To do this we use the formula of Sherman–Morrisson or of Woodbury (cf. A.S. HOUSEHOLDER) :

$$\text{(I.90)} \qquad (K_0+u\tilde{v})^{-1} = [I - \sigma K_0^{-1} u\tilde{v}]K_0^{-1}$$

with $\sigma = (1+\tilde{v}\, K_0^{-1}u)^{-1}$ where u and v are column– vectors with $\tilde{v}$ the transpose of v, so that $u\tilde{v}$ is a matrix of rank 1, and I is the identity matrix. Formula (I.90) implies that if we seek X such that

$$(K_0 + u\tilde{v})X = B$$

we compute Z such that $K_0 Z = B$ and w such that $K_0 w = u$ and then we have :

$$X = Z - \sigma(\tilde{v}Z)w$$

with $\sigma = (1 + \tilde{v}w)^{-1}$.

To invert the system associated to $K_0^n + K_1^n$ we use this factorization process four times, successively adding to K_0^n the two columns associated to the points M_0^n and M_1^n, the last row and the last column of the matrix. Once the matrix K_0^n is decomposed into $LD\tilde{L}$, we have to perform five solutions of linear systems associated to K_0^n. The majority of the computation time for each iteration is devoted to the decomposition $LD\tilde{L}$ of the matrix K_0^n.

I.4.4.2 *Quasi–Newton Algorithm* *AQN*

The algorithm for the model problem (I.76) takes the following form for $\mathscr{L}$ linear :

(I.91) $$(\mathscr{L} - F'(\psi^0))\psi^{n+1} = F(\psi^n) - F'(\psi^0)\psi^n$$

and for $\mathscr{L}$ nonlinear

(I.92)
$$(\mathscr{L}^0 + \mathscr{L}'^0 - F'(\psi^0))\psi^{n+1} = F(\psi^n) - \mathscr{L}^n\psi^n + (\mathscr{L}^0 + \mathscr{L}'^0 - F'(\psi^0))\psi^n.$$

This consists of taking the linearization at the initial value of ψ in order that the operator $(\mathscr{L}^0 + \mathscr{L}'^0 - F'(\psi^0))$ which has to be inverted should be the same throughout all the iterations, but in this case the quadratic convergence of Newton's method is replaced by a superlinear convergence.

For the problem which interests us, we take $(\psi^0, \lambda^0) \in V_\ell \times R$ satisfying the hypotheses H_1', H_2', and H_3'.

From $u^n = (\psi^n, \lambda^n)$ we compute $u^{n+1} = (\psi^{n+1}, \lambda^{n+1}) \in V_\ell \times R$ by :

(I.93)

$$\begin{cases} b_{u^0}(\psi^{n+1},\varphi) - \lambda^{n+1} c_{u^n}(\varphi) = \sum_{i=1}^{k} \frac{I_i}{S_i} \int_{B_i} \varphi \, dS + \int_{\Omega_{cv}} j_{cv} \, \varphi \, dS \\ \qquad + b_{u^0}(\psi^n,\varphi) - a_{\mu^n}(\psi^n,\varphi), \; \forall\varphi \in V_\ell \\ d_{u^0}(\psi^{n+1},1) + \lambda^{n+1} c_{u^n}(1) = I_p + d_{u^0}(\psi^n,1) \; . \end{cases}$$

This algorithm (I.93) is intermediate between the Newton algorithm (I.89) and a standard quasi– Newton algorithm of type (I.92), in that the term $c_{u^n}(\varphi)$ is recalculated at each iteration.

The linear system (I.93) is solved by using the factorization (I.90), as in the Newton algorithm *AN*, the symmetric matrix K_0 being computed and triangularized once and for all at the first iteration. Only the last column of the matrix K, namely the vector $(c_{u^n}(\varphi_i)\ (i \leqslant i \leqslant N_\ell),\ c_{u^n}(1))$ is recalculated at each iteration. In practice the matrix K_0 has been recalculated every five iterations.

These various algorithms (*AP1, AP2, AMW, AN, AQN*) will now be compared on the following examples of equilibria in the Tokamaks TFR, JET, TORE Supra and INTOR.

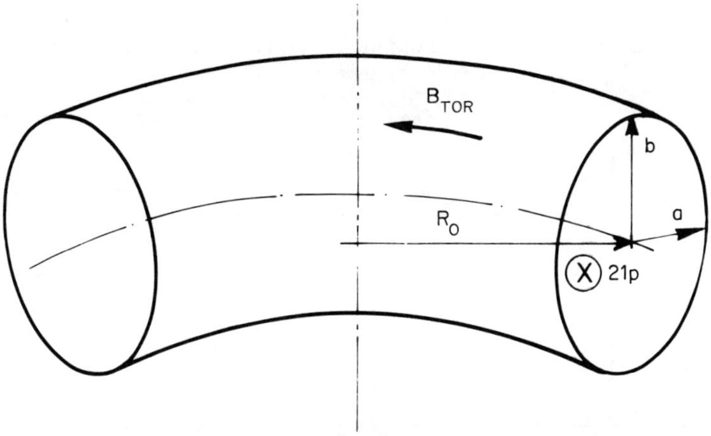

Figure I.8 : Characteristic parameters for the plasma in a Tokamak

I.5 NUMERICAL RESULTS. APPLICATIONS TO TFR, JET, TORE supra AND INTOR :

The equilibrium configurations studied in this section concern four Tokamaks :

. The Tokamak TFR of Fontenay aux Roses.

. The JET (Joint European Torus) at Culham (Great Britain).

. TORE supra, under construction at Cadarache.

. INTOR (International Tokamak Reactor), the international reactor project.

The following test cases were obtained by means of the code SCED (Self–Consistent for Equilibrium and Diffusion) using the numerical methods of I.4 (cf. J. BLUM, J. LE FOLL, B. THOORIS [1]).

I.5.1 Characteristic parameters for a Tokamak and its plasma :

The parameters which characterize a Tokamak are (see Fig I.8).

. the major radius R_0 of the torus.

. the minor horizontal radius a of the plasma and the minor vertical radius b

. the toroidal field B_0 at the centre of the plasma

. the total plasma current $(2I_p)$.

From these we derive the aspect ratio R_0/a of the machine, describing its more or less compact character, and the elongation e = b/a of the plasma.

The characteristics of the four Tokamaks considered here are given in the table below:

	R_0(m)	a(m)	b(m)	R_0/a	e = b/a	B_0 (T)	$2I_p$(MA)
TFR	0.98	0.24	0.24	4.1	1.	6.	0.6
JET	2.96	1.25	2.10	2.4	1.7	2.8	3.8
TORE Supra	2.25	0.75	0.75	3.	1.	4.5	1.7
INTOR	5.20	1.30	2.10	4.	1.6	5.5	6.4

Table I.1

The poloidal system is made up of a set of coils creating with the plasma the poloidal field B_p and playing the following roles: producing magnetizing current, controlling the plasma position, optimizing its shape and possibly creating a divertor. The coupling between these poloidal field coils, which constitute the primary of the transformer, and the plasma which is its secondary can be ensured either by a magnetic circuit (Fig I.2a) or through the air. Iron has the advantage of being more economical in Ampère–turns of magnetization and of channelling the flux into well–defined regions of space, but it makes controlling the plasma more difficult, as we shall see later. Air allows a more compact machine, which is favourable from the theoretical point of view, but couples the primary and the plasma less efficiently and creates parasitic fields. In the case of iron transformers the variation of available flux in the iron core is a determining factor for the long–term evolution of equilibria. The characteristics of the poloidal field system of the four Tokamaks studied here are as follows :

	Nature of the transformer	Variation of flux in the core	Number of types of poloidal coils	Divertor
TFR	iron	2 Wb	2	No
JET	iron	25 Wb	4	No
TORE Supra	iron	14 Wb	5	No
INTOR	air	--	11	Yes

Table I.2

The data necessary for simulating an equilibrium are, apart from the geometry of the Tokamak and the function $\bar{\mu}(B_p^2)$, the currents I_i in the coils B_i, the total plasma current I_p, the parameter β and the function g characterizing the plasma current density.

In this Section the current density j_{cv} in the vacuum vessel will be taken equal to zero. *The currents I_i and I_p correspond to the integral of the current density over the part of the coils and plasma situated in the upper half of the torus; to obtain the total currents in the coil and plasma these data should be multiplied by two.*

I.5.2 Characteristic Results of a Simulation :

The results of the simulation are the poloidal flux $\psi(r,z)$, and the boundary Γ_p of the plasma. Knowing the function $\psi(r,z)$ we can calculate a certain number of the characteristic parameters of the equilibrium configuration, namely :

. the level of saturation of the iron

. the parameters β_p and ℓ_i characterizing the plasma pressure and its current density profile respectively

. the radial position of the plasma, and its elongation.

The level of saturation of the iron can be characterized by the flux in the iron core:

$$\psi_F = 2\pi\psi(F) \tag{I.94}$$

where F is the point in the median plane marking the limit of the iron core (see Fig I.4). The quantity ψ_F is a result of each simulation, but we shall see in Chapter VII that ψ_F depends mainly on the Ampère– turns, i.e. the algebraic sum of the currents in the coils, the vacuum vessel and the plasma. Given the algebraic sum of the Ampère– turns we can decide the approximate level of saturation of the iron for the case under consideration.

The mean "beta poloidal" coefficient β_p and the internal inductance ℓ_i of the plasma will enter in a fundamental way into the calculation of equilibrium fields in Chapter VII. The quantity β_p is defined as the ratio of the mean kinetic pressure $\bar{p}$ of the plasma to its magnetic pressure $\bar{B}_p^2/2\mu_0$, where $\bar{B}_p$ denotes the mean poloidal field on the boundary of the plasma :

$$\beta_p = \frac{2\mu_0 \bar{p}}{\bar{B}_p^2} \tag{I.95}$$

with

$$\bar{p} = \frac{\int_{\Omega_p} p(\psi) r \, dS}{\int_{\Omega_p} r \, dS} , \quad \bar{B}_p = \frac{\int_{\Gamma_p} B_p d\ell}{\int_{\Gamma_p} d\ell} = \frac{\mu_0 I_p}{L_o} ,$$

where $p(\psi)$ is given by (I.24) and L_o is the length of the arc Γ_p . The quantity β, identified in Remark I.2 as the ratio of the kinetic pressure to the magnetic pressure at $r = R_o$, is a first approximation to β_p, and the two quantities, one given a priori and the other calculated a posteriori, will be compared for each case.

The internal inductance ℓ_i of the plasma per unit length is defined from its magnetic energy by :

$$\ell_i = \frac{4\pi}{\mu_0^2 \bar{R} I_p^2} \int_{\Omega_p} B_p^2 r \, dS \tag{I.96}$$

where $\bar{R}$ is the mean major radius of the plasma defined by :

$$\bar{R} = \frac{\int_{\Omega_p} r \, dS}{\int_{\Omega_p} dS} . \tag{I.97}$$

The quantity ℓ_i characterizes the "peakedness" nature of the current density profile (for a flat current $\ell_i = 0.5$, and ℓ_i is greater the more the current profile is spiked).

We next define the radial position Δ_H of the plasma. The boundary Γ_p cuts the Or– axis in general at two points M_i and M_e with coordinates R_i and R_e ; let I be the mid– point of the segment M_iM_e (see Fig I.9). The intersection of the limiter D with the Or– axis in general consists likewise of two points J_o and K_o ; let I_o be the mid– point of the segment J_oK_o ; and R_o the abscissa of I_o. The algebraic measure $\overline{I_oI}$ on the Or– axis represents the horizontal displacement of the plasma relative to the limiter and is denoted by Δ_H :

$$\Delta_H = \frac{R_i+R_e}{2} - R_o \ . \tag{I.98}$$

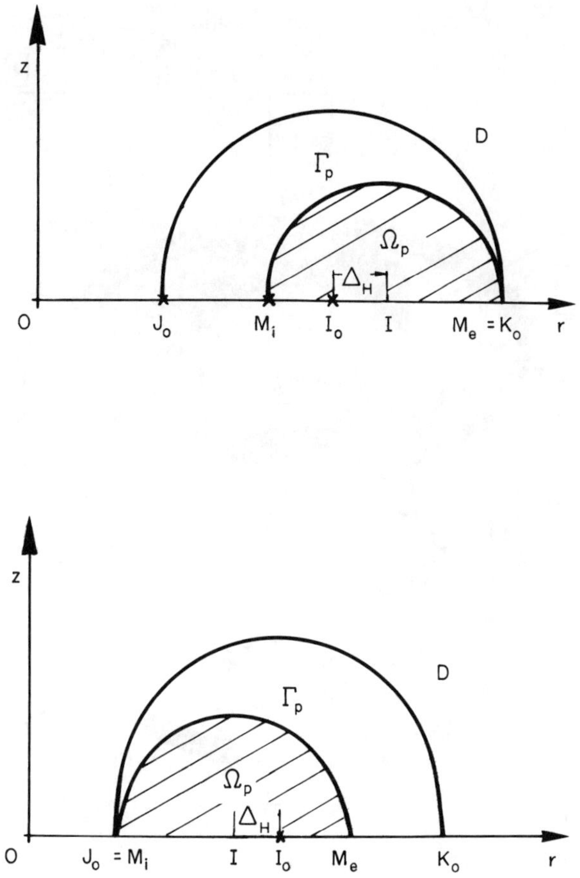

Figure I.9 : *Definition of the radial position* Δ_H *of the plasma*

If the point of contact between Γ_p and D is in the median plane, the quantity Δ_H is positive if the plasma touches D at the exterior point K_0 (Fig.1.9a) and is negative if it touches D at the interior point J_0 (Fig. I.9b). The major radius R of the plasma and is minor radii a (horizontal) and b (vertical) as well as its elongation e are then defined as follows :

$$
(I.99) \quad \begin{cases} a = \dfrac{R_e - R_i}{2} \quad , \quad b = \sup\limits_{M \in \Gamma_p} z(M) \\[2ex] R = \overline{OI} = R_0 + \Delta_H \; , \; e = \dfrac{b}{a} \quad . \end{cases}
$$

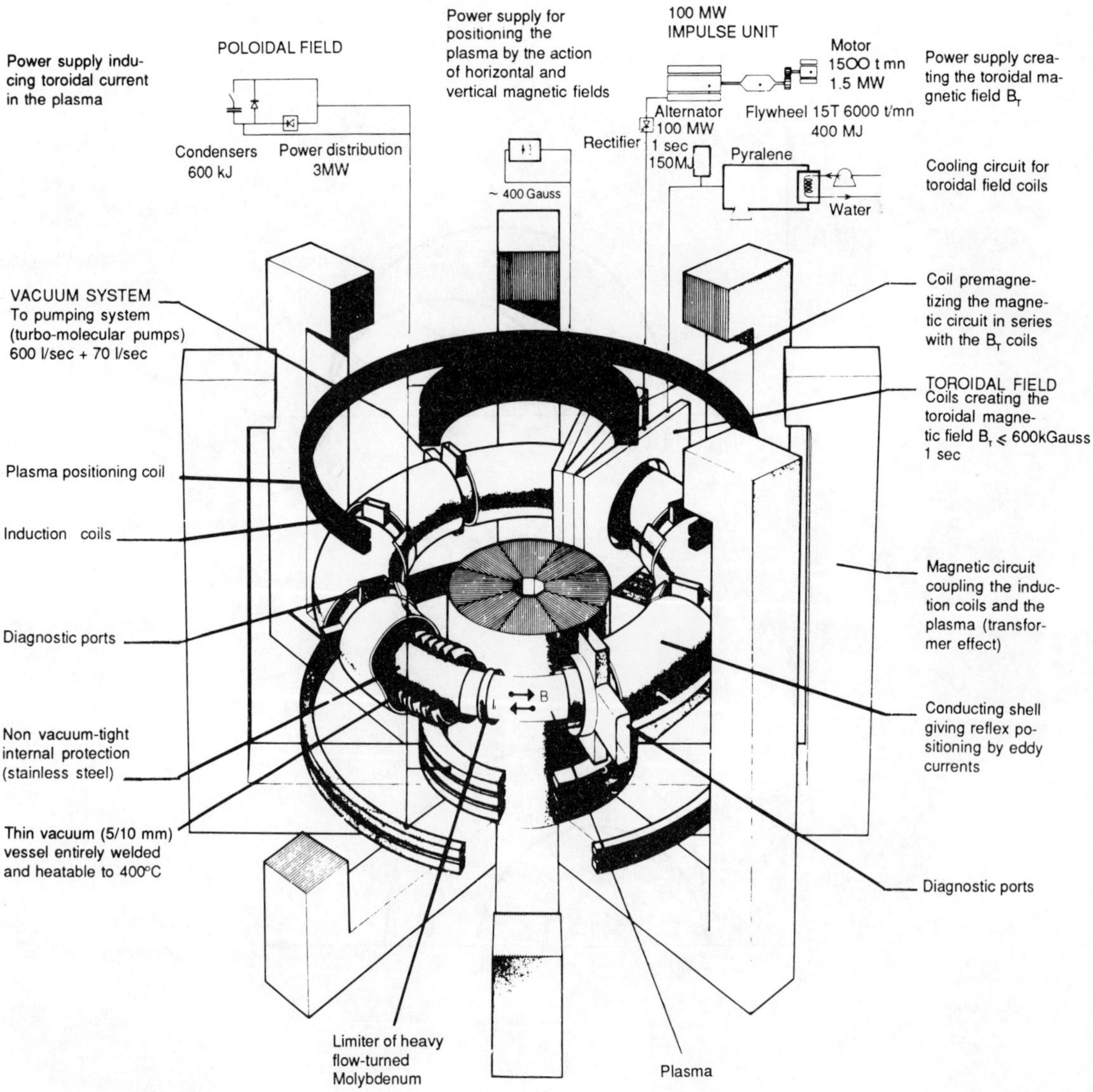

Figure I.10 : The Tokamak TFR.

I.5.3 The Tokamak TFR (cf. EQUIPE TFR)

Figure I.10 represents a model of the Tokamak TFR at Fontenay aux Roses. The magnetic circuit consists of a cylindrical iron core with circular section and eight return outer limbs. These arms destroy the axisymmetric character of the configuration. The coils of the poloidal circuit are of two types : interior coils positioned against the iron core and exterior coils positioned against the outer limbs. These coils play the role of primary coils (to induce the plasma current), serve to pre program the vertical field to ensure equilibrium of the plasma, and effect a feedback to displacements of the plasma. The fact that two types of coil alone have to satisfy so many demands is the origin of difficulties in controlling the equilibrium and displacements of the plasma in TFR that we study in Chapter VII.

Figure I.11 represents the triangulation of a meridian section of TFR containing 1600 nodes, 3000 triangles and used for solving problem (P_I') by the finite element method. The axisymmetric equivalent of the iron outer limbs is represented here. There are two air gaps, i.e. zones of air that lie between two pieces of iron, one in the core and the other in the arms. The limiter is represented by a semicircle of radius 24 cm. The function $\mu_r(B_P)$ representing the relative magnetic permeability $(\mu = \mu_0\mu_r)$ as a function of the modulus of the poloidal field is shown in Figure I.12. The function $\bar{\mu}(B_P^2)$ can easily be deduced from this.

We shall now compare the various numerical methods defined in the preceding section for a certain number of equilibrium configurations of TFR. We consider four configurations whose data are represented in Table I.3.

Configuration No	I_p (kA)	I_1 (kA)	I_2 (kA)	α	β	γ	δ	ζ
1	100	-2	-100	1.5	1.	3.	0	1.
2	100	3	-105	1.5	1.	3.	0	1.
3	100	-41.5	- 73.5	1.5	0.25	3.	0	1.
4	100	-23.6	- 78.4	1.	1.	1.	1.	0

Table I.3

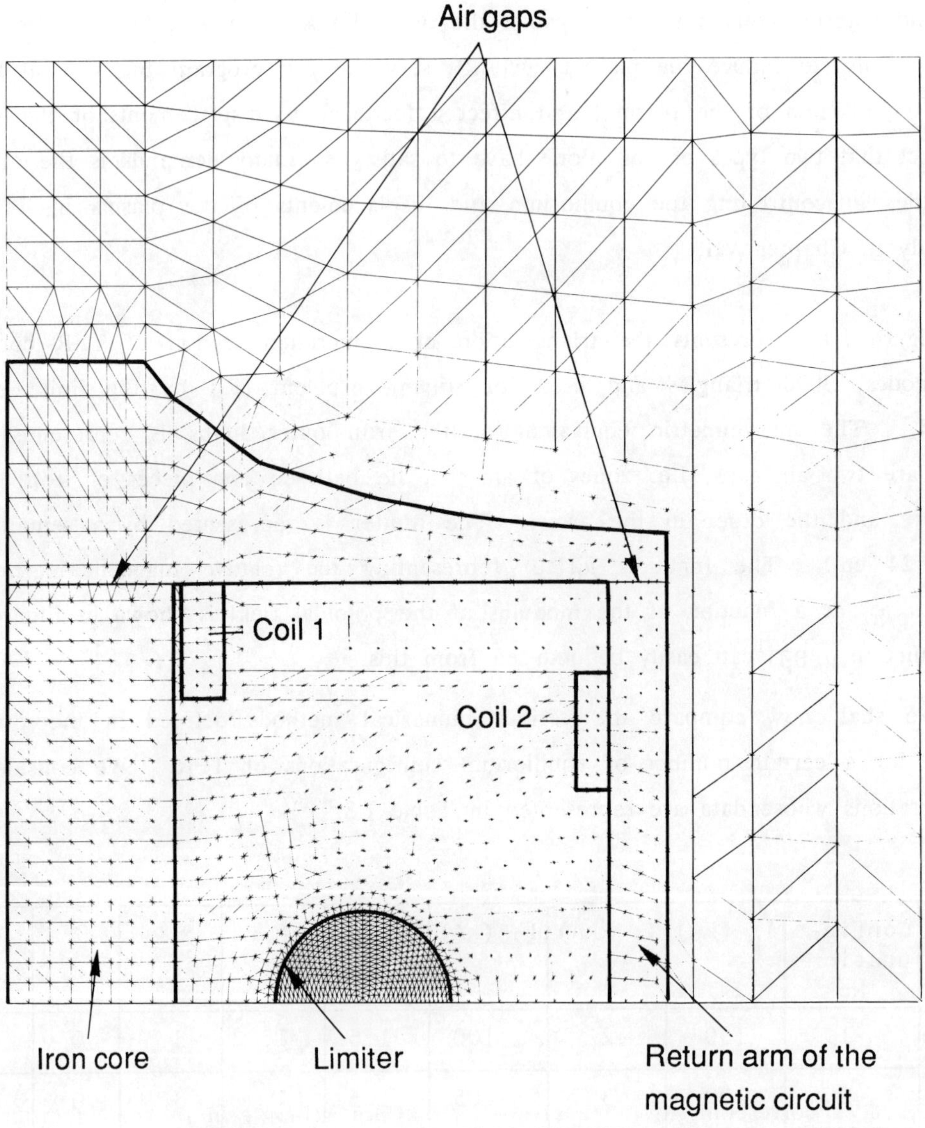

Figure 1.11 : Triangulation of a meridian section of TFR.

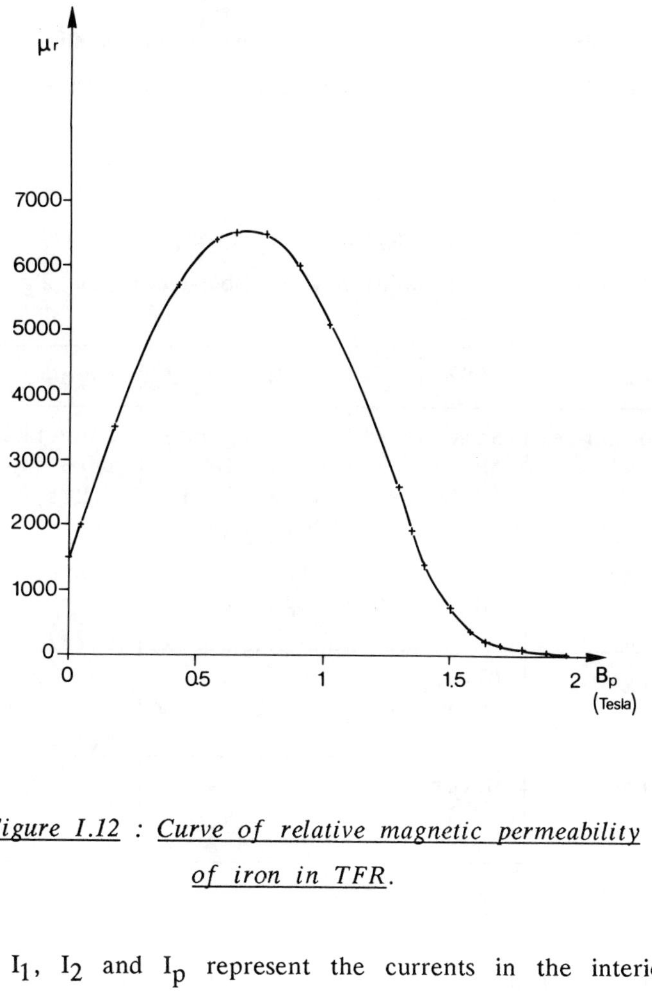

Figure I.12 : Curve of relative magnetic permeability of iron in TFR.

The currents I_1, I_2 and I_p represent the currents in the interior coils, exterior coils and in the plasma, respectively, for the upper half of the torus. The coefficients α, β, γ, δ, ζ correspond to the expression (I.32) for $g(\psi_N)$. In the configurations 1, 2 and 3 the function $g(\psi_N)$ is thus

$$g(\psi_N) = (1-\psi_N^{1.5})^3 .$$

This expression is chosen as a generalization of the dependence of the current density on $(1-\rho^2/a^2)^3$ deduced from experimental observations (cf. Remark I.3).

In configuration 4 we have the case of a "flat" current density profile not vanishing on Γ_p and with $g(\psi_N)$ equal to 1.

The test for convergence of the various numerical algorithms is :

$$\varepsilon_o = \frac{\sum_{j=1}^{N_\ell} \left| \psi^{n+1}(M_j) - \psi^n(M_j) \right|}{\sum_{j=1}^{N_\ell} \left| \psi^n(M_j) \right|} < 10^{-5} .$$

The algorithms are initiated starting from a plasma which is slightly displaced inwards ($\Delta_H < 0$) for cases 1 and 4, and slightly displaced outwards ($\Delta_H > 0$) in cases 2 and 3.

Table I.4 gives the number of iterations necessary for convergence according to the numerical method in the case considered, with corresponding calculation time on the CDC 7600, as well as the spectral radius $\rho(M_u)$ associated to the algorithm *AP2*.

Case no.	AP1	AP2	AMW	AN	AQN
1	Convergence very slow after 12 it. (55s) $\varepsilon_o=10^{-2}$	Same as AP1 $\rho(M_u)=0.992$	Convergence very slow after 25 it. (3mm) $\varepsilon_o=10^{-3}$	Converges in 6 it. (27s)	Di-verges
2	Diverges	Diverges $\rho(M_u)= 1.08$	after 25 it $\varepsilon_o = 10^{-3}$	Converges in 5 it. (23s)	Di-verges
3	Diverges	Diverges $\rho(M_u) = 1.1$	Same as cases 1 and 2	Converges in 5 it. (23s)	Di-verges
4	Diverges	Diverges $\rho(M_u)=1.005$	Same as cases 1 and 2	Converges in 6 it. (27s)	Di-verges

Table I.4

Observe that the algorithms of Picard type (*AP1* and *AP2*) either diverge or converge extremely slowly (this being explained by the fact that $\rho(M_u)$ is very close to 1). The quasi– Newton algorithm *AQN* diverges because the starting solution is too far from the sought– for equilibrium. Only the Newton algorithm gives rapid convergence in all cases. In Chapter IV we shall see a classification of the equilibria of TFR by the ratio $I_1/(I_1+I_2)$. For the present we can describe the principal characteristics of the equilibria obtained by Table I.5 :

Case No.	ψ_F(Wb)	β_P	ℓ_i	Δ_H(cm)	a (cm)	b (cm)	e
1	0.069	0.997	1.54	-9.63	14.37	13.9	0.965
2	-0.014	0.999	1.55	+3.	21.	19.5	0.927
3	-0.86	0.25	1.57	+0.1	23.9	21.8	0.913
4	-0.08	0.995	0.496	-5.	19.	17.3	0.91

Table I.5

The configurations 1, 2 and 4 correspond to an essentially zero flux in the iron core; this is due to the magnetizing current $(I_e + I_i + I_p)$ which is very weak and is -2kA in the three cases. On the other hand, in case 3 the value of $(I_e + I_i + I_p)$ is -30kA which gives rise to the flux of -0.86 Wb in the iron core.

Figure I.13a shows the flux lines corresponding to equidistant ψ values, as well as the boundary Γ_p of the plasma for case 1. Figures I.14a and I.15a show these flux lines for configurations 2 and 3 respectively.

Comparing Tables I.3 and I.5 we observe that β_P is very close to the given value of β, which shows that β is a good approximation to β_P and allows a simple dependence of $j_T(r,\psi)$ on β. The quantity ℓ_i is close to 1.5. in cases 1, 2 and 3 and this corresponds to a sharply peaked current density ; in case 4 the value of ℓ_i is close to 0.5, corresponding to a "flat" current. Figures I.13b, I.14b and I.15b represent the limiter D, the boundary Γ_p of the plasma and 20 equipotentials of ψ inside the plasma and corresponding to equidistant flux, for the respective cases 1, 2 and 3. Observe that in case 1 the plasma is much displaced off centre towards the interior, while in case 2 it is displaced outwards and in case 3 it is almost centred. This corresponds to the Δ_H values given in Table I.5. In the four cases the plasma is quasi-circular with a slight ellipticity tending to flatten it; this corresponds to e values given in Table I.5 and lying between 0.91 and 0.965.

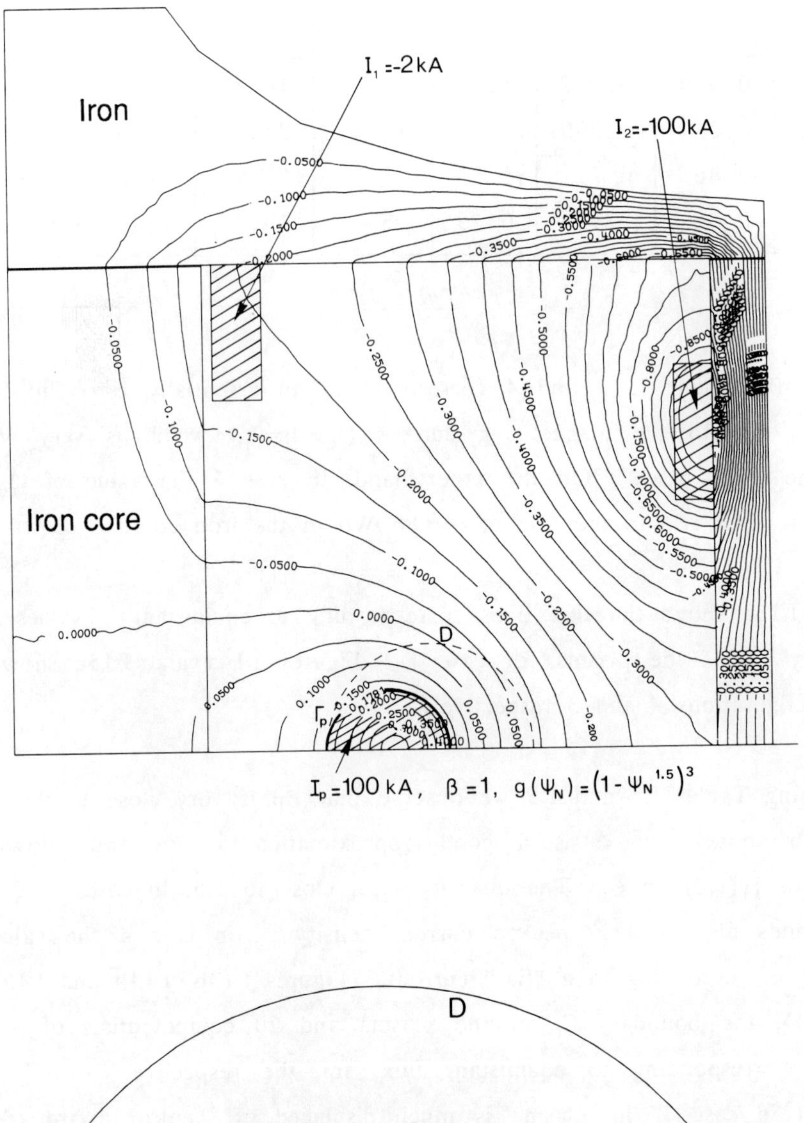

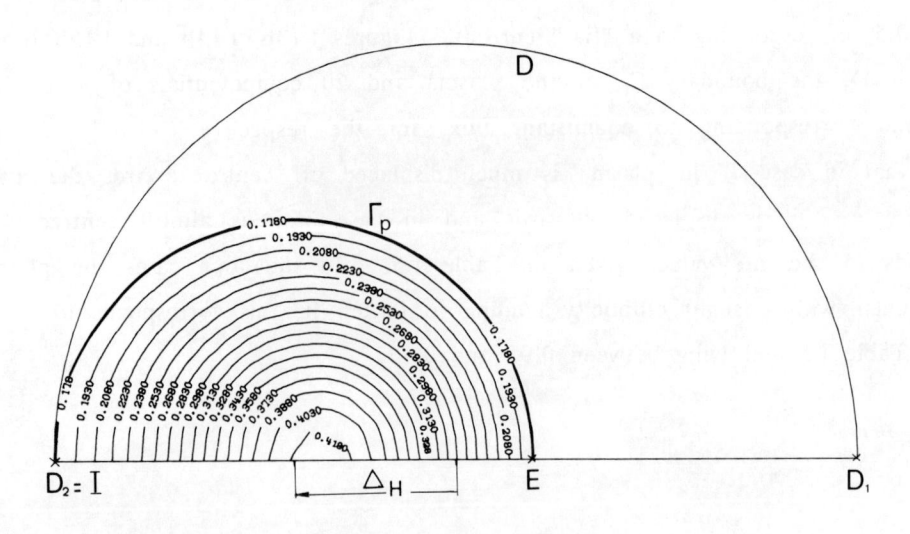

Figure I.13 :

Equilibrium configuration No.1 in TFR

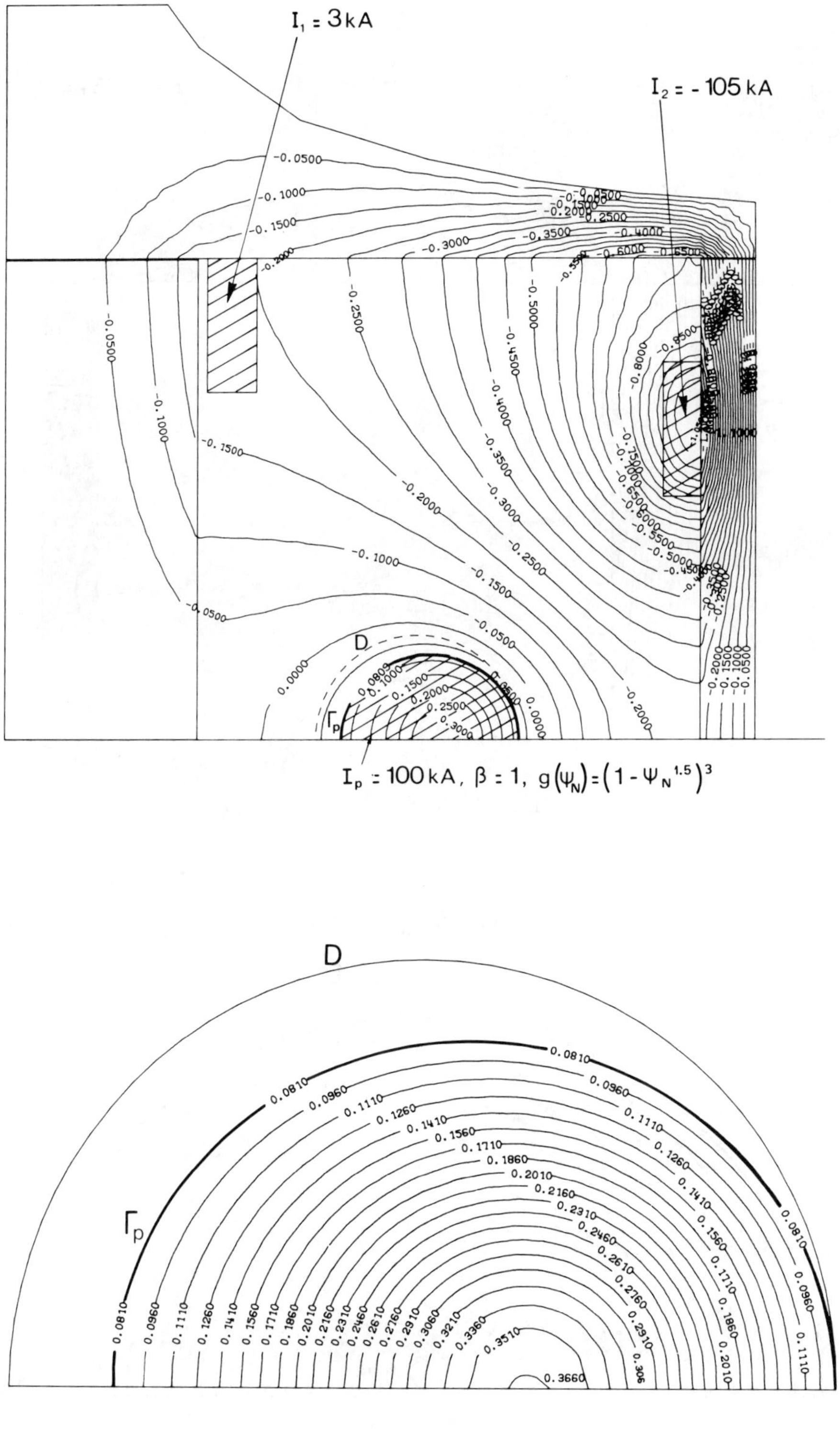

<u>*Figure I.14*</u> :

<u>*Equilibrium configuration No.2 in TFR*</u>

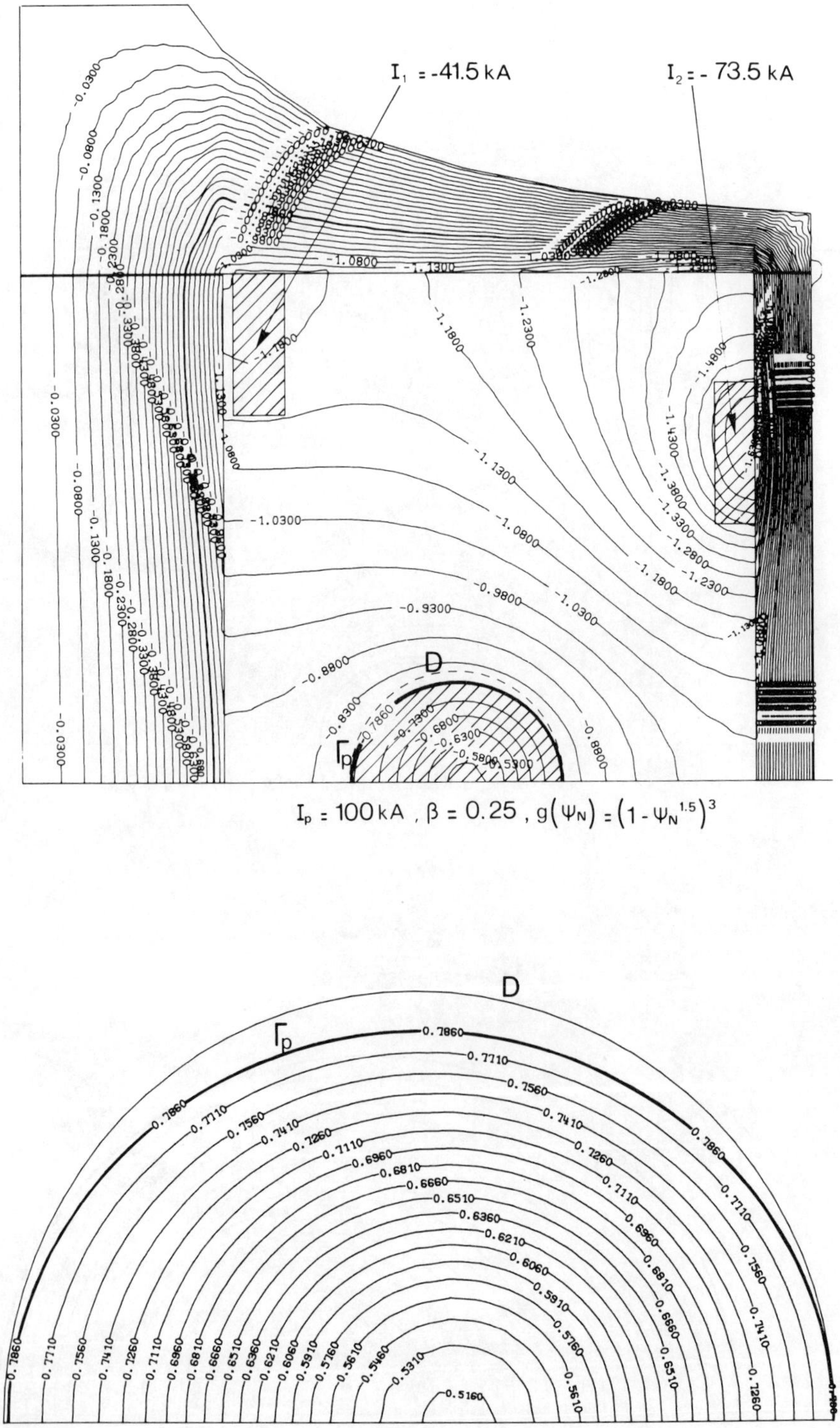

<u>*Figure I.15*</u> :

<u>*Equilibrium configuration No.3 in TFR*</u>

I.5.4 *The JET* *(Joint European Torus) (cf. JET PROJECT) :*

The JET is the European Tokamak functioning at Culham near Oxford, and destined to approach and even attain ignition. Figure I.16 represents a drawing of JET in which it can be seen that, as in TFR, the magnetic circuit consists of an iron core and eight return arms. Figure I.17 shows a meridian section of the torus. The poloidal field system is made up of four types of coils : coil 1, which surrounds the iron core, produces the main magnetizing current ; the sum of the currents in coils 3 and 4 ensures the centring of the radial position of the plasma, while the ratio of their currents imposes a certain shape (circular or D–shaped) on the plasma ; coil 2, whose current is weak for the size of the coil, is divided into two parts : one serves to control the vertical displacements of the plasma and thus does not enter into our problem since we consider only plasmas which are symmetric relative to the horizontal plane, while the other part (the "shaping coil") refines the optimization of the plasma shape. The vacuum chamber Ω_{cv} is D–shaped. The limiter is made of an exterior plate (D_1), but the protections (D_2) tend to prevent the plasma from touching the inside of the vacuum vessel ; therefore we have taken a limiter to consist of the two points D_1 and D_2. The function $\mu_r(B_P)$ describing the relative magnetic permeability of the iron is represented in Figure I.18. An axisymmetric equivalent for the eight iron return arms was found by referring to a three dimensional magnetostatic code.

We now consider the results obtained by the various numerical methods for two test–cases, one corresponding to a plasma with elliptic shape and the other to a circular plasma. The data for the two cases are collected in Table I.6

Test case	I_p(MA)	I_1(MA)	I_2(MA)	I_3(MA)	I_4(MA)	α	β	γ	δ	ζ
1	1.9	-6.14	0	0.105	-1.384	1.	1.5	1.	0	1.
2	1.3	-4.17	0	-0.394	-0.808	1.	2.	1.	0	1.

<u>*Table I.6*</u>

The function $g(\psi_N)$ defined by (I.32) is in these two cases :

$$g(\psi_N) = 1-\psi_N \quad .$$

In both cases the algorithm is started from a plasma slightly off–centre towards the outside and for a convergence test of $\varepsilon_0 < 10^{-5}$ the results obtained are in Table I.7 :

Test case	*AP1*	*AP2*	*AMW*	*AN*	*AQN*
1	diverges	diverges $\rho(M_u)=1.09$	after 25 it. (3mn) $\varepsilon_o=10^{-3}$ very slow convergence	convergence in 4 it. (17s)	diverges
2	converges in 35 it. (2mn)	converges in 35 it. $\rho(M_u)=0.8$	same as case 1	converges in 5 it. (21s)	converges in 6 it. (9s)

Table I.7

Observe that, as in the case of *TFR*, only Newton's algorithm turns out to be convergent and rapid. The characteristics of the equilibrium configurations obtained are given in Table I.8.

Test case	ψ_F(Wb)	β_p	ℓ_i	Δ_H(cm)	a(cm)	b(cm)	e
1	-7.68	1.494	0.907	9.1	117.2	170.	1.45
2	-6.19	1.96	0.986	3.	123.2	126.5	1.027

Table I.8

From the values of ψ_F we observe that the iron is much more saturated than in *TFR* ; β_P is still very close to β, and ℓ_i is of the order of 1. In case 1 the plasma is highly elliptical, while in case 2 it is quasi– circular.

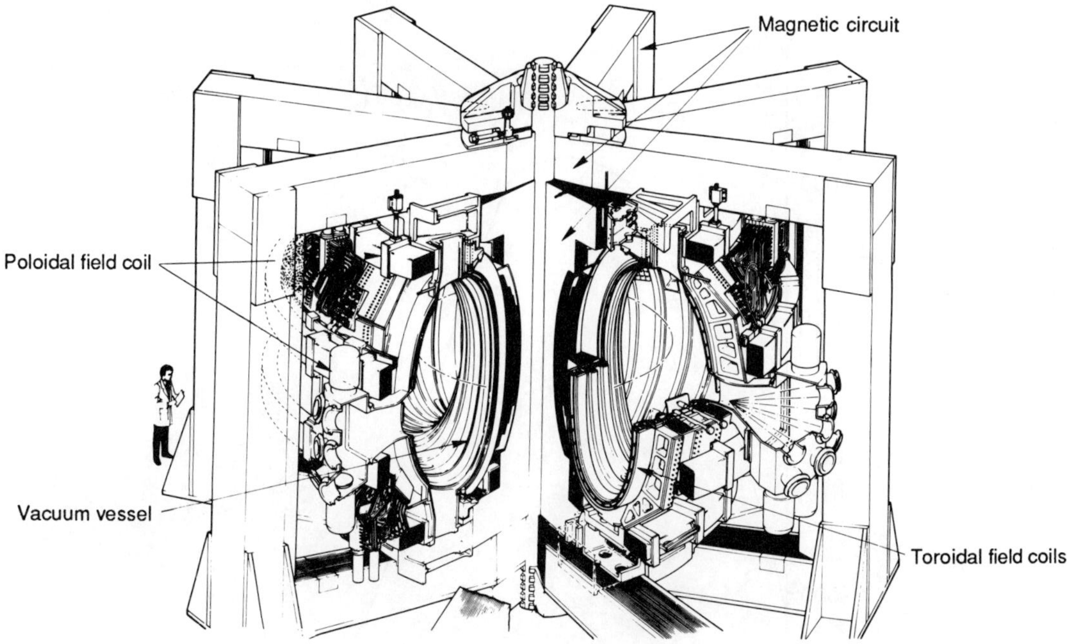

Fig.I.16 : The Tokamak JET

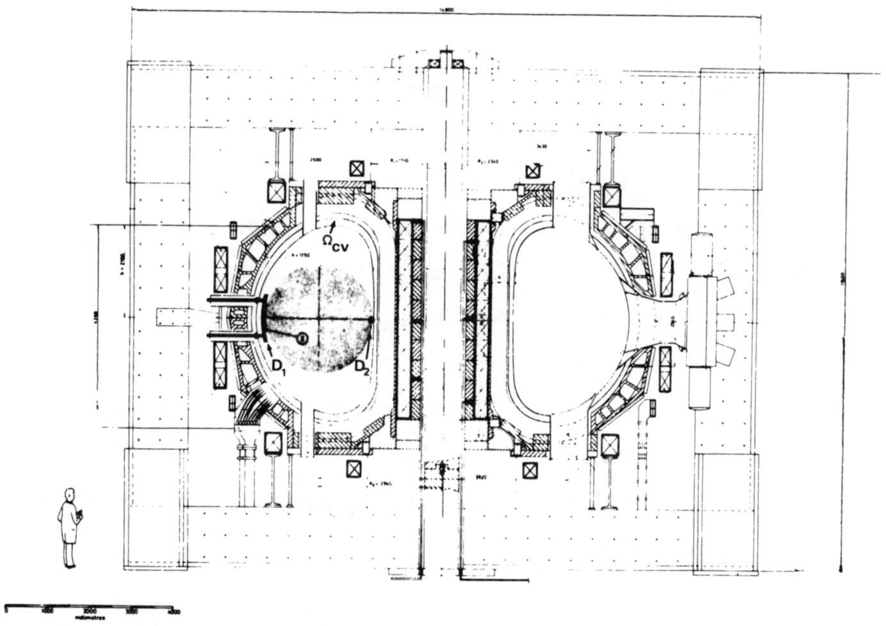

JET : elevation

Figure I.17 : Meridian cross−section of JET.

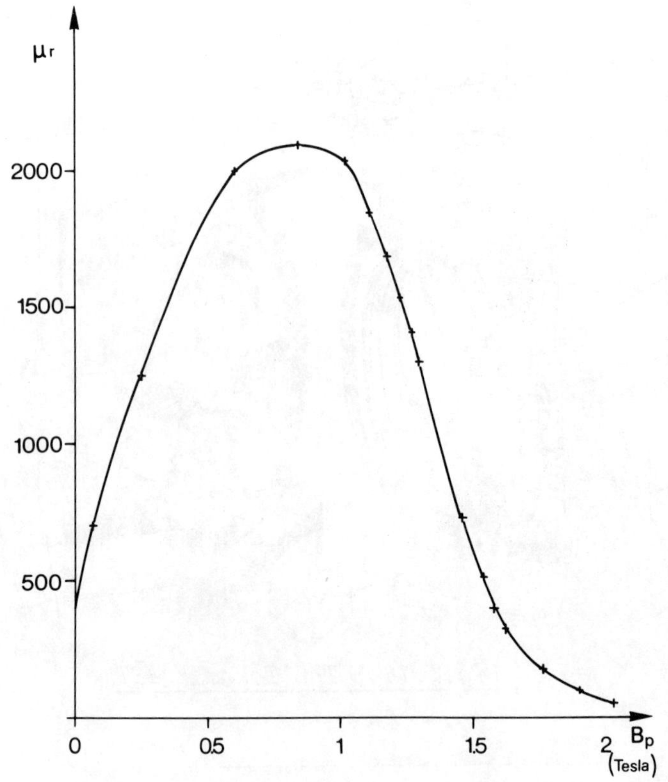

Figure I.18 : *The $\mu_r(B_p)$ curve for JET*

Figures I.19a and I.20a represent the flux lines and the plasma boundary Γ_p in cases 1 and 2 respectively; Figures I.19b and I.20b represent the free boundary Γ_p and 20 lines of flux inside the plasma in the domain Ω_v ; Figures I.19c and I.20c represent the poloidal field B_P, with each arrow proportional to the intensity of B_P. Note that in case 1 the plasma has an elliptical shape, and in case 2 a circular shape. A typical discharge for JET corresponds initially to the formation of a small circular plasma touching the exterior limiter which then grows as long as gas is injected into the chamber and finally fills it, taking a D shape.

a)

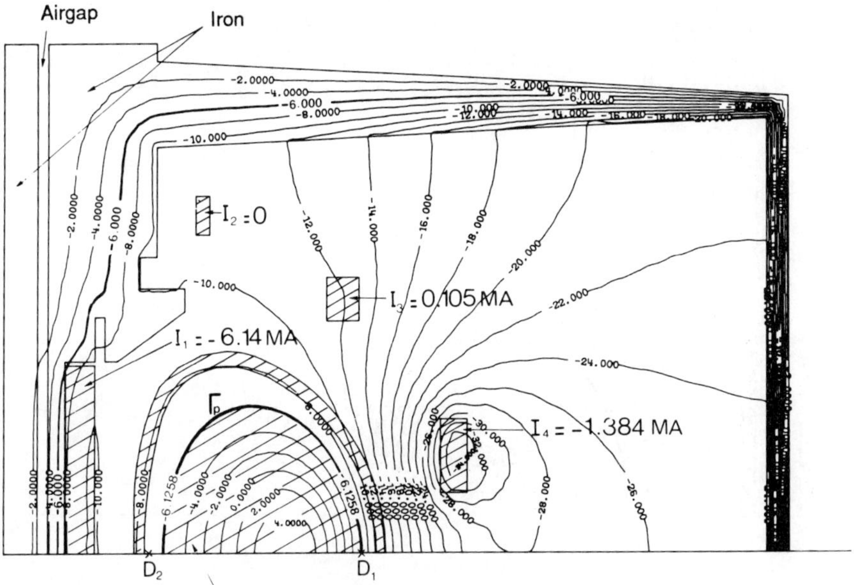

$I_p = 1.9$ MA , $\beta = 1.5$, $g(\Psi_N) = 1 - \Psi_N$

b)

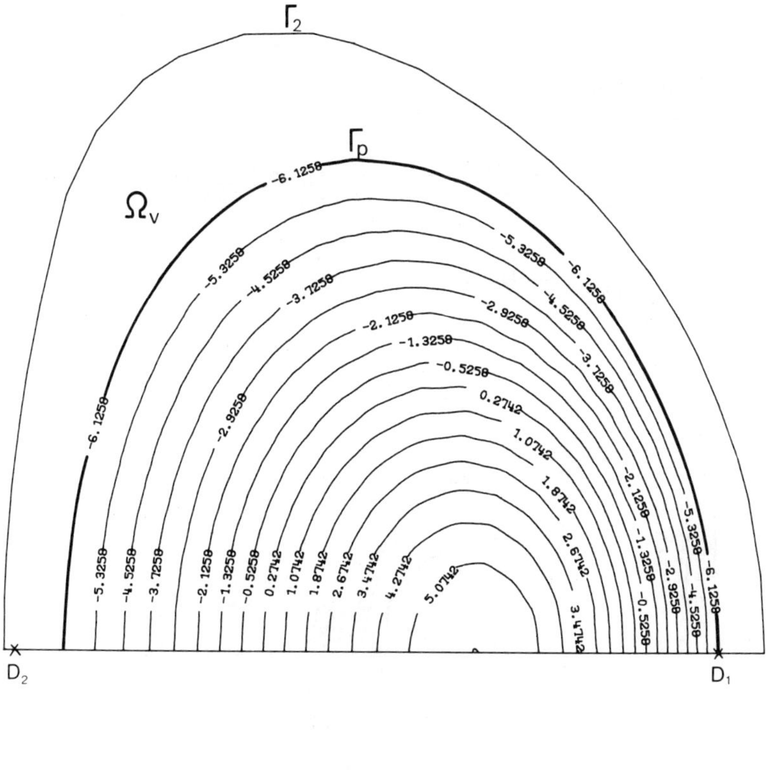

c)

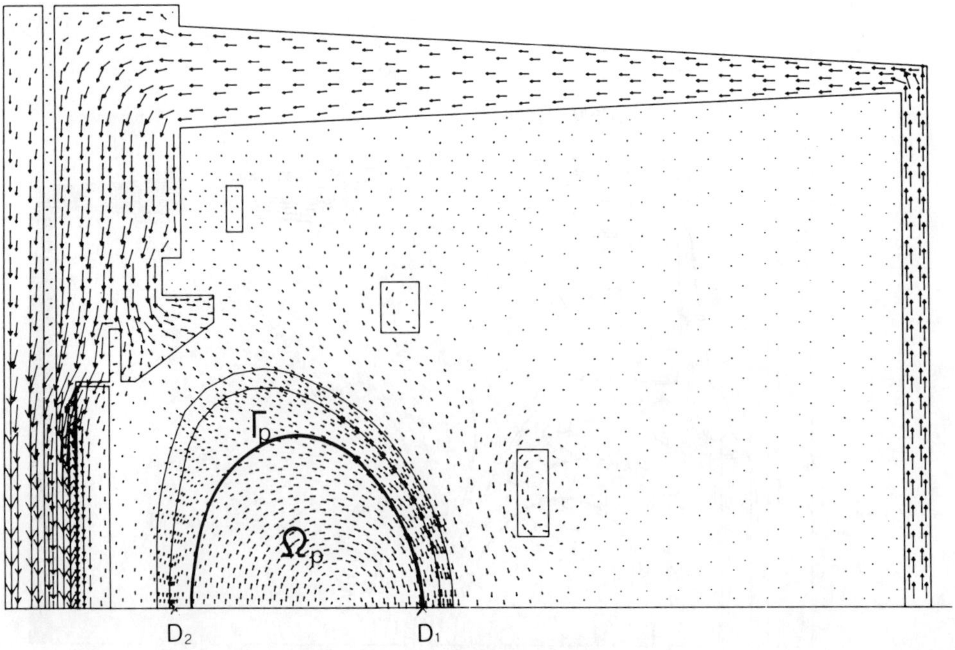

Figures I.19: Equilibrium configuration 1 in JET (elliptical plasma)

a)

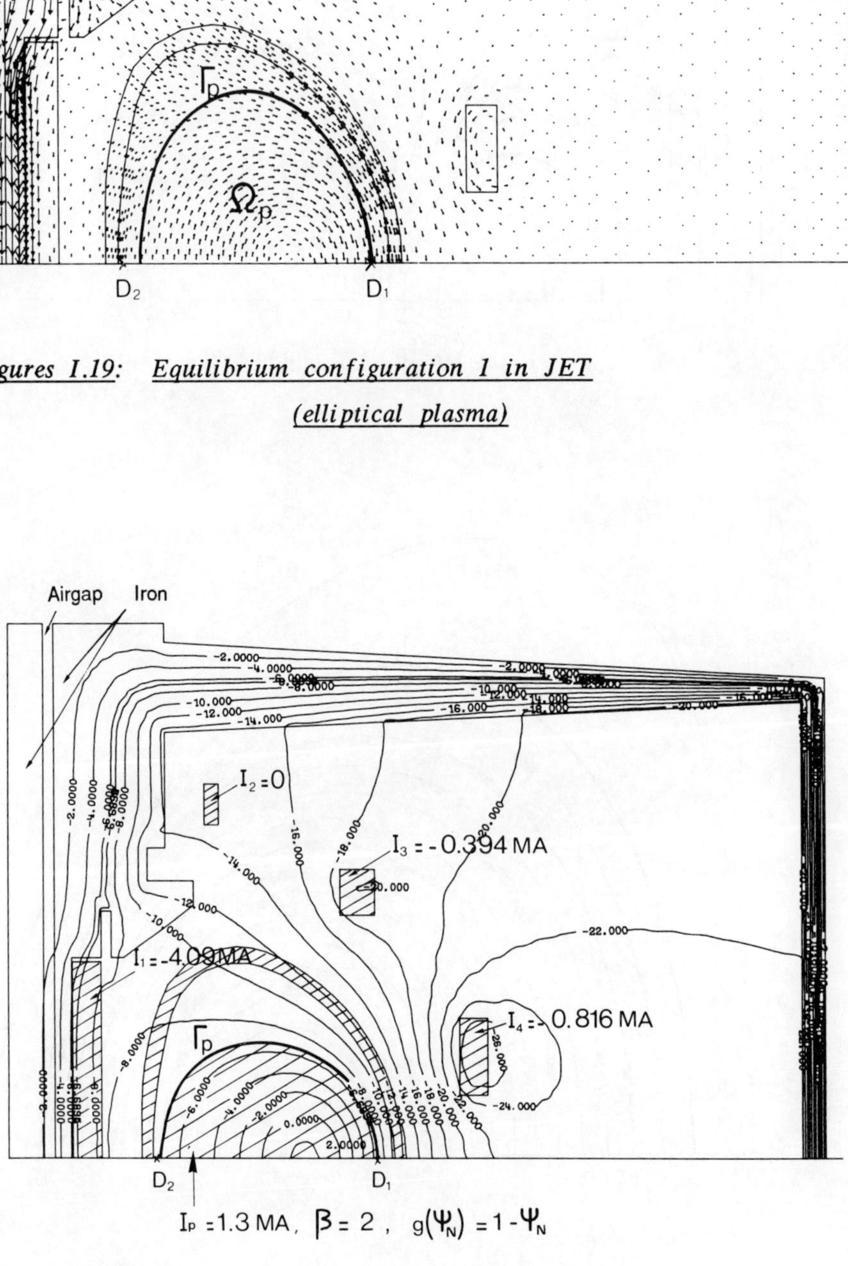

$I_p = 1.3$ MA, $\beta = 2$, $g(\Psi_N) = 1 - \Psi_N$

b)

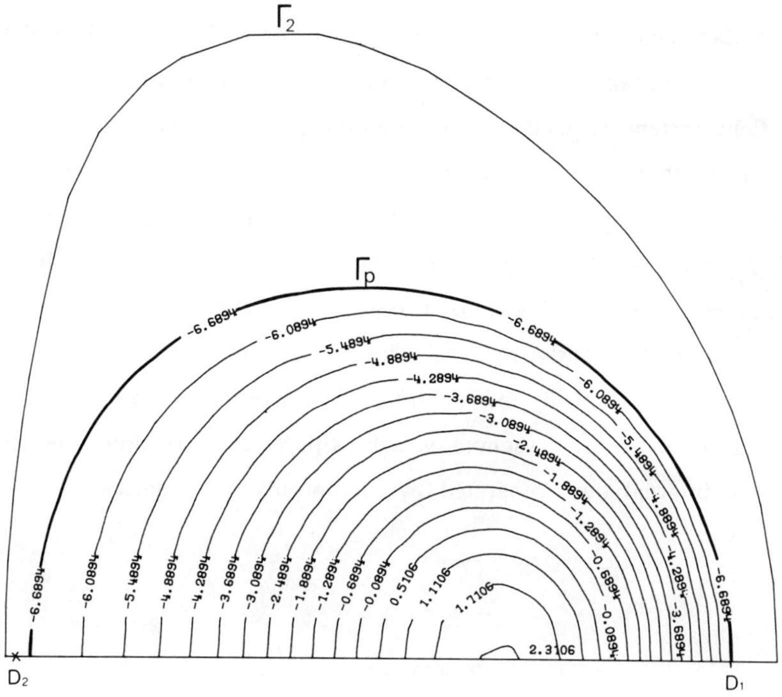

c)

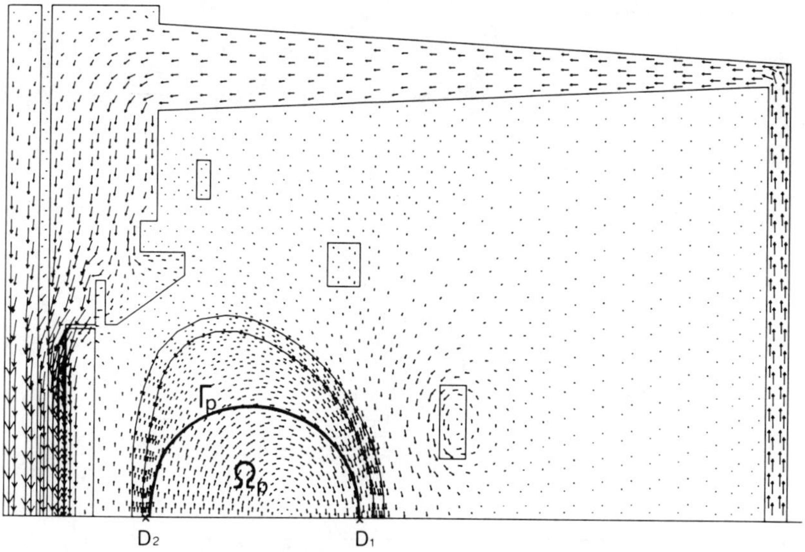

<u>*Figure I.20*</u> : <u>*Equilibrium configuration 2 in JET*</u>
<u>*(circular plasma)*</u>

1.5.5 TORE Supra (cf. R. AYMAR et al [11]) :

TORE Supra is the new Tokamak of the association EURATOM–CEA, under construction at Cadarache, and has superconducting toroidal field coils which permit long (30s) discharges, an indispensible property for the reactor of the future. In contrast, the poloidal field system is made up of traditional copper coils and is represented in Figure I.21. The magnetic circuit consists of an iron core which can reach saturation, together with six return arms. There are five poloidal field coils: coil 1 against the core, coil 2 against the core extension, coils 3, 4 and 5 surrounding the external vacuum vessel, and coils placed symmetrically to the latter four. These coils, mounted in parallel, serve simultaneously for the premagnetization, induction of plasma current, adjustment of the vertical field and control of horizontal displacements of the plasma. In TORE Supra there are two vacuum vessels with circular section, the external one or cryostat, and the internal one, separated by the toroidal field coils.

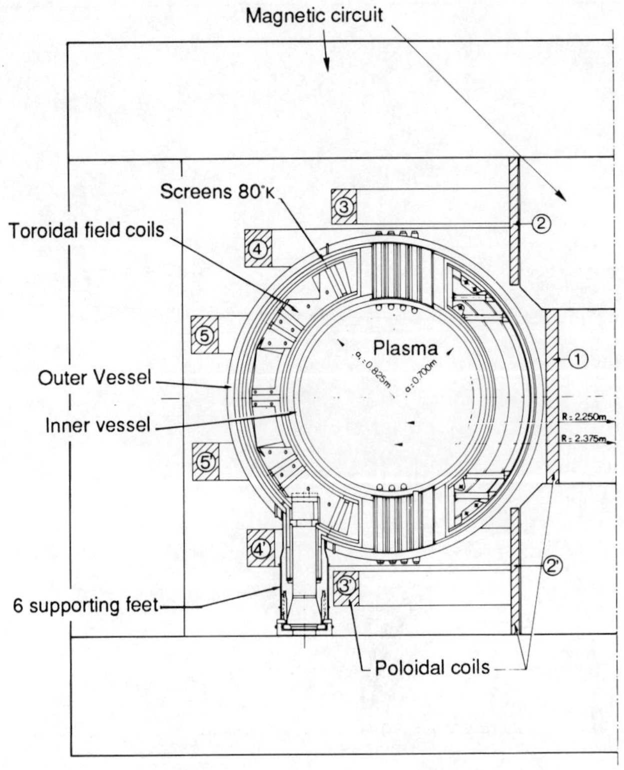

Figure I.21 : Meridian section of TORE Supra.

A typical discharge consists generally of a phase of plasma current increase and then a current plateau. We consider here two test-cases, one corresponding to the beginning of the plateau (case 1) and the second to the end of the plateau (case 2).

The data for the two cases are collected in Table I.9:

Test case	I_p(MA)	I_1(MA)	I_2(MA)	I_3(MA)	I_4(MA)	I_5(MA)	α	β	γ	δ	ζ
1	0.85	0.5	-0.15	-0.197	-0.197	-0.197	1.	0.87	1.	0	1.
2	0.85	-1.2	-0.25	-0.207	-0.207	-0.207	1.	0.8	1.	0	1.

Table I.9

The characteristic outputs in the two cases, whose convergence is obtained by Newton's method, are given in Table I.10:

Test case	ψ_F (Wb)	β_p	ℓ_i	Δ_H(cm)	a(cm)	b(cm)	e
1	3.84	0.86	0.98	-10.9	74.1	73.4	0.99
2	-4.61	0.77	0.9	- 9.5	75.5	76.1	1.01

Table I.10

Figures I.22 represent the equilibrium configuration corresponding to case 1 (beginning of the plateau).

1.5.6 INTOR (cf. INTOR REPORT)

The dimensions given in Table I.1 show clearly that we are concerned here with a reactor project. The coils of the poloidal and toroidal field systems are of superconducting material. The large dimensions of the machine make the cost of an iron transformer too substantial, which is why INTOR is designed with an air transformer. One of its peculiarities is that it has a poloidal divertor. This device is able to prevent the impurities situated at the plasma boundary from penetrating the plasma, and conducts them away for a certain distance at which point they are canalized by a collector ; to do this an electromagnetic device is used which creates a separatrix on the boundary of the plasma, i.e. a flux line possessing a point of zero field, also called a stagnation point or hyperbolic point. The meridian section of INTOR represented in Figure I.23 shows such a device.

Here we are interested exclusively in the case of an asymmetric divertor which has the advantage of requiring only one collecting device instead of two situated symmetrically with respect to the Or axis, as in the case of a symmetric divertor. In this case the assumptions of symmetry relative to the equatorial plane made in (I.25) are no longer valid and we are obliged to work in a complete domain Ω. Moreover, the fact that there is no magnetic circuit canalizing the flux lines means that we have to solve the problem throughout the half-plane $r > 0$, or at least in a domain Ω whose boundaries are very far from the experimental device (here we have taken $r =$ 100m for AB and $z =$ 80m for BC). The plasma boundary Γ_p is no longer defined by its contact with the limiter, but as being the separatrix, i.e. the flux line passing through the hyperbolic point. To be more precise, referring to Figure I.24, we observe that there are two hyperbolic points in Ω_v that we call X_1 and X_2 and which are not symmetrical relative to the median plane because of currents in the coils imposed on the outside. We consider the one of these two points that corresponds to the greater value of ψ, and to fix the ideas we suppose that this point X_1 corresponds to the negative z coordinate. Then Γ_p and Ω_p are defined as follows:

$$\Gamma_p = \{M \in \Omega_v | \psi(M) = \psi(X_1) \text{ and } z(M) > z(X_1)\}$$

(I.100)

$$\Omega_p = \{M \in \Omega_v | \psi(M) > \psi(X_1) \text{ and } z(M) > z(X_1)\} .$$

a)

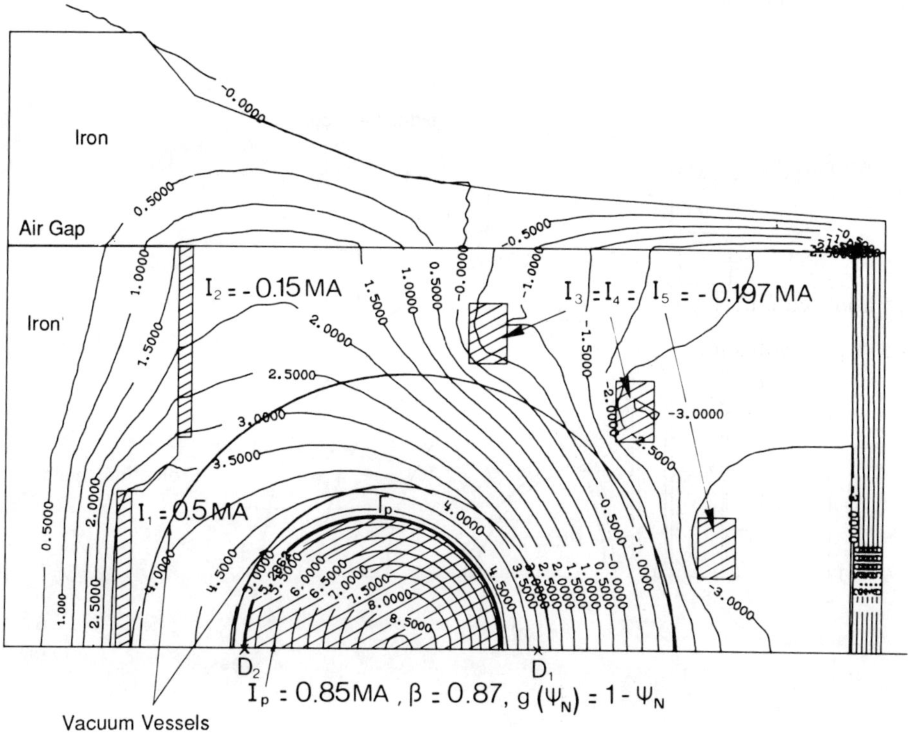

b)

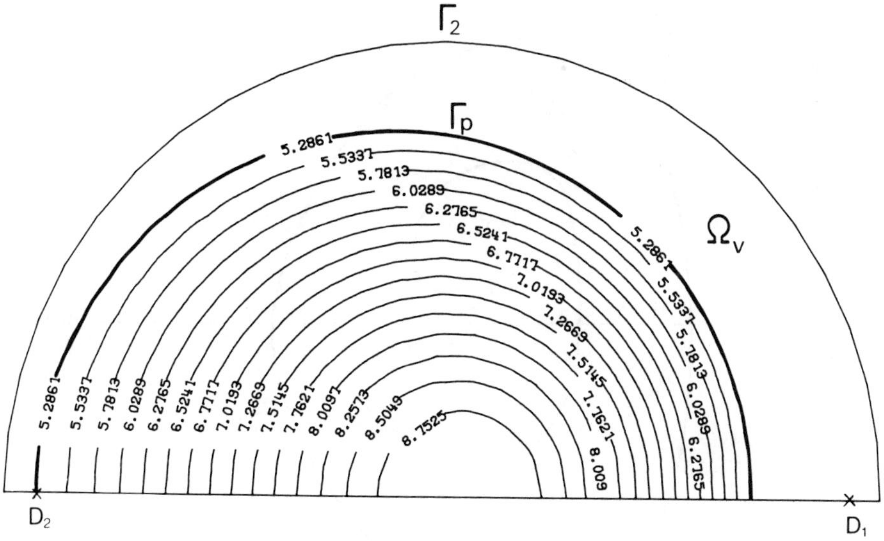

Figure I.22 : Equilibrium configuration 1 in TORE Supra (beginning of flat top).

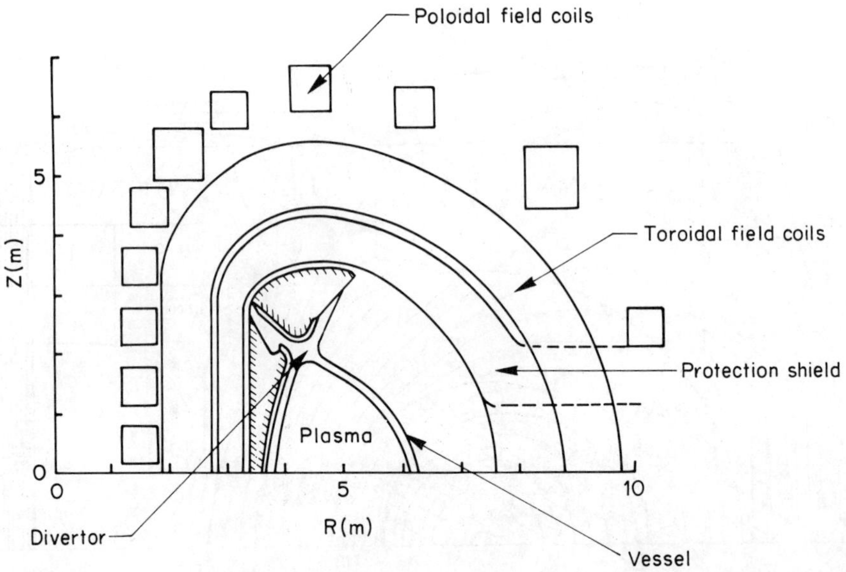

Figure I.23 : *Meridian section of INTOR*.

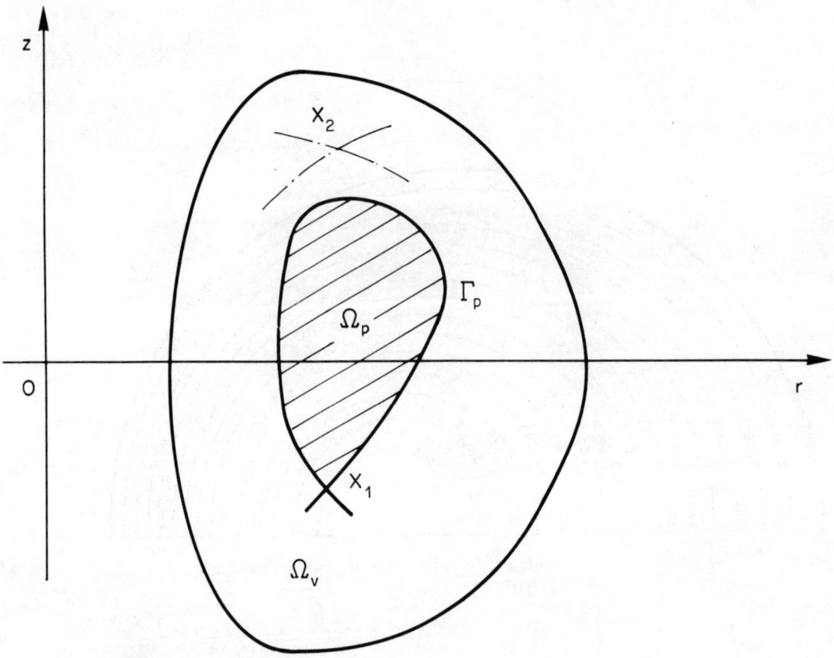

Figure I.24 : *Definition of the plasma boundary in the case of a divertor*.

The equations of Sections I.2, I.3 and I.4 extend easily to the case when there is no symmetry relative to the Or– axis. In the linearization it is the point X_1 which replaces the point M_0.

We propose to study the configuration corresponding to $I_p = 4.7$ MA (total plasma current), $\beta = 2.65$, with $g(\psi_N) = 1 - \psi_N$ and the currents I_j in the upper coils and I_j' in the lower coils given in the following Table I.11. :

j	1	2	3	4	5	6	7	8	9	10	11
I_j (MA)	-5.63	-5.63	-3.6	-3.6	-3.6	6.28	3.14	3.14	-0.15	-3.02	-3.39
I_j (MA)	-6.02	-6.02	-3.6	-3.6	-3.6	8.28	4.14	4.14	-0.15	-5.27	-2.13

Table I.11

Figure I.25a represents the lines of flux for the equilibrium configuration corresponding to these data and also the plasma boundary Γ_p, while Figure I.25b shows ten flux lines between Γ_p and a point G_1 situated at 30cm from the plasma in the median plane. Note the hyperbolic points X_1 and X_2, to which there corresponds a flux difference of 1.4Wb. Figure I.25b allows the throats of the divertor to be easily visualized; observe in particular that to canalize the impurities situated up to 30cm from the plasma it is necessary to have a divertor throat of more than 1.50m around the point X_1 (this is due to the fact that B_P is zero at X_1 and very small nearby).

The optimization of the position of the point X_1 and its control will be the object of a section in Chapter II.

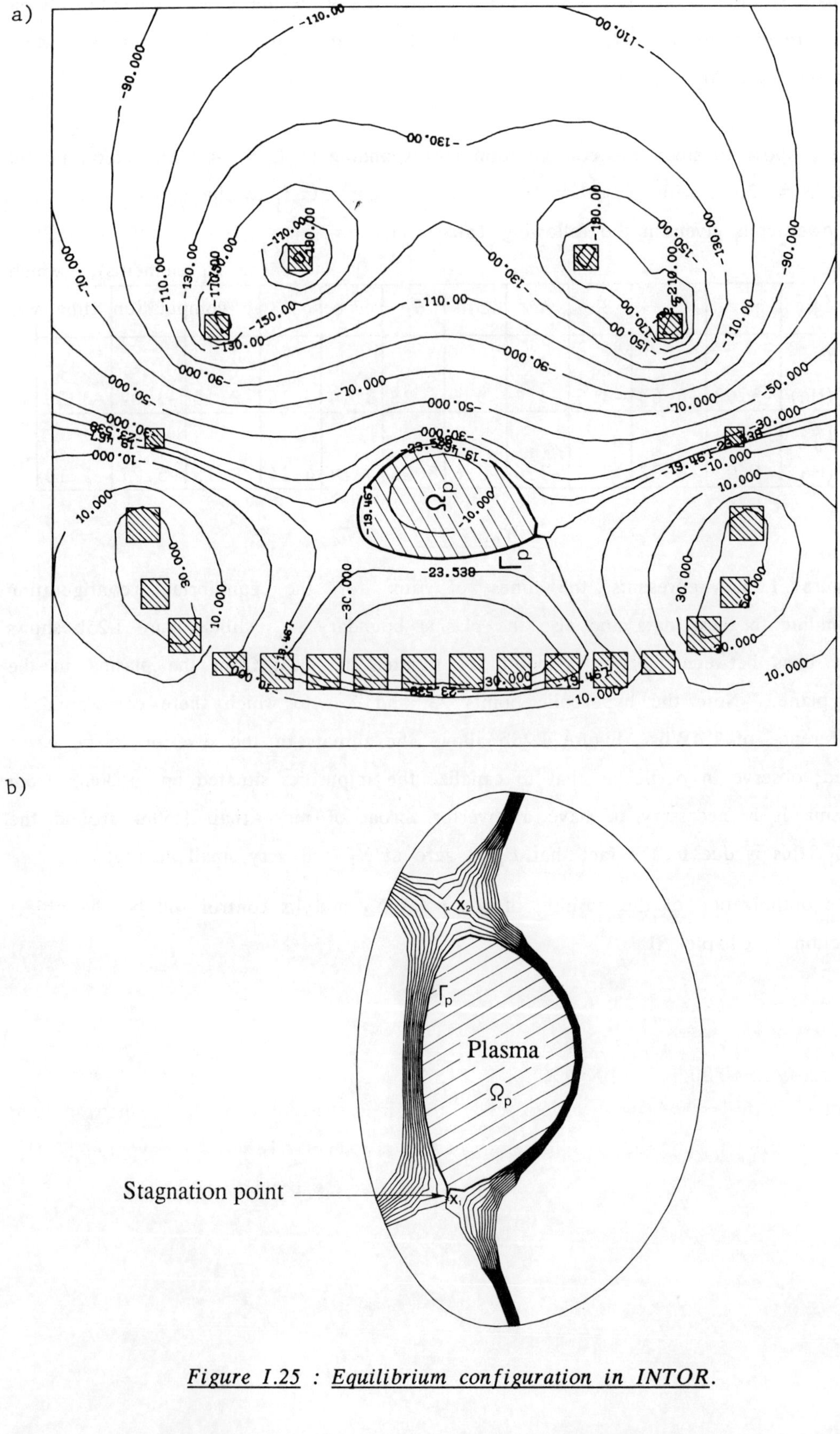

Figure I.25 : Equilibrium configuration in INTOR.

I.6 SOME COMMENTS ON THE NUMERICAL METHODS OF I.4, MOTIVATED BY THE TEST CASES OF I.5 :

I.6.1 The Finite Element Method

The finite elements employed in the test-cases of I.5 are the finite elements P_1 defined in I.4.2. One test was carried out for TFR using finite elements P_2 with a coarser mesh : to obtain the same type of precision for the solution it was appropriate to use a mesh consisting of 800 triangles (instead of 1500 for the P_1 elements), which corresponds approximately to the same number of vertices. The computation time was multiplied by 2 because of the greater number of terms in the matrix to be calculated, and the greater band width for the profile method used. This is explained by the fact that the geometry is quite complicated (iron, air gap, coils...) and that it is not possible to reduce very significantly the number of triangles of the mesh. Moreover, since the function ψ is not in $H^2(\Omega)$ because of the discontinuity of its normal derivative at the air-iron interface, there is no error bound for the magnetostatic part of the problem and we have only strong $H^1(\Omega)$ convergence of the solution of the discrete problem to the solution of the continuous problem (cf. R. GLOWINSKI – A. MARROCCO). All this leads us to think that the finite elements P_1 are the ones best adapted to this problem. We could envisage a method with P_1 elements in the air and iron, and P_2 elements in the plasma, with intermediate elements with four vertices then being necessary in order to make the junction. Finite elements of P_2 type in the plasma would allow the magnetic axis to be determined with greater precision; in fact in the case of P_1 elements it is necessarily a vertex of the grid.

Quadrangles and elements of type Q_1 or Q_2 could equally well be used to solve this problem.

I.6.2 Newton's Method

From the test-cases of I.5 it is clear that it is Newton's method which is here best adapted for treating the nonlinearities. We saw in I.4.4 that the hypotheses H_1', H_2', H_3' have to be satisfied by the solution of the discrete problem (P_I').

Hypothesis H_1' requires uniqueness of the point of contact between Γ_p and D ; in fact when the plasma is centred ($\Delta_H = 0$) it touches simultaneously the interior and exterior points of the limiter and the derivative problem no longer exists. In these cases we observe that Newton's algorithm diverges, with the plasma oscillating between the interior and exterior points of the limiter and its linearization changing at each iteration of the algorithm.

Hypothesis H_2' is the uniqueness of the magnetic axis; all the cases treated satisfy this hypothesis.

Hypothesis H_3' is the non-vanishing of the poloidal field $\mathbf{B_P}$ on Γ_p. Now this hypothesis fails in the case of the divertor because the X-point is a point of zero field for the continuous problem. However, in the discrete problem this hyperbolic point is necessarily a vertex of the mesh since ψ is linear on each triangle; the boundary Γ_p is a polygonal line and the value of $\nabla\psi$ has to be considered on each segment of it. Unless ψ has the same value at the three vertices of a triangle, $\nabla\psi$ is nonzero on each segment of Γ_p. Newton's algorithm thus works perfectly even in the case of a divertor.

I.6.3 *Method of Solution of the Linear Systems* :

A direct method, namely the Crout method with profile combined with the factorization method presented in I.4.4.1, was used to solve the linear system at each iteration of Newton's method. An iterative method of over-relaxation was compared with this direct method, but it requires a considerably longer computation time since the optimal value of the over-relaxation parameter is greater than 1.99 and this leads to a spectral radius for the over-relaxation matrix very close to 1. This has to do with the presence of air-gaps, which require a very fine mesh relative to the mesh-size for the iron, and with the fact that the relative permeability μ_r of iron is of the order of 10^3. It would appear to be necessary to optimize the over-relaxation parameter in different ways according to the region concerned (iron, air, air-gaps) and to the cases of iron saturated or not. All these considerations have led to the use of a direct method compatible with the memory capacity available on CDC 7600.

COMMENTS ON BIBLIOGRAPHY

The numerical results presented in this chapter were obtained using the stationary version of the SCED (Self–Consistent for Equilibrium and Diffusion) code : cf. J. BLUM – J. LE FOLL – B. THOORIS [1]. To our knowledge, the only other code which treats the problem of equilibrium with free boundary in a Tokamak with iron is the one presented in G. CENACCHI – E. SALPIETRO – A. TARONI : this code likewise uses a finite element method. In the absence of iron, the Green's functions for the operator L can be exhibited: these functions are used to calculate the equilibrium with free boundary for Tokamaks without iron in K. LACKNER, L.E. ZAKHAROV, J.L. JOHNSON et al. Finite difference methods are used in J.L. JOHNSON et al, M.S. CHU et al. Various algorithms (Picard, Marder–Weitzner,...) have been tried out for the treatment of nonlinearities in K. LACKNER.

Instead of calculating the function $\psi(r,z)$ one can look for the functions $r(\psi,\chi)$ and $z(\psi,\chi)$, where χ is the orthogonal coordinate of ψ; this "inverse" method is used by L. DELUCIA – S.C. JARDIN – A.M. TODD to solve the free boundary equilibrium problem. A review of these "inverse" methods is given by H.R. HICKS – R.A. DORY – J.A. HOLMES and by L.M. DEGTYAREV – V.V. DROZDOV.

The method of moments is used by L.L. LAO to calculate the equilibrium of the plasma.

The equilibrium of rotating plasmas has been studied by S. SEMENZATO – R. GRUBER – H.P. ZEHRFELD and by W. KERNER – O. JANDL, using finite element methods with variable mesh following the lines of flux.

For equilibrium problems in a perfect superconducting shell treated by R. TEMAM [1] and [2], finite element methods have been used to obtain numerical solutions in C. GUILLOPE, J. LAMINIE and M. SERMANGE. Picard and Marder–Weitzner algorithms were used to treat by M. SERMANGE the nonlinearities. Finite element analysis of related problems has been studied by F. KIKUCHI – T. AIZAWA and G. CALOZ – J. RAPPAZ.

In Chapter IV, when studying equilibrium solution branches, we shall return to the convergence of the various Picard, Marder–Weitzner and Newton–Raphson algorithms.

2. Static control of the plasma boundary by external currents

In Chapter I the problem consisted of finding the equilibrium configuration, and the free boundary of the plasma in particular, the currents in the coils being known. In this chapter it is the inverse problem we intend to solve: that of finding the currents in the coils which allow a desired magnetic configuration to be obtained in the "best" way possible, and in particular such that the plasma boundary should have certain desired characteristics: radial position, elongation or shape.

Many authors have already applied themselves to the problem of determining currents over a continuous sheet or in coils that give the plasma a certain form, but in the absence of a magnetic circuit. The approach used in L.E. ZAKHAROV, J.P. BOUJOT – J.P. MORERA – R. TEMAM or C. MERCIER–SOUBBARAMAYER is the following:

(i) solve the equilibrium equations in the interior of the plasma, with its boundary being the desired boundary Γ_d;

(ii) solve the external "ill–posed" problem of determining those currents in the coils that are consistent with extension of the interior solution, i.e. the continuity of the flux and its normal derivative on Γ_d;

(iii) verify (possibly with the help of a free boundary code) that the boundary Γ_p obtained with these currents is "close" to Γ_d.

This approach requires using successively several codes to solve one and the same problem and, in certain particular cases, the boundary Γ_p obtained in (iii) can be very different from Γ_d since the free boundary is not directly involved in the determination of the currents in the coils.

Another approach has been used in K. LACKNER, J.L. JOHNSON et al., A. OGATA – H. NINOMIYA which brings in the free boundary problem but relies on the notion of additivity of the flux, which is not satisfied here because of the nonlinearity due to the presence of the magnetic circuit.

We propose here a self-consistent approach which takes account simultaneously of the nonlinearity due to the iron and of the existence of a free boundary. This approach which was presented in J. BLUM – R. DEI CAS and J. BLUM – J. LE FOLL – B. THOORIS [2], consists of formulating the problem as an optimal control problem for a system governed by the partial differential equations of Chapter I, and using methods of solution for such problems (cf. J.L LIONS. [2]). Practical applications are given to the various Tokamaks considered (TFR, TORE Supra, JET and INTOR).

II.1 FORMULATION OF THE VARIOUS CONTROL PROBLEMS

We aim to determine the currents in certain coils in order that, according to the problem in hand, the plasma should have a given radial position, height or shape. The problem is formulated in the terminology of optimal control with control parameters, equations of state and cost-function (cf. J.L. LIONS [2]).

II.1.1 The Control Parameters

The k coils in the Tokamak are divided into two categories:

- the m coils $B_1,\ldots,B_m$ whose currents $I_1,\ldots,I_m$ are the optimization parameters;
- the (k– m) coils $B_{m+1},\ldots,B_k$ whose currents are given and kept fixed in the optimization.

Define I_S and I_T by

$$I_S = \sum_{i=1}^{m} I_i$$

$$I_T = \sum_{i=1}^{k} I_i \,.$$

In all the control problems that we shall consider, the total plasma current I_p and the algebraic sum I_T of the Ampère–turns will be given data, fixed in the optimization. Consequently, since the currents $I_{m+1},\ldots,I_k$ are likewise fixed, the sum I_S of the currents involved in the optimization is a given quantity for the problem.

Let Λ_i denote the percentages of the currents I_i in the m coils B_i taken relative to their sum I_S :

$$\text{(II.1)} \quad \begin{cases} \Lambda_i = \dfrac{I_i}{I_S} = \dfrac{I_i}{\sum\limits_{j=1}^{m} I_j} \;, \quad i \in \{1,\ldots,m\} \\[2ex] \Lambda = (\Lambda_1,\ldots,\Lambda_{m-1}) \in \mathbb{R}^{m-1} \end{cases}$$

with Λ_m deduced trivially from Λ by:

$$\Lambda_m = 1 - \sum_{i=1}^{m-1} \Lambda_i$$

The vector Λ is the control vector; its components are the $(m-1)$ control parameters Λ_i. This vector may be subject to certain constraints (positivity of certain components, for example). Let U_{ad} denote the closed convex subset of $\mathbb{R}^{m-1}$ consisting of the admissible controls.

II.1.2 The State Equations

The state vector is the pair $(\psi,\lambda) \in V^{p_0}(\Omega) \times \mathbb{R}$ where V^{p_0} is defined by (I.35). The state equations are the equations (I.37) of the problem (P_I') which can be written in terms of the Λ_i as follows:

$$\text{(II.2)} \quad \begin{cases} a_\mu(\psi,\varphi) = \sum\limits_{i=1}^{m} \dfrac{\Lambda_i I_S}{S_i} \displaystyle\int_{B_i} \varphi dS + \sum\limits_{i=m+1}^{k} \dfrac{I_i}{S_i} \displaystyle\int_{B_i} \varphi dS \\[2ex] \qquad + \lambda \displaystyle\int_{\Omega_p} h(r)\, g(\psi_N)\varphi dS \;, \quad \forall\varphi \in V^{p_o'}(\Omega) \\[2ex] I_p = \lambda \displaystyle\int_{\Omega_p} h(r)\, g(\psi_N) dS \end{cases}$$

with $\Omega_p = \{ M \in \Omega_v | \psi(M) > \sup_D \psi \}$ and where the current density j_{cv} in the vacuum chamber has been taken to be zero.

II.1.3 *The Cost Function*

We distinguish the following three problems :

II.1.3.1 *Control of the Radial Position*

Here we assume that the point of contact between the plasma and the limiter is situated in the median plane. To impose on the plasma a certain radial position Δ_H we require its boundary to pass through two given points I and E in the median plane, at least one of these points being a point of the limiter (see Fig II.1). To do this we minimize the cost function :

$$J_1 = \tfrac{1}{2}[\psi(E) - \psi(I)]^2 \tag{II.3}$$

with respect to Λ defined by (II.1).

If the minimum of J_1 is zero and if the plasma touches the limiter at E or at I then the plasma boundary will indeed pass through both the points E and I.

Remark II.1

It is natural to consider two coils (m= 2) in this optimization; in fact, since the current I_S is fixed, one single control parameter should enable us to make the plasma pass through two points (I and E) one of which is the point of contact with the limiter. The minimum of J_1 then has to equal zero. □

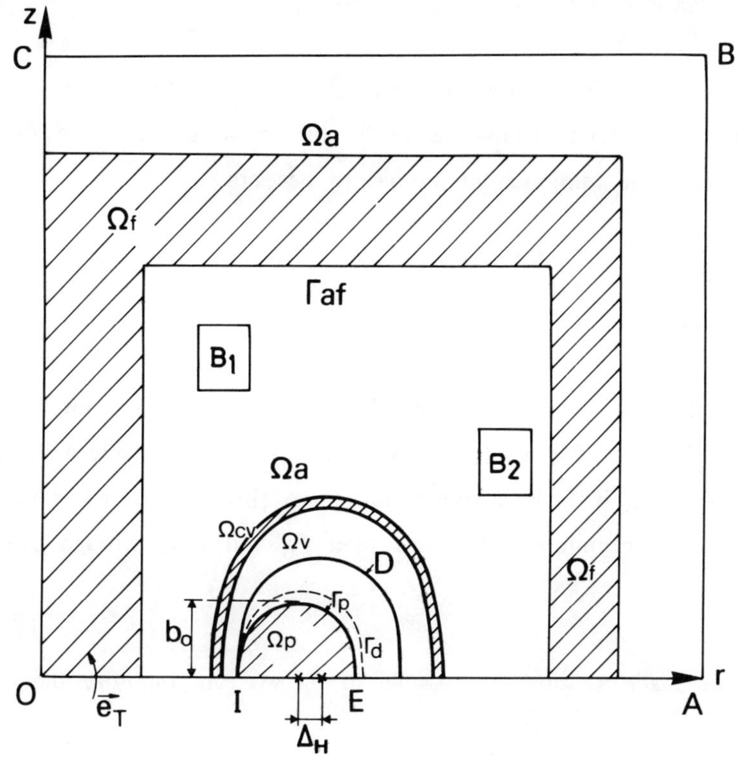

Figure II.1 : *Meridian section of the torus*

II.1.3.2 Control of the Height

In the machines such as JET where the plasma is non-circular, we require it to have a given elongation as well as a certain radial position. To do this we could make the plasma boundary pass through a third point H (as well as the points I and E), supposed to be the topmost point of the plasma. For this we can minimize

$$J_2 = J_1 + \tfrac{1}{2}[\psi(E) - \psi(H)]^2 \; . \tag{II.4}$$

However, even if the minimum of J_2 is zero and the plasma boundary passes through the three points I, E and H, the point H is not necessarily the stipulated highest point of the plasma. We prefer to impose on the plasma a certain height b_0 (see Fig II.1) and minimize with respect to Λ the following cost function :

$$J_3 = J_1 + \frac{K}{2}\,[\sup_{M\in\Gamma_p} z(M) - b_0]^2 \tag{II.5}$$

where K is a constant designed to give equal weight to the two terms in the cost function J_3.

Remark II.2

It is natural to consider three coils (m= 3) in this optimization. In fact the two free control parameters should enable us to fix the radial position and the height of the plasma. In this case the minimum of J_3 has to be zero. □

II.1.3.3 Control of the Shape

We require the free boundary Γ_p of the plasma to be as "close" as possible to a desired boundary Γ_d (see Fig II.1); Γ_d will be taken to be internally tangent to the limiter D at a point F_o if D is a continuous curve (in Fig II.1 we have F_o = I), and if D consists of a finite number of points then Γ_d will pass through one of the points also denoted F_o.

For Γ_p to be made as close as possible to Γ_d we can minimize the cost function

$$J_4 = \frac{1}{2} \int_{\Gamma_d} [\psi(M) - \psi(F_o)]^2 \, d\Gamma \tag{II.6}$$

with respect to Λ.

If J_4 were zero then Γ_d would be the line of flux passing through F_o and would thus be identified with Γ_p.

In the particular case of a configuration with divertor defined in Section I.5.6, there is a hyperbolic point X_1 on the boundary Γ_p and we require that this point should be the stagnation point of the divertor. If we wish for Γ_p, defined by equation (I.100), to be as close as possible to Γ_d then it is natural to minimize

$$J_5 = \frac{1}{2} \int_{\Gamma_d} [\psi(M) - \psi(X_1)]^2 \, d\Gamma \tag{II.7}$$

with respect to Λ.

In what follows we shall denote by F_1 the point F_o in the case of a boundary defined by its contact with a limiter and the point X_1 in the case of a divertor.

As we have only a finite number of control parameters at our disposal, the minima of J_4 or J_5 cannot be zero and we achieve "$\Gamma_p = \Gamma_d$" only as best we can.

Remark II.3

If we have the currents in m coils as control parameters we might naturally expect to be able to make the plasma boundary pass through m given points. This is not true in general; the m points have to be conveniently positioned in order for the plasma boundary to pass through them. □

II.1.3.4 The General Case

To synthesize the cost functions (II.3), (II.5), (II.6) and (II.7) into one and the same functional we define J by :

$$J = \frac{1}{2}\{K_1 \, [\psi(E) - \psi(I)]^2 + K_2[\sup_{M \epsilon \Gamma_p} z(M) - b_0]^2 + K_3 \int_{\Gamma_d} [\psi(M) - \psi(F_1)]^2 \, d\Gamma + K_4 \sum_{i=1}^{m} I_i^2\} \tag{II.8}$$

where the coefficients K_1, K_2, K_3 and K_4 can be taken equal to zero or not according to the problem being considered, and where the term $K_4 \Sigma \, I_i^2$ represents the energy cost of the configuration. This regularizing term allows configurations which present current dipoles and are very unfavourable from the energy point of view to be avoided.

II.1.4 The Control Problem (P_{II}) :

The optimal control problem (P_{II}) can now be formulated as follows : find $\Lambda \epsilon U_{ad}$ and $(\psi,\lambda) \; \epsilon \; V^{P_0}(\Omega) \times R$ satisfying the equations of state (II.2) and such that :

$$\left\{ \begin{array}{l} J(\Lambda,\psi) = \inf\limits_{\substack{\Lambda' \epsilon U_{ad} \\ (\psi',\lambda') \epsilon V^{P_0}(\Omega) \times R}} J(\Lambda',\psi') \end{array} \right. \tag{II.9}$$

where (ψ',λ') are related to Λ' by the equations of state (II.2) and where J is given by the expression (II.8) with the currents I_i related to Λ_i by (II.1).

In Chapter III we shall give conditions sufficient for the existence of a solution to the control problem for the shape of the plasma in a simplified model.

II.2 INTRODUCTION OF THE LAGRANGIAN AND OPTIMALITY SYSTEM

II.2.1 Definition of the Lagrangian

Since the system of state equations is nonlinear, and the cost function J non-convex, we shall define a Lagrangian in order to derive formally the conditions necessary for optimality for the problem (P_{II}). This comes down to regarding the state equations (II.2) as constraints, and introducing a Lagrange multiplier $(\chi,\nu) \in V^{p'}_{0}(\Omega) \times R$ by the Lagrangian N defined thus :

$$
\begin{aligned}
N(\Lambda,(\psi,\lambda),(\chi,\nu)) &= J(\Lambda,\psi) + a_{\mu}(\psi,\chi) \\
&- \sum_{i=1}^{m} \frac{\Lambda_i I_S}{S_i} \int_{B_i} \chi \, dS - \sum_{i=m+1}^{k} \frac{I_i}{S_i} \int_{B_i} \chi \, dS \\
&- \lambda \int_{\Omega_p} h(r)\, g(\psi_N)\chi \, dS + \nu\left[\lambda \int_{\Omega_p} h(r) g(\psi_N) dS - I_p\right] ,
\end{aligned}
\tag{II.10}
$$

$$
\Lambda \in \mathcal{U}_{ad},\ (\psi,\lambda) \in V^{p}_{0}(\Omega) \times R,\ (\chi,\nu) \in V^{p'}_{0}(\Omega) \times R .
$$

The pair (χ,ν) is the Kuhn–Tucker vector (cf. KUHN–TUCKER) for the problem (P_{II}) if :

$$
J(\Lambda,\psi) = \inf_{\substack{\Lambda' \in \mathcal{U}_{ad} \\ (\psi',\lambda') \in V^{p}_{0}(\Omega)\times R}} J(\Lambda',\psi') = \inf_{\substack{\Lambda' \in \mathcal{U}_{ad} \\ (\psi',\lambda') \in V^{p}_{0}(\Omega)\times R}} N(\Lambda',\psi',\lambda',\chi,\nu)
\tag{II.11}
$$

related to Λ' by (II.2)

A necessary and sufficient condition for $(\Lambda,(\psi,\lambda))$ to be a solution to the problem (P_{II}) and (χ,ν) to be the associated Kuhn–Tucker vector is that $(\Lambda,(\psi,\lambda),(\chi,\nu))$ should be a saddle–point for N, i.e.

$$(II.12)\qquad \begin{cases} N(\Lambda,\psi,\lambda,\chi',\nu') \leqslant N(\Lambda,\psi,\lambda,\chi,\nu) \leqslant N(\Lambda',\psi',\lambda',\chi,\nu) \\ \forall(\Lambda',\psi',\lambda',\chi',\nu') \in \mathcal{U}_{ad} \times V^{p_o}(\Omega) \times R \times V^{p'_o}(\Omega) \times R. \end{cases}$$

Since the Lagrangian N is non–convex the existence of saddle–points is not guaranteed, but we shall deduce from (II.12) conditions necessary for optimality of the problem (P_{II}).

II.2.2 *First Order Necessary Conditions for Optimality*

Let H4 be the following hypothesis : the plasma boundary Γ_p has a unique point M_2 where the z–coordinate attains its maximum, which is assumed to be nondegenerate i.e. :

$$(II.13)\qquad \begin{cases} z(M_2) = \sup\limits_{M\in\Gamma_p} z(M) \\ \dfrac{\partial\psi}{\partial r}(M_2) = 0 \\ \dfrac{\partial^2\psi}{\partial r^2}(M_2) \neq 0\ . \end{cases}$$

<u>Proposition II.1</u>

If the hypotheses H1, H2, H3 of Paragraph I.3 and the hypothesis H4 above are satisfied, then a necessary condition for $(\Lambda,(\psi,\lambda),(\chi,\nu)) \in \mathcal{U}_{ad} \times V^{p_o}(\Omega) \times R \times V^{p'_o}(\Omega) \times R$ to be a saddle point for N is that (Λ,ψ,λ) satisfy the state equations (II.2) and

$$
\text{(II.14)}\quad
\begin{cases}
b_u(\varphi,\chi-\nu) = - K_1[\psi(E) - \psi(I)][\varphi(E) - \varphi(I)] \\[2ex]
\qquad - K_2[z(M_2) - b_o][\dfrac{\varphi(M_o) - \varphi(M_2)}{\dfrac{\partial\psi}{\partial z}(M_2)}] \\[2ex]
\qquad - K_3 \displaystyle\int_{\Gamma_d} [\psi(M) - \psi(F_1)][\varphi(M) - \varphi(F_1)]d\Gamma, \\
\qquad\qquad \forall \varphi \in V^{p_o}(\Omega) \\[2ex]
c_u(\chi - \nu) = 0
\end{cases}
$$

$$
\sum_{i=1}^{m-1} I_S(\Lambda_i'-\Lambda_i)[-\frac{1}{S_i}\int_{B_i}\chi\, dS + \frac{1}{S_m}\int_{B_m}\chi\, dS + K_4 I_S(\Lambda_i-\Lambda_m)] \geqslant 0, \quad \forall \Lambda \in \mathcal{U}_{ad}
$$

(II.15)

where the forms b_u and c_u are defined by equations (I.50) and (I.51) and where M_o is the point of contact of Γ_p and the limiter D. □

The (formal) proof of Proposition II.1 uses the following lemma.

<u>Lemma II.1</u>

If the hypotheses H1, H3 and H4 are satisfied the mapping which associates $\sup_{M\in\Gamma_p} z(M)$ to $\psi \in V^{p_o}(\Omega)$ is Gâteaux differentiable and its derivative is the map which to $\tilde{\psi} \in V^{p_o}(\Omega)$ associates $\dfrac{\tilde{\psi}(M_o)-\tilde{\psi}(M_2)}{\dfrac{\partial\psi}{\partial z}(M_2)}$ □

<u>Proof of Lemma II.1</u> :

From the hypotheses H1 and H3 of Section I.3 and from hypothesis H4 it is clear that for θ sufficiently small the point M_2' with maximum z-coordinate on the curve with equation

$$
(\psi+\theta\tilde{\psi})(M) = \sup_D(\psi+\theta\tilde{\psi})
$$

approaches M_2 as θ tends to 0. Let M_0' be a point of D where $\sup_D(\psi + \theta\tilde{\psi})$ is attained. We have :

$$(II.16) \qquad (\psi+\theta\tilde{\psi})(M_2') = (\psi + \theta\tilde{\psi})(M_0') .$$

Using the fact that M_0' approaches M_0 as θ tends to zero (see the proof of Lemma I.2 in Section I.3) we expand (II.16) to first order and obtain

$$(II.17) \qquad \begin{aligned} &\psi(M_2) + \nabla\psi(M_2)\cdot M_2 M_2' + o(\|M_2 M_2'\|) \\ &+ \theta\tilde{\psi}(M_2) + o(\theta) = \psi(M_0) + \theta\tilde{\psi}(M_o) + o(\theta) . \end{aligned}$$

The vector $\nabla\psi$ is nonzero at M_2 by hypothesis H3 and, using the fact that $\partial\psi/\partial r\ (M_2) = 0$ by (II.13) we have

$$(II.18) \qquad \nabla\psi(M_2)\cdot M_2M_2' = \frac{\partial\psi}{\partial z}(M_2)\times[z(M_2')-z(M_2)] .$$

By (II.17) and (II.18), and using the fact that $\psi(M_2) = \psi(M_0)$ by definition of Γ_p we obtain

$$\lim_{\theta\to 0} \frac{z(M_2')-z(M_2)}{\theta} = \frac{\tilde{\psi}(M_0)-\tilde{\psi}(M_2)}{\frac{\partial\psi}{\partial z}(M_2)} ,$$

where $\partial\psi/\partial z\ (M_2)$ is different from zero by the hypothesis H3. □

<u>Proof of Proposition II.1</u> :

The inequality on the left hand side of (II.12) can be written :

$$a_\mu(\psi,\chi'-\chi) - \sum_{i=1}^{m} \frac{\Lambda_i I_S}{S_i}\int_{B_i}(\chi'-\chi)\,dS - \sum_{i=m+1}^{k}\frac{I_i}{S_i}\int_{B_i}(\chi'-\chi)dS$$

$$-\lambda\int_{\Omega_p} h(r)g(\psi_N)(\chi'-\chi)dS + (\nu'-\nu)[\lambda\int_{\Omega_p} h(r)g(\psi_N)dS - I_p] \leqslant 0,$$

$$\forall\ (\chi',\nu') \in V^{p_0'}(\Omega)\times R .$$

Taking $\chi' = \chi + \varphi$ and $\chi' = \chi - \varphi$ with $\nu' = \nu$ we obtain the first equation of (II.2); taking $\nu' = \nu + 1$ and $\nu' = \nu - 1$ we obtain the second equation of (II.2).

From the inequality on the right in (II.12) we deduce

$$\text{(II.19)} \qquad \begin{cases} (N'_\psi(\Lambda,\psi,\lambda,\chi,\nu),\varphi) = 0 \quad , \quad \forall\varphi \in V^{p_o}(\Omega) \\ N'_\lambda(\Lambda,\psi,\lambda,\chi,\nu) = 0 \end{cases}$$

where N'_ψ denotes the Gâteaux derivative of N with respect to ψ and N'_λ is the derivative of N with respect to λ. Let us calculate the Gâteaux derivative of J with respect to ψ in the direction of φ ; using Lemma II.1 we obtain :

$$(J'_\psi(\Lambda,\psi),\varphi) = K_1[\psi(E) - \psi(I)][\varphi(E) - \varphi(I)]$$

$$\text{(II.20)} \qquad + K_2[z(M_2) - b_o] \frac{\varphi(M_o) - \varphi(M_2)}{\frac{\partial\psi}{\partial z}(M_2)}$$

$$+ K_3\int_{\Gamma_d} [\psi(F_1) - \psi(M)][\varphi(F_1)-\varphi(M)], \forall\varphi\in V^{p_o}(\Omega).$$

Using the derivations carried out in Section I.3 we can write the relations (II.19) in the form of the equations (II.14) defining χ and ν.

Finally, from the equation on the right of (II.12) we deduce that

$$\text{(II.21)} \qquad (N'_\Lambda(\Lambda,\psi,\lambda,\chi,\nu),\Lambda'-\Lambda) \geqslant 0 \quad , \quad \forall\Lambda' \in \mathcal{U}_{ad}.$$

After computing N'_Λ, we can write this inequality in the form of (II.15), which formally proves the Proposition. □

__Remark II.4__

From the general properties of the Lagrangian and from Proposition II.1 we deduce that the gradient G of the cost function J is the vector in R^{m-1} with components

$$\text{(II.22)} \qquad G_i = \frac{1}{S_m}\int_{B_m} \chi\, dS - \frac{1}{S_i}\int_{B_i} \chi\, dS + K_4 I_S(\Lambda_i - \Lambda_m)$$

$$i \in \{1,\ldots,m-1\}.$$

This gradient will be used in the numerical method for optimization of J. □

II.2.3 Interpretation of the Adjoint State in Terms of Partial Differential Equations

Define the space $V_c^{p_0}(\Omega)$ as the space of functions of $V^{p_0}(\Omega)$ defined by (I.35) with a constant added, namely :

$$V_c^{p_0}(\Omega) = \{\varphi \in L^{p_0}(\Omega) \text{ such that } \frac{1}{r} \nabla \varphi \in [L^{p_0}(\Omega; r\,dr\,dz)]^2 \quad \text{(II.23)}$$

$$\text{and } \varphi = \text{constant on } \Gamma_0 \}.$$

In equations (II.14) we put :

$$\chi' = \chi - \nu .$$

The variable χ', which belongs to $V_c^{p_o'}(\Omega)$, is the real unknown quantity in equations (II.14).

Recall that δ_x denotes the Dirac measure at a point x of Ω, and that δ_{Γ_p} is the measure on Γ_p defined by

$$(\delta_{\Gamma_p}, v) = \int_{\Gamma_p} v \, d\Gamma . \quad \text{(II.24)}$$

We can now establish formally the following proposition :

<u>Proposition II.2</u>

If the variable $\chi' = \chi - \nu$ defined by (II.14) and belonging to $V_c^{p_o'}(\Omega)$ exists then it satisfies the following partial differential equations :

$$
\text{(II.25)}\quad
\left\{
\begin{array}{l}
(L + L')\chi' - \lambda \left\{ \dfrac{h(r)g'(\psi_N)}{\psi(M_o)-\psi(M_1)} \chi' 1_{\Omega_p} \right. \\[2ex]
\quad + \delta_{M_1} \displaystyle\int_{\Omega_p} \dfrac{h(r)g'(\psi_N)[\psi(M)-\psi(M_o)]\chi'}{[\psi(M_o)-\psi(M_1)]^2} \, dS \\[2ex]
\quad + \delta_{M_o} \left[\displaystyle\int_{\Omega_p} \dfrac{h(r)g'(\psi_N)[\psi(M_1)-\psi(M)]\chi'}{[\psi(M_o)-\psi(M_1)]^2} \, dS \right. \\[2ex]
\quad \left.\left. + \displaystyle\int_{\Gamma_p} h(r)g(1)\left(\dfrac{\partial\psi}{\partial n}\right)^{-1} \chi' \, d\Gamma \right] - \left(\delta_{\Gamma_p}, h(r)g(1)\left(\dfrac{\partial\psi}{\partial n}\right)^{-1} \chi'\right) \right\} \\[2ex]
\quad = - K_1[\psi(E) - \psi(I)](\delta_E - \delta_I) \\[2ex]
\quad - K_2[z(M_2)-b_o]\left[\left(\dfrac{\partial\psi}{\partial z}\right)(M_2)\right]^{-1} (\delta_{M_o} - \delta_{M_2}) \\[2ex]
\quad - K_3\{[\delta_{\Gamma_d}, \psi - \psi(F_1)] - \delta_{F_1} \displaystyle\int_{\Gamma_d} [\psi-\psi(F_1)] d\Gamma\} \\[2ex]
\displaystyle\int_{\Omega_p} h(r)g(\psi_N) \, \chi' \, dS = 0 \\[2ex]
\dfrac{\partial\chi'}{\partial z} = 0 \quad \text{on} \quad \Gamma_1
\end{array}
\right.
$$

where

M_o is the point of contact between Γ_p and D,

M_1 is the magnetic axis of Ω_p,

M is a generic point of Ω_p,

L is defined by (I.13),

and L' is the following operator :

$$L'. = \nabla\cdot\left[\frac{2\mu'}{\mu^2 r^3} (\nabla\psi.\nabla.) \nabla\psi\right] . \qquad \square$$

The (formal) proof is carried out by considering in (II.14) a function φ belonging to D(Ω) (the set of C^∞ functions with compact support in Ω) and applying Green's fomula. We obtain in this way the first equation of (II.25). If we multiply this equation by a function φ in $V^{P0}(\Omega)$ and integrate over Ω we obtain by applying Green's formula and subtracting the first equation of (II.14) :

(II.26) $$\int_{\Gamma_1} \frac{1}{\mu r} \frac{\partial \chi'}{\partial z} \varphi \, d\Gamma = 0, \qquad \forall \varphi \in V^{p_o}(\Omega)$$

and hence the symmetry condition on Γ_1. □

In Chapter III we shall see conditions which are sufficient for the existence of the adjoint state χ' for a simplified problem.

__*Remark II.5*__:

We note in the variational formulation (II.14) as well as with the partial differential equations (II.25) that the adjoint state (χ,ν) is in fact the adjoint state for the linearized problems (I.52) or (I.71). This is explained by the fact that the adjoint state is defined by the vanishing of the derivative of the Lagrangian N with respect to the pair (ψ,λ), and this involves the derivatives of the equations of state with respect to (ψ,λ). □

__*Remark II.6*__:

We observe the duality between $\psi \in V^{p_o}(\Omega)$ and $\chi \in V^{p_o'}(\Omega)$, a familiar notion for control problems (cf. J.L. LIONS [2]). We chose p_o greater than 2 in order that ψ should be continuous; this gives a meaning to the cost function J given by (II.8), and corresponding to pointwise observations. The adjoint state χ then belongs to a space of functions which are less regular ($p_o' < 2$); this χ is not continuous and does not even belong to $H^1(\Omega)$, because of the Dirac measures belonging to $W^{-1,p_o'}(\Omega)$ and featuring on the right hand side of the equations (II.25). □

II.2.4 __*Cases Without Constraints or with Inequality Constraints*__

II.2.4.1 __*The Case Without Constraints*__

If $U_{ad} = R^{m-1}$ the inequality (II.15) becomes the following equality:

(II.27) $$\frac{1}{S_i} \int_{B_i} \chi \, dS - K_4 I_S \Lambda_i = \frac{1}{S_m} \int_{B_m} \chi \, dS - K_4 I_S \Lambda_m, \forall i \in \{1,\ldots,m-1\} .$$

In the case when there is no regularization ($K_4 = 0$) the optimality condition (II.27) means that the averages of χ (or of χ') over each of the m coils B_i are equal.

II.2.4.2 Inequality Constraints

We take the case where

$$\mathcal{U}_{ad} = \{\Lambda \in R^{m-1} | a_i \leqslant \Lambda_i \leqslant b_i, \quad \forall i \in \{1,\ldots,\ell_o\} \text{ with } \ell_o<m\} .$$

This condition means that for technological reasons the currents in certain coils are bounded below and above (from the size of the coils and the maximum admissible current density); in general

$$b_i = \frac{I_i^{max}}{I_S}, \quad a_i = -\frac{I_i^{max}}{I_S}$$

where I_i^{max} is the maximum admissible current in the coil B_i. In certain cases a_i will be taken equal to zero because technological constraints may prevent the current from changing sign in a coil.

If we denote

$$c_i = -\frac{1}{S_i}\int_{B_i} \chi \, dS + \frac{1}{S_m}\int_{B_m} \chi \, dS + K_4 I_S (\Lambda_i - \Lambda_m)$$

the inequalities (II.15) then become

. for $i \in \{1,\ldots,\ell_o\}$ there are the following alternatives :

$$\text{(II.28)} \quad \begin{cases} c_i > 0 \text{ and } \Lambda_i = b_i \\ \text{or} \\ c_i < 0 \text{ and } \Lambda_i = a_i \\ \text{or} \\ c_i = 0 \end{cases}$$

. for $i \in \{\ell_o+1,\ldots,m\}$ $\quad c_i = 0.$ □

In what follows we shall consider exclusively the case with no constraints. This applies in particular to the projected Tokamaks, because for them it is necessary to calculate the currents in the coils and from hence the tensions to apply to the generators that will create the desired configurations "as well as possible", in order to calibrate the required power levels. The constraints are applied once the experimental device has been realized and the power of the generators fixed.

II.3 NUMERICAL METHODS OF SOLUTION

The problem (P_{II}) is an optimal control problem with nonlinear state equations and non–quadratic functional. We shall solve it by a sequential quadratic method (cf. J.P. YVON, F. BONNANS – C..SAGUEZ and references) consisting of the solution of a sequence of control problems with linear equations of state and quadratic functional. In Chapter I it was Newton's method that turned out to be the most efficient for solving the equations of state; the sequential quadratic method is derived from Newton's method, with the difference that Λ is modified at each iteration in order to minimize a functional close to J. It is for this reason that this method has been chosen here for solving problem (P_{II}).

To begin with, we specify that we are going to solve the control problem for the state equations (I.74) for the discrete problem in the space $V_\ell \times R$. The algorithm consists of external iterations, each one being the solution of a linear quadratic control problem and consisting of a sequence of internal conjugate gradient iterations. More precisely, the nth external iteration of the algorithm consists of defining for the linearized problem in $(\psi^n,\lambda^n) \in V_\ell \times R$ a control problem ($P^n{}_{II}$) with a quadratic functional J^n "close" to J. This linear quadratic control problem ($P^n{}_{II}$) will itself be solved by a succession of internal iterations of conjugate gradient, and $(\Lambda^{n+1},\psi^{n+1},\lambda^{n+1})$ will thus be computed as being the optimum for the problem ($P^n{}_{II}$).

II.3.1 External Iterations of the Sequential Quadratic Algorithm

These are defined as follows :

. Let $(\Lambda^o,\psi^o,\lambda^o) \in R^{m-1} \times V_\ell \times R$;

. at the nth iteration of the algorithm we consider the linearized problem in $u^n = (\psi^n,\lambda^n)$ for the system (I.74); then $(\psi,\lambda) \in V_\ell \times R$ is related to the control vector $\Lambda \in R^{m-1}$ by equations (I.89), supposing that hypotheses H_1', H_2' and H_3' of Section I.4.5 are satisfied :

$$
\text{(II.29)} \quad \begin{cases} b_{u^n}(\psi,\varphi) - \lambda c_{u^n}(\varphi) = \sum_{i=1}^{m} \frac{\Lambda_i I_S}{S_i} \int_{B_i} \varphi \, dS \\ + \sum_{i=m+1}^{k} \frac{I_i}{S_i} \int_{B_i} \varphi \, dS + a'_{u^n}(\psi^n,\varphi) \;, \; \forall \varphi \in V_\ell \\ d_{u^n}(\psi,1) + \lambda c_{u^n}(1) = I_p \end{cases} .
$$

The functional J defined by (II.8) is not quadratic with respect to ψ because of the term in sup z(M), $M \in \Gamma_p$. Suppose hypothesis H'_4 (the discrete analog of H4 in Section II.2.2) is satisfied, namely

H'_4) $\sup_{M \in \Gamma_p} z(M)$ is attained at a unique point of Γ_p.

We then replace sup z(M), $M \in \Gamma_p$ by its first order expansion in ψ given by Lemma (II.1).

We define the following quadratic functional :

$$
\text{(II.30)} \quad \begin{cases} J^n(\Lambda,\psi) = \frac{1}{2} \{K_1 [\psi(E)-\psi(I)]^2 \\ \quad + K_2 \left[\sup_{M \in \Gamma_p^n} z(M) - b_o + \dfrac{\psi(M_o^n)-\psi(M_2^n)}{\dfrac{\partial \psi^n}{\partial z}(M_2^n)} \right] \\ \quad + K_3 \int_{\Gamma_d} [\psi(F_1)-\psi(M)]^2 \, d\Gamma + K_4 I_S^2 \sum_{i=1}^{m} \Lambda_i^2 \} \end{cases}
$$

and look for $(\Lambda^{n+1},\psi^{n+1},\lambda^{n+1}) \in R^{m-1} \times V_\ell \times R$ satisfying (II.29) and such that :

$$
(P_{II}^n) \qquad J^n(\Lambda^{n+1},\psi^{n+1}) = \inf_{\substack{\Lambda \in R^{m-1} \\ (\psi,\lambda) \in V_\ell \times R \\ \text{related to } \Lambda \text{ by (II.29)}}} J^n(\Lambda,\psi) .
$$

Remark II.7

In the sequential quadratic method proposed by S.P. HAN, the second derivative of the equations of state is introduced into the cost function in order to ensure super– linear convergence of the algorithm. In the algorithm defined above we have only linear convergence provided that the gradient of J with respect to ψ is sufficiently small (cf. J.C. GILBERT.). □

The following proposition is easily proved :

<u>Proposition II.3</u>

The optimality system for the linear quadratic control problem ($P^n{}_{II}$) can be written

$$\text{(II.31)}\quad \begin{cases} b_{u^n}(\psi^{n+1},\varphi) - \lambda^{n+1} c_{u^n}(\varphi) = \sum_{i=1}^{m} \dfrac{\Lambda_i^{n+1} I_S}{S_i} \displaystyle\int_{B_i} \varphi \, dS \\ \quad + \sum_{i=m+1}^{k} \dfrac{I_i}{S_i} \displaystyle\int_{B_i} \varphi \, dS + a'_{u^n}(\psi^n,\varphi) \ , \quad \forall \varphi \in V_\ell \\ d_{u^n}(\psi^{n+1},1) + \lambda^{n+1} c_{u^n}(1) = I_p \end{cases}$$

$$\text{(II.32)}\quad \begin{cases} b_{u^n}(\varphi,\chi^{n+1}-\nu^{n+1}) = - K_1[\psi^{n+1}(E)-\psi^{n+1}(I)][\varphi(E)-\varphi(I)] \\ - K_2\left[\sup\limits_{M\in\Gamma_p^n} z(M)-b_o+\dfrac{\psi^{n+1}(M_o^n)-\psi^{n+1}(M_2^n)}{\dfrac{\partial\psi^n}{\partial z}(M_2^n)}\right]\left[\dfrac{\varphi(M_o^n)-\varphi(M_2^n)}{\dfrac{\partial\psi^n}{\partial z}(M_2^n)}\right] \\ - K_3 \displaystyle\int_{\Gamma_d} [\psi^{n+1}(F_1)-\psi^{n+1}(M)][\varphi(F_1)-\varphi(M)] \, d\Gamma, \quad \forall\varphi \in V_\ell \\ c_{u^n}(\chi^{n+1} - \nu^{n+1}) = 0 \end{cases}$$

$$\text{(II.33)}\quad \frac{1}{S_i}\int_{B_i} \chi^{n+1} \, dS - K_4 I_S \Lambda_i^{n+1} = \frac{1}{S_m}\int_{B_m} \chi^{n+1} \, dS - K_4 I_S \Lambda_m^{n+1}$$

$$\forall i \in \{1,\ldots,m-1\}$$

for $(\Lambda^{n+1},\psi^{n+1},\lambda^{n+1},\chi^{n+1},\nu^{n+1}) \in R^{m-1} \times V_\ell \times R \times V_\ell \times R$.

II.3.2 <u>Internal Conjugate Gradient Iterations</u>

To solve problem ($P^n{}_{II}$) we use a conjugate gradient algorithm (cf. J..CEA, E. POLAK). We start off with

$$\Lambda^n_o = \Lambda^n .$$

The jth iteration is described as follows :

. calculate $(\psi^n_j, \lambda^n_j) \in V_\ell \times R$ by

$$\text{(II.34)} \quad \begin{cases} b_{u^n}(\psi^n_j,\varphi) - \lambda^n_j c_{u^n}(\varphi) = \sum_{i=1}^{m} \frac{(\Lambda^n_j)_i I_S}{S_i} \int_{B_i} \varphi \, dS \\ + \sum_{i=m+1}^{k} \frac{I_i}{S_i} \int_{B_i} \varphi \, dS + a'_{u^n}(\psi^n,\varphi) , \quad \forall\varphi \in V_\ell \\ d_{u^n}(\psi^n_j,1) + \lambda^n_j \, c_{u^n}(1) = I_p \end{cases}$$

. define $(\chi^n_j, \nu^n_j) \in V_\ell \times R$ by

$$\text{(II.35)} \quad \begin{cases} b_{u^n}(\varphi,\chi^n_j - \nu^n_j) = - K_1[\psi^n_j(E)-\psi^n_j(I)][\varphi(E)-\varphi(I)] \\ - K_2 \left[\sup_{M\in\Gamma^n_p} z(M) - b_o + \dfrac{\psi^n_j(M^n_o)-\psi^n_j(M^n_2)}{\frac{\partial\psi^n}{\partial z}(M^n_2)} \right] \left[\dfrac{\varphi(M^n_o)-\varphi(M^n_2)}{\frac{\partial\psi^n}{\partial z}(M^n_2)} \right] \\ - K_3 \int_{\Gamma_d} [\psi^n_j(F_1)-\psi^n_j(M)][\varphi(F_1)-\varphi(M)] \, d\Gamma, \quad \forall\varphi \in V_\ell \\ c_{u^n}(\chi^n_j - \nu^n_j) = 0 . \end{cases}$$

The gradient G^n_j of the functional J^n at the jth iteration has the following components $(G^n_j)_i$:

$$\text{(II.36)} \quad (G^n_j)_i = - \frac{1}{S_i} \int_{B_i} \chi^n_j \, dS + \frac{1}{S_m} \int_{B_m} \chi^n_j \, dS + K_4 I_S[(\Lambda^n_j)_i - (\Lambda^n_j)_m] .$$

Put

$$\text{(II.37)} \quad \begin{cases} w^n_o = G^n_o \\ w^n_j = G^n_j + \kappa^n_j \, w^n_{j-1} \quad \text{for } j > 0 . \end{cases}$$

Define the sequence $(\Lambda_j^n, \psi_j^n, \lambda_j^n) \in R^{m-1} \times V_\ell \times R$ by

(II.38)
$$\begin{cases} \Lambda_{j+1}^n = \Lambda_j^n - \rho_j^n w_j^n \\ \psi_{j+1}^n = \psi_j^n - \rho_j^n \hat{\psi}_j^n \\ \lambda_{j+1}^n = \lambda_j^n - \rho_j^n \hat{\lambda}_j^n \end{cases}$$

where $(\hat{\psi}_j^n, \hat{\lambda}_j^n) \in V_\ell \times R$ are defined by

(II.39)
$$\begin{cases} b_{u^n}(\hat{\psi}_j^n, \varphi) - \hat{\lambda}_j^n \, c_{u^n}(\varphi) = I_S \sum_{i=1}^{m-1} (w_j^n)_i \left[\frac{1}{S_i} \int_{B_i} \varphi \, dS - \frac{1}{S_m} \int_{B_m} \varphi \, dS\right], \\ \qquad \forall \varphi \in V_\ell \\ d_{u^n}(\hat{\psi}_j^n, 1) + \hat{\lambda}_j^n \, c_{u^n}(1) = 0 \end{cases}$$

and where ρ_j^n and κ_j^n are such that

(II.40)
$$J^n(\Lambda_j^n - \rho_j^n w_j^n, \psi_j^n - \rho_j^n \hat{\psi}_j^n) = \inf_{\rho>0} J^n(\Lambda_j^n - \rho w_j^n, \psi_j^n - \rho \hat{\psi}_j^n)$$

and

(II.41)
$$\begin{cases} K_1[\hat{\psi}_j^n(E) - \hat{\psi}_j^n(I)][\hat{\psi}_{j-1}^n(E) - \hat{\psi}_{j-1}^n(I)] \\ +K_2 \dfrac{[\hat{\psi}_j^n(M_o^n) - \hat{\psi}_j^n(M_2^n)]\,[\hat{\psi}_{j-1}^n(M_o^n) - \hat{\psi}_{j-1}^n(M_2^n)]}{\left[\frac{\partial \psi^n}{\partial z}(M_2^n)\right]^2} \\ +K_3 \int_{\Gamma_d} [\hat{\psi}_j^n(F_1) - \hat{\psi}_j^n(M)][\hat{\psi}_{j-1}^n(F_1) - \hat{\psi}_{j-1}^n(M)]\,d\Gamma \\ +K_4 I_S^2 \{ \sum_{i=1}^{m-1} (w_j^n)_i (w_{j-1}^n)_i + [\sum_{i=1}^{m-1} (w_j^n)_i][\sum_{i=1}^{m-1} (w_{j-1}^n)_i] \} = 0. \end{cases}$$

As the number of control parameters for problem $(P^n{}_{II})$ is $(m-1)$, the conjugate gradient algorithm converges in at most $(m-1)$ iterations and therefore

$$(II.42)\quad \begin{cases} \Lambda^n_{m-1} = \Lambda^{n+1} \\ \psi^n_{m-1} = \psi^{n+1} \\ \lambda^n_{m-1} = \lambda^{n+1} \end{cases}.$$

Then the matrix for the system (II.29) can be recalculated for the next external iteration (P_{II}^{n+1}).

II.3.3 *Structure and Convergence of the Algorithm*

The algorithm is summarized in the flow diagram on the following page.

In the cases studied, this algorithm generally converged in 5 external iterations for $\varepsilon_0 = 10^{-5}$. A typical case of control of plasma shape for a mesh of 1500 vertices and 3000 triangles requires 10 seconds of calculation on CRAY 1. Given the limited number of control parameters, almost all the execution time is devoted to the computation and factorization of the matrix of the system (II.29), with the internal conjugate gradient iterations representing only a small part of the calculation time. The execution time for solving problem (P_{II}) by this sequential quadratic method is thus hardly any longer than that of Newton's method for solving the direct problem. This justifies the choice of such an algorithm rather than a method of steepest descent or nonlinear conjugate gradient (Fletcher– Reeves or Polak– Ribière), which would require numerous solutions of nonlinear state equations in order to determine the parameter of optimal descent.

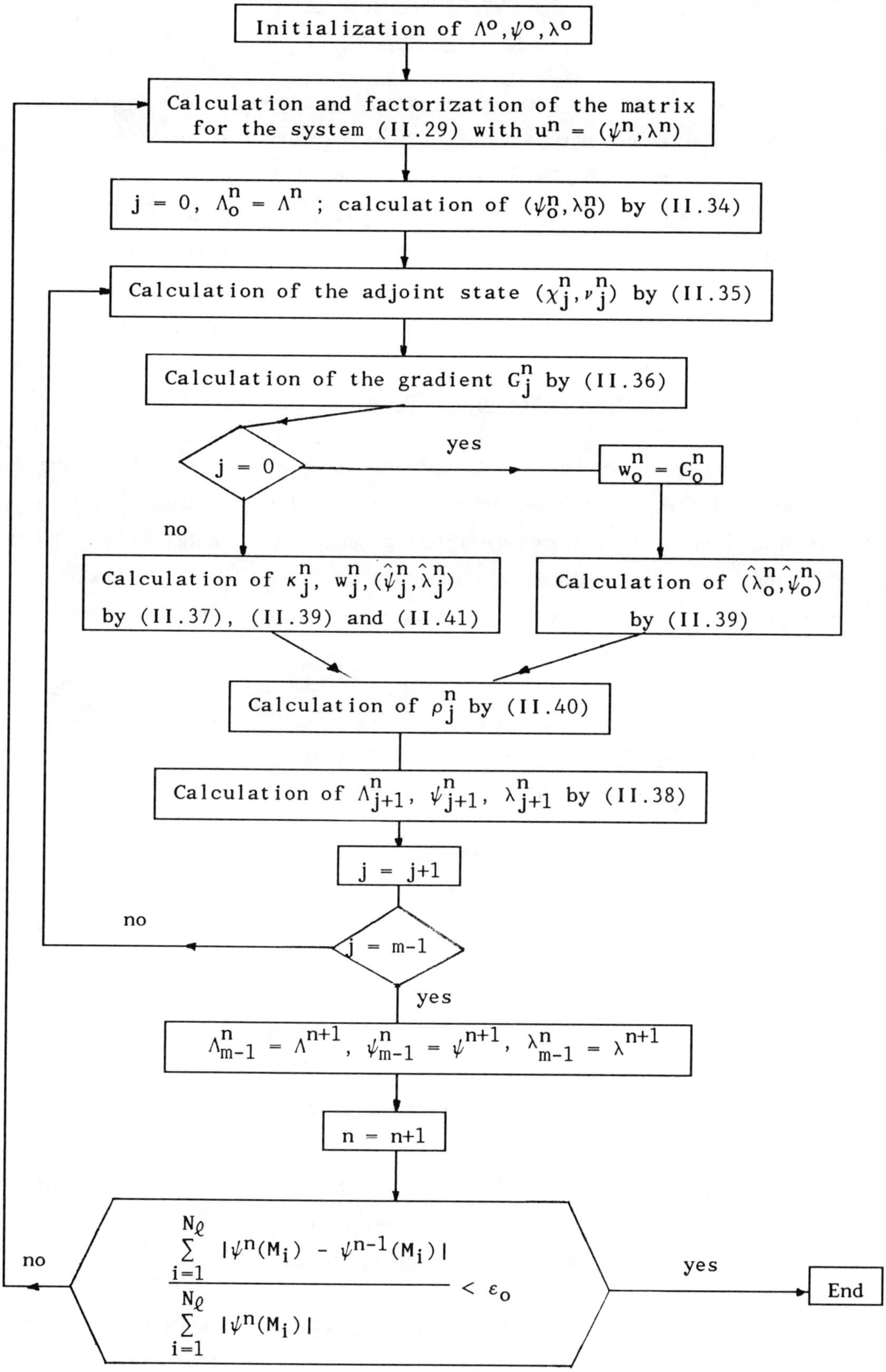
Initialization of $\Lambda^o, \psi^o, \lambda^o$
Calculation and factorization of the matrix for the system (II.29) with $u^n = (\psi^n, \lambda^n)$
$j = 0$, $\Lambda_o^n = \Lambda^n$; calculation of (ψ_o^n, λ_o^n) by (II.34)
Calculation of the adjoint state (χ_j^n, ν_j^n) by (II.35)
Calculation of the gradient G_j^n by (II.36)
$j = 0$
yes
$w_o^n = G_o^n$
no
Calculation of κ_j^n, w_j^n, $(\hat{\psi}_j^n, \hat{\lambda}_j^n)$ by (II.37), (II.39) and (II.41)
Calculation of $(\hat{\lambda}_o^n, \hat{\psi}_o^n)$ by (II.39)
Calculation of ρ_j^n by (II.40)
Calculation of Λ_{j+1}^n, ψ_{j+1}^n, λ_{j+1}^n by (II.38)
$j = j+1$
no
$j = m-1$
yes
$\Lambda_{m-1}^n = \Lambda^{n+1}$, $\psi_{m-1}^n = \psi^{n+1}$, $\lambda_{m-1}^n = \lambda^{n+1}$
$n = n+1$
no
$\dfrac{\sum_{i=1}^{N_\ell} |\psi^n(M_i) - \psi^{n-1}(M_i)|}{\sum_{i=1}^{N_\ell} |\psi^n(M_i)|} < \varepsilon_o$
yes
End

II.4 NUMERICAL RESULTS: CONTROL OF RADIAL POSITION, ELONGATION AND SHAPE OF THE PLASMA

We saw in Section II.1 that two coils allow the radial position of the plasma to be stipulated, whereas it is necessary to have at least three coils in order to give it a definite elongation as well. For control of the shape of the plasma a much greater number of coils is needed. Recall that, as in Chapter I, the currents in the coils and the plasma correspond to the upper half of the torus and have to be multiplied by two to obtain the total (upper and lower) currents; this applies to the TFR, TORE Supra and JET configurations which are here symmetric with respect to the equatorial plane.

II.4.1 Control of the Radial Position of the Plasma

For the optimization we involve the currents in the innermost coil B_1 and the outermost coil B_k. These two coils create a vertical field in the plasma region which allows control of the radial position Δ_H of the plasma, defined in Section I.5.2. The control vector Λ then has a single component $\Lambda_1 = I_1/(I_1+I_k)$. We minimize the functional (J_1) given by (II.3), where determining the points I and E corresponds to fixing a desired radial position :

$$\Delta_H = \tfrac{1}{2}[r(I) + r(E)] - R_o \ . \qquad \text{(II.43)}$$

If the constants K_2, K_3 and K_4 are taken to be zero in the expression (II.8) for J, we are reduced to solving the problem (P_{II}) with $\Lambda \in R$.

In the algorithm of Section II.3, each external iteration consists of a single internal gradient iteration, since it is a one-parameter problem. We consider the Tokamak TFR (cf. Section I.5) for which there are only two coils, the inner one B_1 and the outer one B_2. We put $I_p = 100$kA, $I_S = I_1 + I_2 = -102$kA.

Figure II.2 shows the plasma boundary Γ_p as well as 20 internal lines of flux corresponding to equidistant ψ values, for 6 particular configurations distinguished by their values for β_p, ℓ_i and Δ_H (see Section I.5.2 for the definition of these quantities). The case $\ell_i = 0.5$ corresponds to a flat current density : $g(\psi_N) = 1$. The case $\ell_i = 1.5$ corresponds to a peaked current density : $g(\psi_N) = (1-\psi_N^{1.5})^3$. The value of Δ_H (in cm) corresponds to the displacement of the plasma relative to the vacuum vessel as is determined once the points I and E are given, as in formula (II.43).

The value of Λ obtained numerically for each of the 6 cases is shown beneath each figure. The quantity e denotes the elongation of the plasma; it varies between 0.85 and

a) $\beta_p=0.25$, $l_i=0.5$, $\Delta_H=0.1$, $e=0.85$, $\Lambda=0.48$

b) $\beta_p=0.25$, $l_i=1.5$, $\Delta_H=0.1$, $e=0.91$, $\Lambda=0.29$.

c) $\beta_p=1$, $l_i=0.5$, $\Delta_H=0.1$, $e=0.87$, $\Lambda=0.21$

d) $\beta_p=1$, $l_i=1.5$, $\Delta_H=0.1$, $e=0.91$, $\Lambda=0.03$

e) $\beta_p=0.25$, $l_i=1.5$, $\Delta_H=-5$, $e=0.94$, $\Lambda=0.31$

f) $\beta_p=0.25$, $l_i=1.5$, $\Delta_H=+5$, $e=0.95$, $\Lambda=0.16$

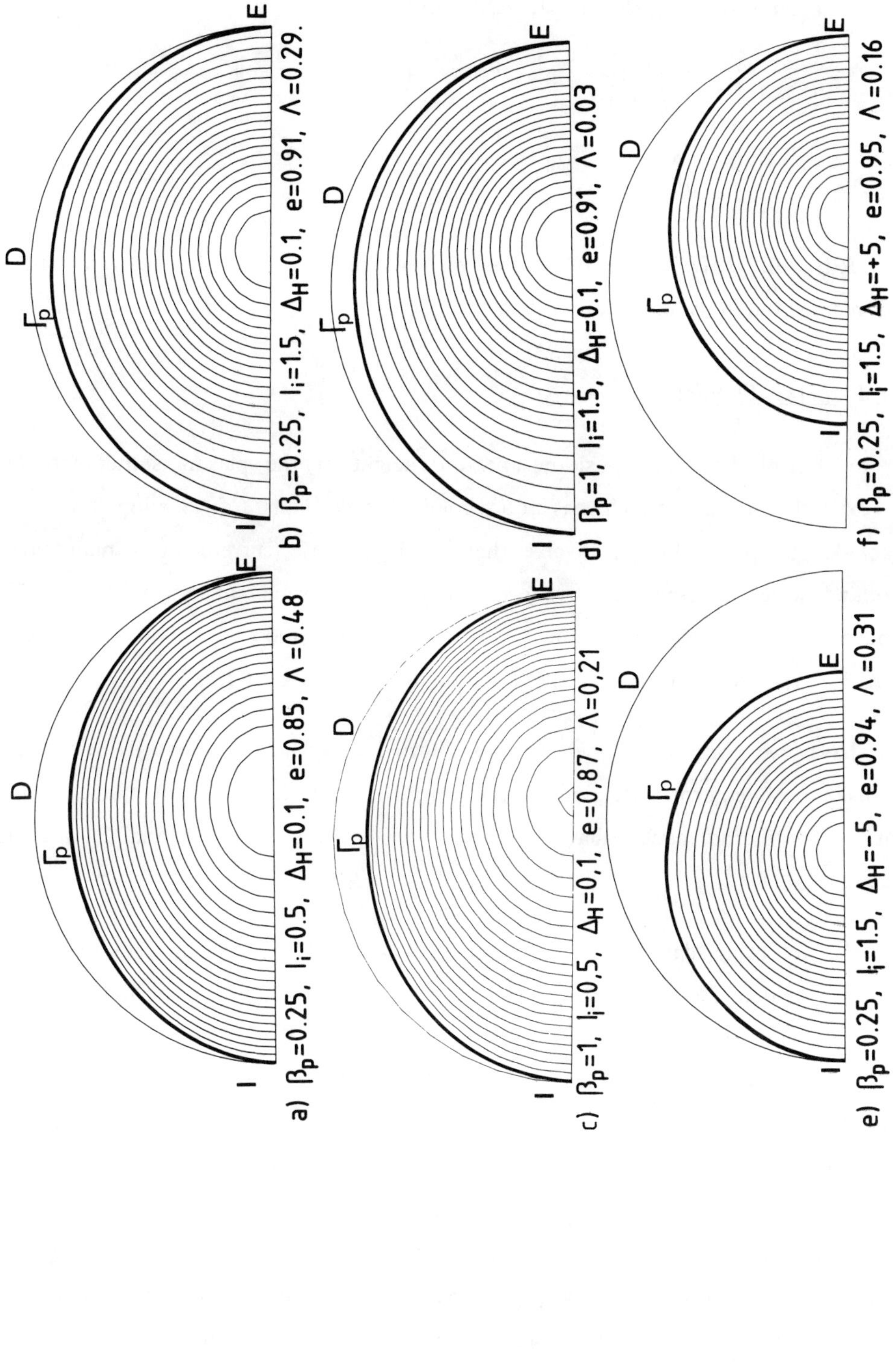

Figure II.2 : Shapes of the plasma in TFR as a function of β_p, l_i and Δ_H.

0.95 according to the case. The plasma boundary is thus slightly flattened relative to a circle. This can be compared with the formula giving the elongation of the plasma by an expansion to second order in a/R (cf. J. ANDREOLETTI) :

$$e = 1 + \frac{a^2}{2R^2}\left[\frac{3}{4}\left(\text{Log}\,\frac{8R}{a} - \frac{\beta_p}{3} - \frac{\ell_i}{6} - \frac{11}{12}\right) - n\left(\text{Log}\,\frac{8R}{a} + \beta_p + \frac{\ell_i}{2} - \frac{3}{2}\right)\right] \tag{II.44}$$

with $n = - R/B_v\ \partial B_v/\partial \Delta_H$, where B_v is the vertical component of the external field in the plasma region and n is its index.

II.4.2 Control of the Elongation

We aim to control the radial position and the height of the plasma simultaneously. Consider the example of JET which has four poloidal field coils (cf. Section I.5). In controlling the elongation we do not involve the coil B_2, whose current I_2 is maintained equal to zero and which is used exclusively to control the plasma shape (see Fig. I.17). We thus have two control parameters $\Lambda_1 = I_1/I_S$ and $\Lambda_2 = I_3/I_S$ with $I_S = I_1 + I_3 + I_4$, and we minimize the functional J_3 given by (II.5).

Since we have two control parameters, two internal conjugate gradient iterations are necessary to make each external iteration of the algorithm converge, and at the optimum we note that $J_3 = 0$ and hence that Γ_p passes through the points I and E and its height is b_0.

Consider the following data : $I_S = -7.42$ MA, $I_p = 1.9$ MA, $\beta = 1.5$, $g(\psi_N) = 1 - \psi_N$. We seek $\Lambda \epsilon R^2$ such that Γ_p passes through the points I (r=1.75m, z=0) and E (r = 4.215m, z = 0) and is of given height b_0 = 2m. We obtain numerically : $I_1 = -6.17$ MA, $I_3 = 0.135$ MA, $I_4 = -1.385$ MA. Figure II.3 (a) shows the equilibrium configuration thus obtained, and Figure II.3 (b) shows Γ_p and 20 lines of flux inside the plasma, corresponding to equidistant ψ. Note that the highest point of the plasma is very close to the inner boundary Γ_2 of the vacuum vessel. Thus this formulation is somewhat inconvenient in that a certain elongation is imposed on the plasma but the point of maximum height may be badly situated on the horizontal line $z = b_0$, with respect to the vacuum vessel in particular. This justifies controlling the shape of the plasma even when there are few control parameters available.

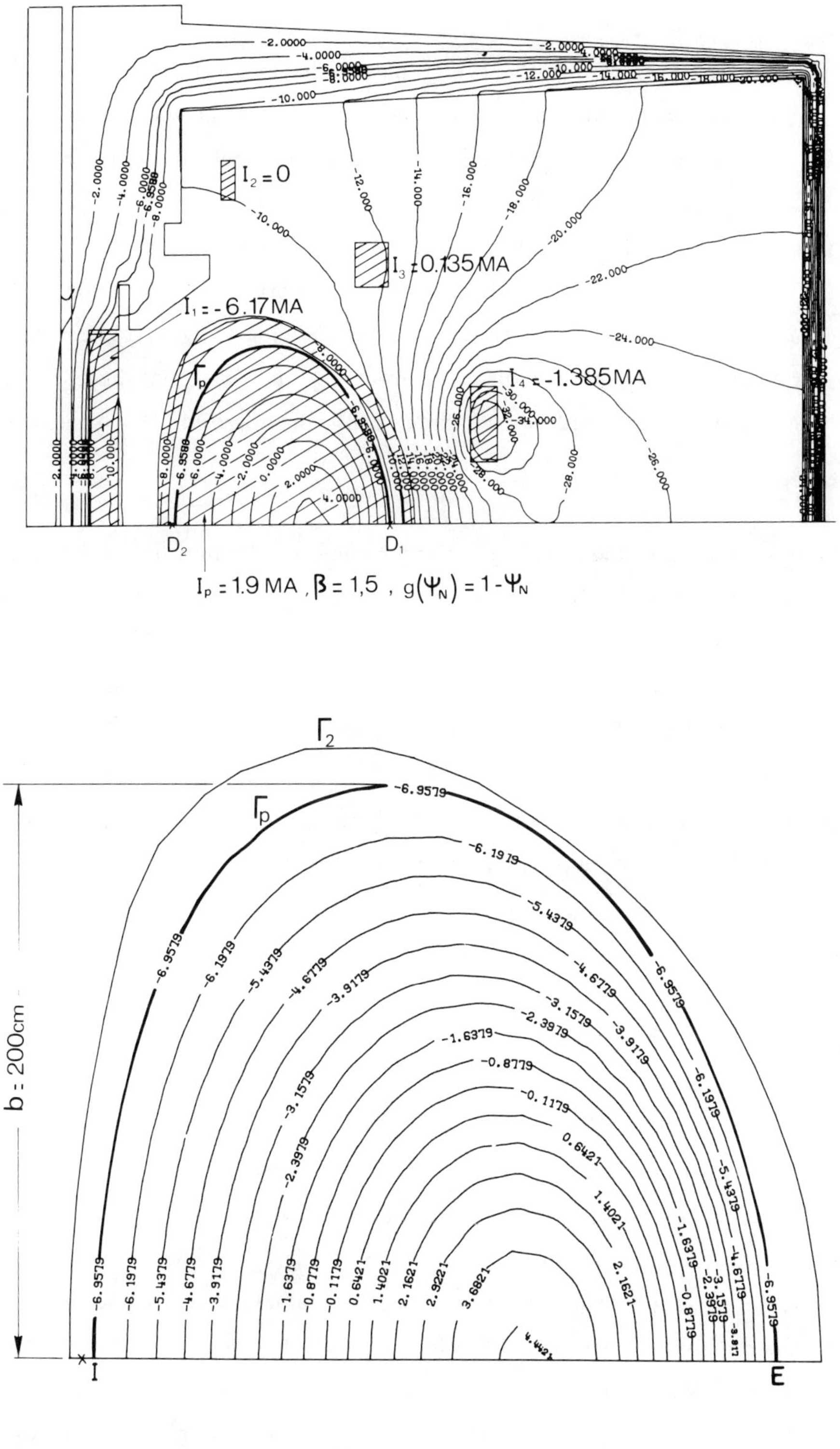

Figure II.3 : *Control of the elongation of the plasma in JET*.

II.4.3 Control of the Shape of the Plasma

II.4.3.1 Application to JET

In the case of JET we wish to realize a plasma having a D shape, with boundary denoted Γ_d. To begin with, we do not involve the coil B_2 in the optimization as in Section II.4.2; the current I_2 is maintained at 0 and we seek the pair $\Lambda = (\Lambda_1,\Lambda_2)$ which minimizes J_4 given by (II.6), where F_0 is the point with coordinates $r = 4.215m$, $z=0$. We take the same data as in Section II.4.2 and obtain numerically $I_1 = -6.135$ MA, $I_3 = 0.078$ MA, $I_4 = -1.362$ MA. The equilibrium configuration obtained is represented in Figs. II.4a and II.4b where the desired boundary Γ_d is represented by the dashed curve. Note that Γ_p is very different from Γ_d and does not have the required D shape. This proves that with only three coils the desired D shape cannot be realized.

We therefore introduce the coil B_2 (the "shaping coil") into the optimization and seek $\Lambda = (\Lambda_1, \Lambda_2, \Lambda_3) \in R^3$ which minimizes J_4. We obtain numerically : $I_1 = -6.185$ MA, $I_2 = 0.285$ MA, $I_3 = -0.196$ MA, $I_4 = -1.322$ MA. Note the appearance of a dipole between the coils B_2 and B_3, and that moreover $I_2 + I_3$ is very close to the value of I_3 obtained previously with three coils only. The configuration obtained is shown in Figs. II.5a and II.5b, where we see that Γ_p is very close to Γ_d and has indeed the required D shape. Thanks to the dipole created between coils 2 and 3 we do obtain the required triangularity for Γ_p, which cannot be obtained with only three coils.

II.4.3.2 Application to Tore Supra.

As was seen in Section I.5, the Tokamak Tore Supra has five poloidal field coils. An initial study was carried out with fixed $I_3 = I_4 = I_5$ in order to distribute the heating and so allow long discharges. Since I_S is likewise fixed, we have to solve a problem with two parameters $\Lambda_1 = I_1/I_S$ and $\Lambda_2 = I_2/I_S$. The aim is to keep the plasma circular during a phase of intensive heating characterized by the increase in β. To do this we minimize J_4 with Γ_d circular and F_0 the point with coordinates $r = 1.4m$, $z = 0$. We solve the control problem with values of β increasing from 1 to 4. The other data are : $I_S = -0.241$ MA, $I_p = 0.85$ MA, $g(\psi_N) = 1-\psi_N$.

a)

b)

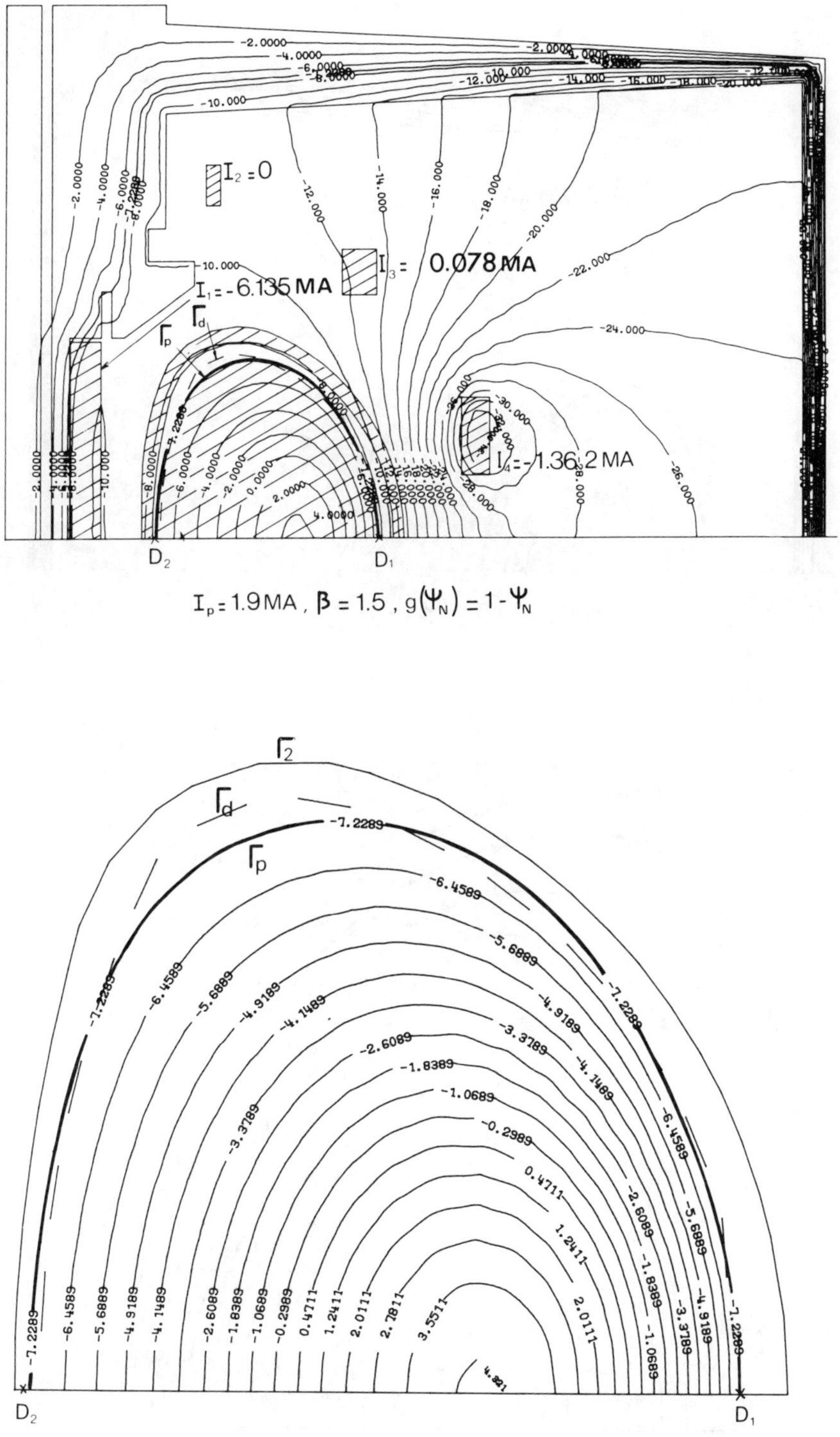

Figure II.4 : *Control of the shape of the plasma in JET (3 coils).*

a)

b)

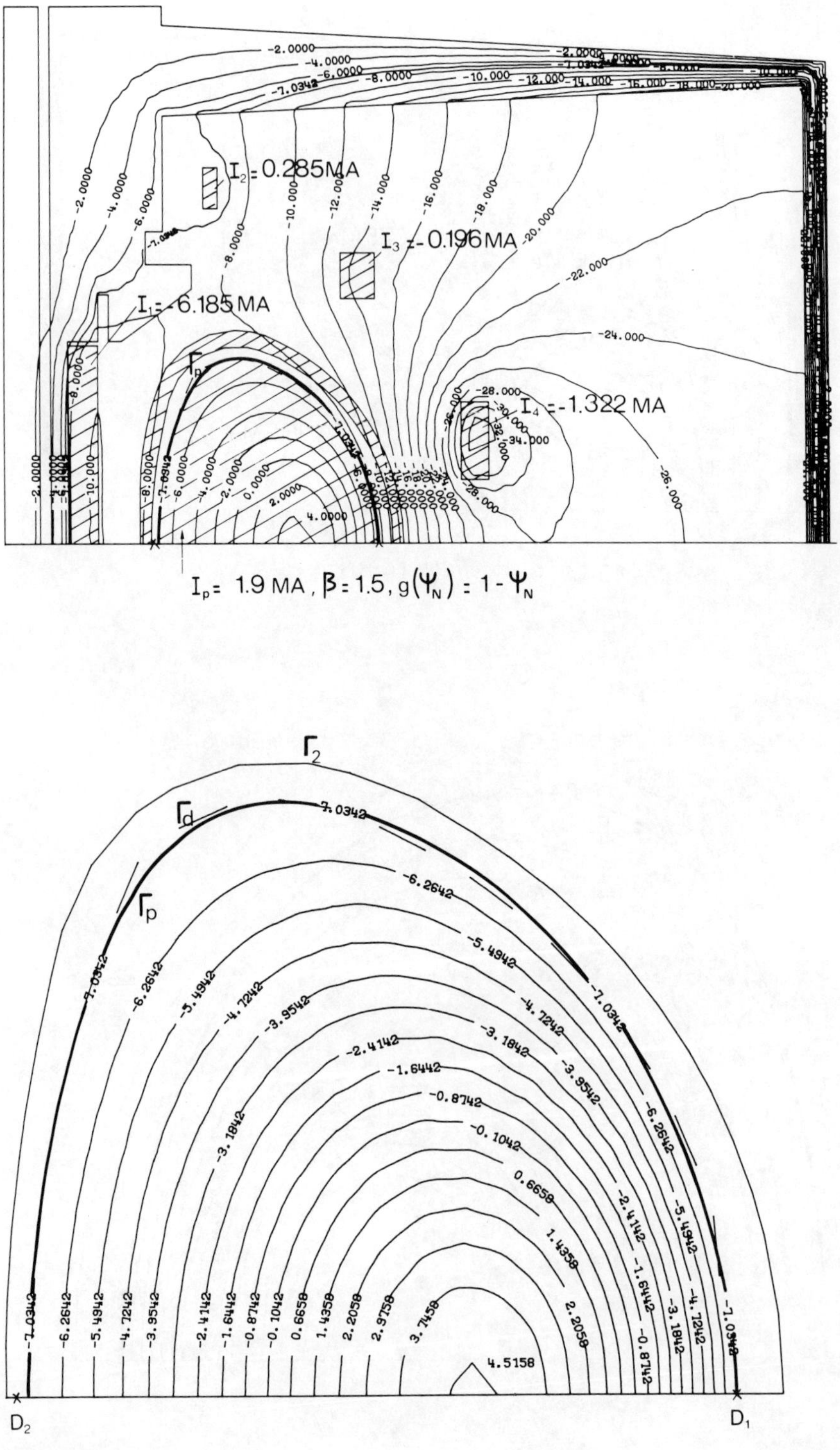

Figure II.5 : *Control of the shape of the plasma in JET (4 coils).*

Fig. II.6 shows the evolution of the currents obtained numerically as a function of β, as well as the variation in J_4. Figs. II.7a and II.7b show the equilibrium configurations thus obtained for β having values of 1, 2 and 2.5, the circular boundary Γ_d being shown as a dashed curve and Γ_2 representing the inner boundary of the vacuum vessel. Note that the plasma is quite circular for $\beta = 1$, but deforms increasingly as β increases. The triangular deformation is such that when β reaches 2.7 the plasma touches Γ_2, which obviously cannot be tolerated. This explains the fact that with the condition $I_3 = I_4 = I_5$ it was not possible to obtain equilibrium for β greater than 2.7.

It is therefore necessary to have a supplementary control parameter. For this we allow freedom for the current I_3 because it is this which, with I_2, determines the shape of the plasma. We still impose the condition $I_4 = I_5$. We then have three control parameters Λ_1, Λ_2 and Λ_3, and the evolution of the currents and of J_4 as a function of β is given in Fig. II.8. Observe that I_3 varies in a way quite different from that of I_4 and I_5, and that for β equal to 2.5 the functional J_4 is 100 times lower than it was with two parameters. Figs. II.9a and II.9b show that the plasma Γ_p is maintained quasi-circular for β going from 1 up to 4. These computations have demonstrated the necessity of proceeding in such a way that I_3 is independent of I_4 and I_5.

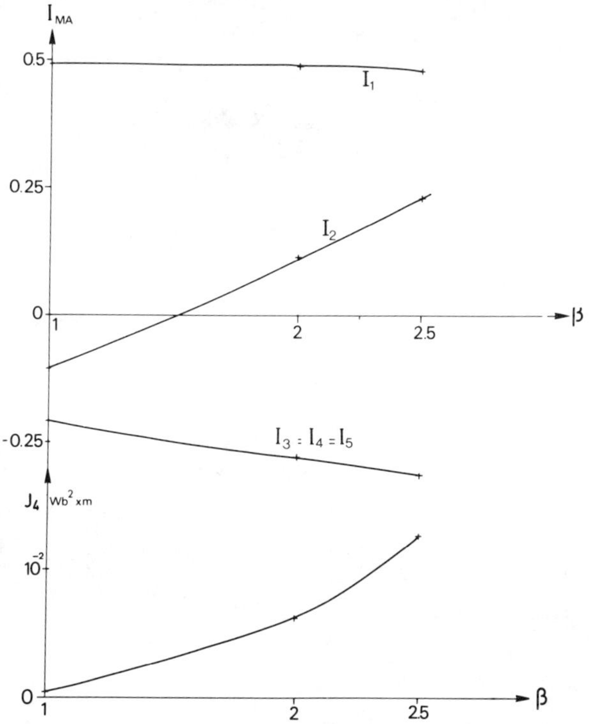

<u>*Figure II.6*</u> : <u>*Optimal currents and minimum of the cost function as a function of β in TORE Supra*</u> *($I_3 = I_4 = I_5$)*

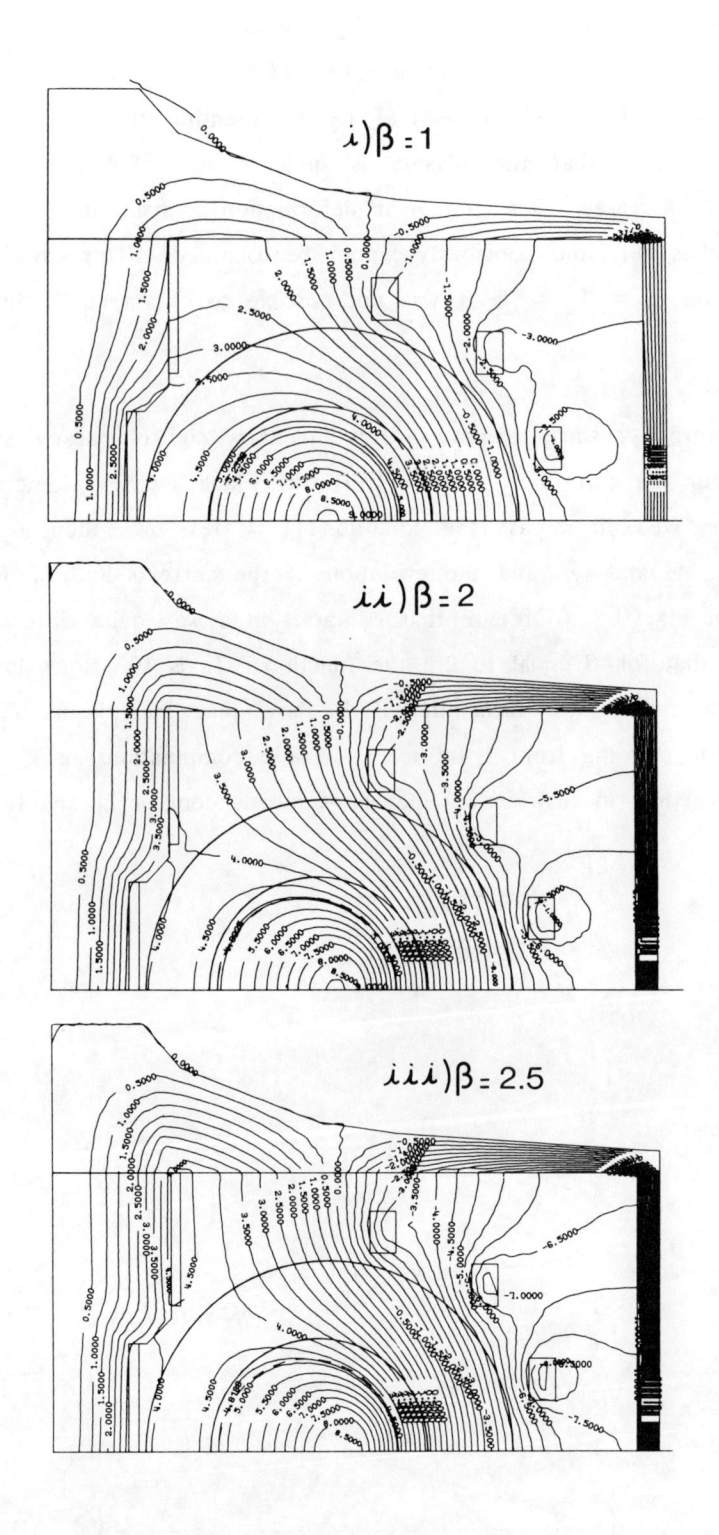

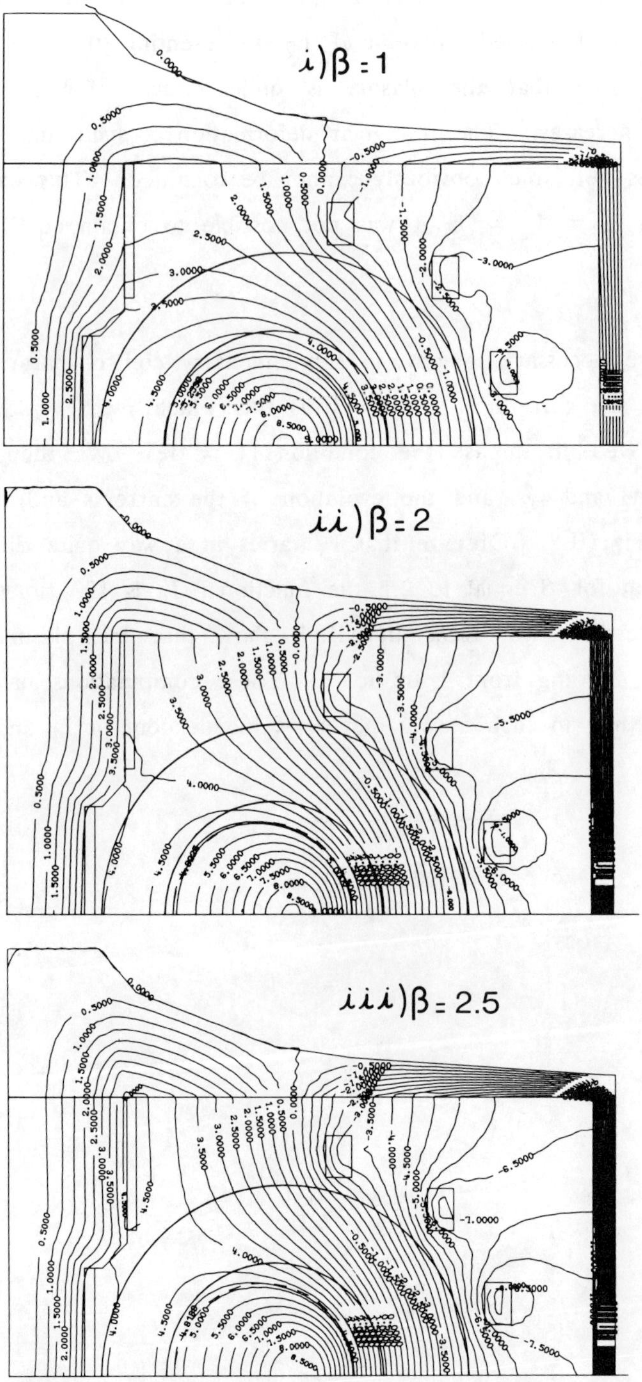

Figure II.7a : *Equilibrium configurations for β values of 1,2 and 2.5 in TORE Supra* ($I_3 = I_4 = I_5$)

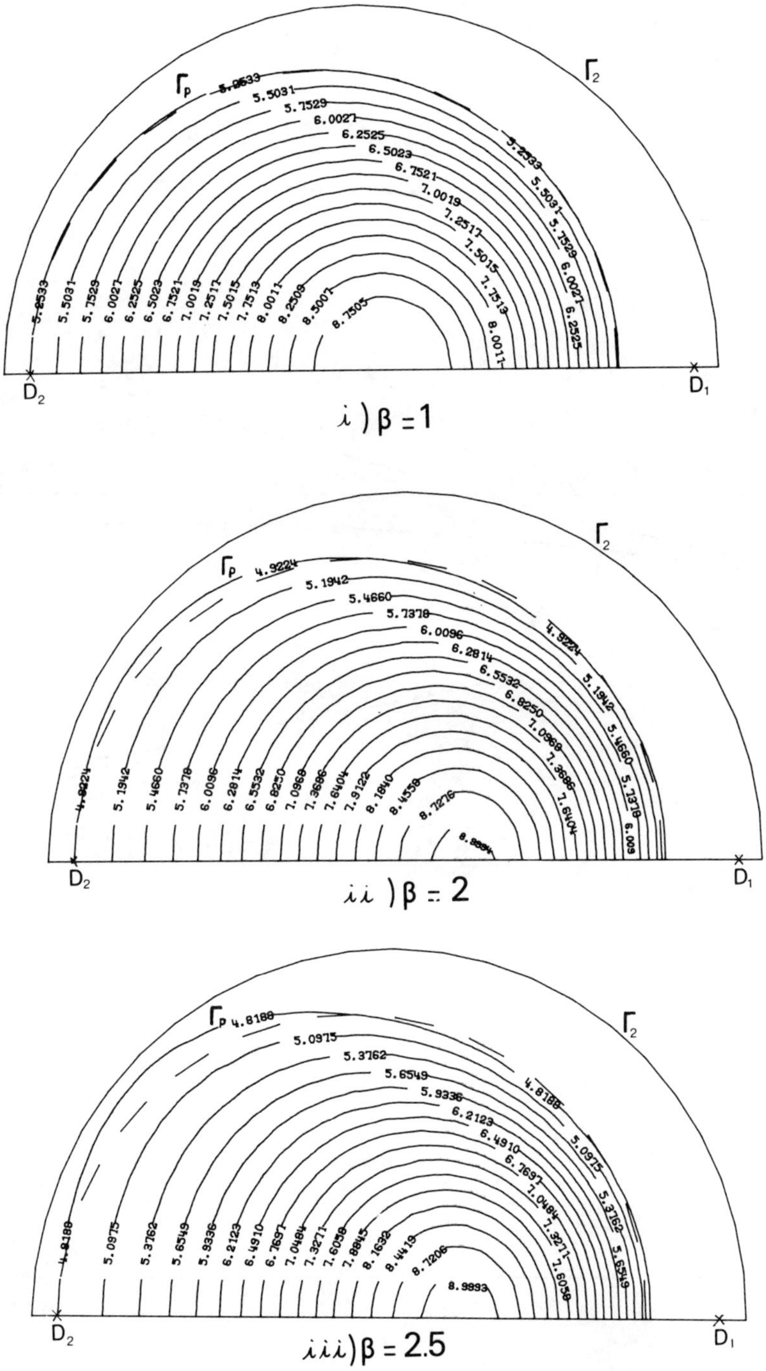

<u>*Figure II.7b*</u> : <u>*Plasma Configurations for β Values of 1, 2, and 2.5 in TORE Supra*</u> ($I_3 = I_4 = I_5$)

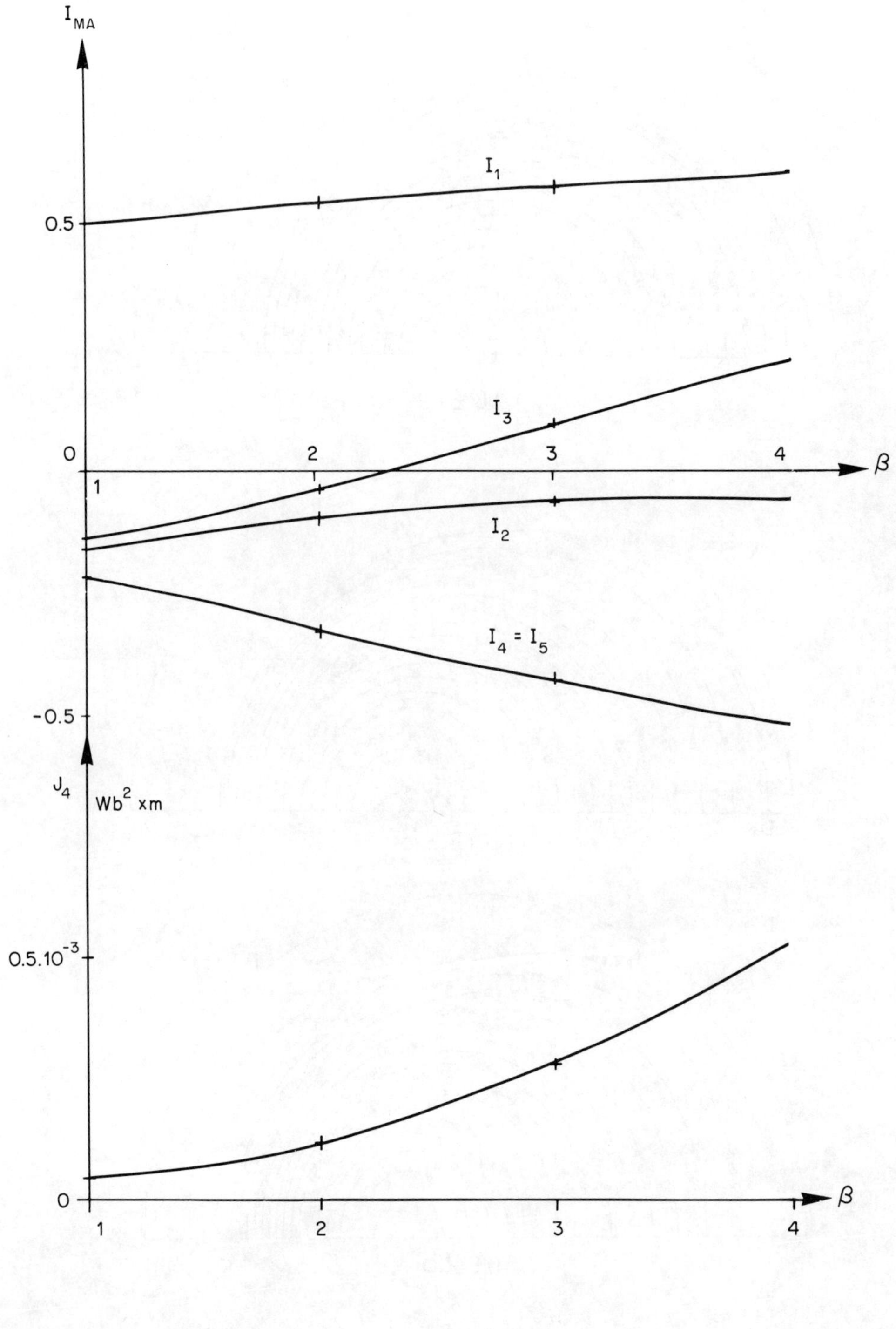

Figure II.8 : Optimal currents and minimum of J_4 *as a function of* β $(I_4 = I_5)$.

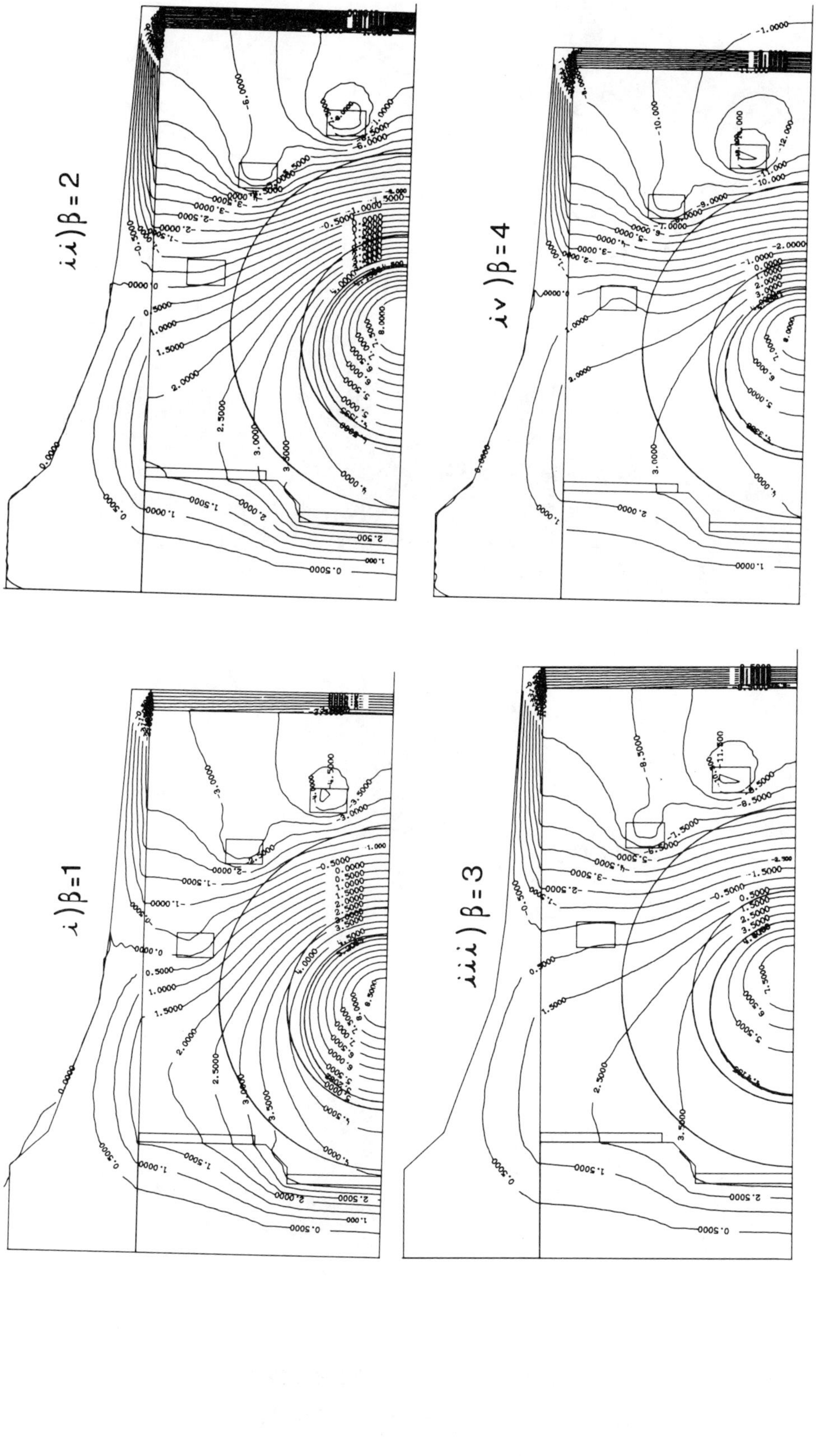

Figure II.9a : *Equilibrium Configurations for* β *Going From 1 to 4 in TORE Supra* ($I_4 = I_5$)

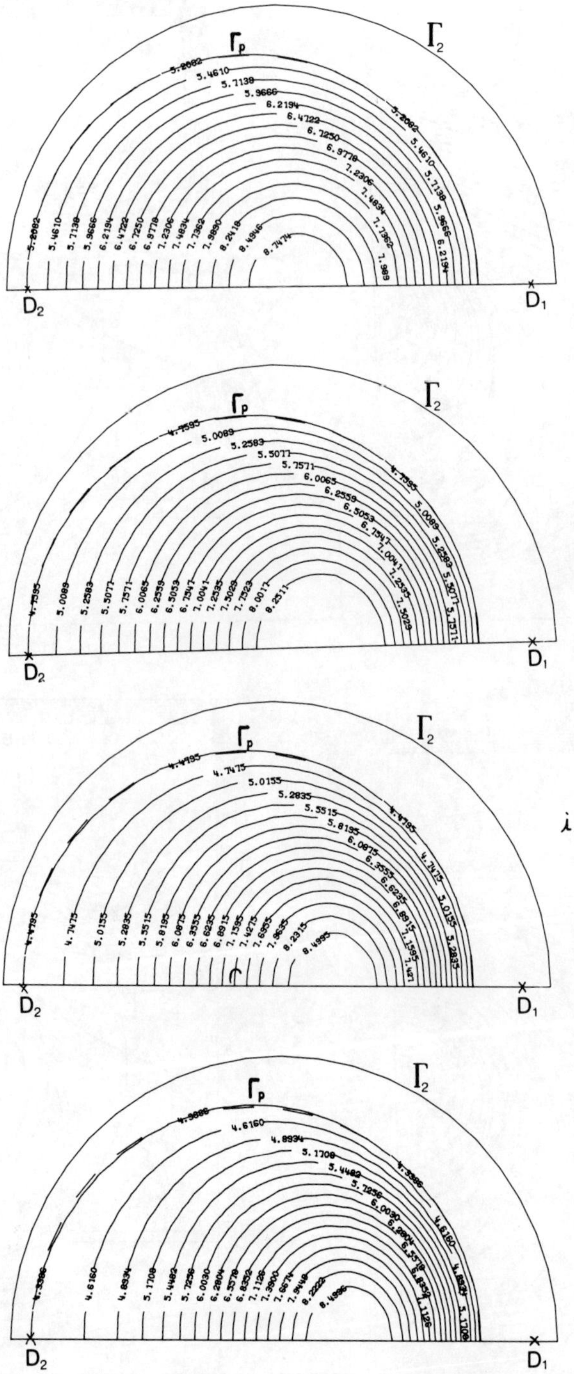

i) $\beta = 1$

ii) $\beta = 2$

iii) $\beta = 3$

iv) $\beta = 4$

Figure II.9b : Equilibrium configurations for β going from 1 to 4 in TORE Supra ($I_4 = I_5$)

II.4.3.3 *Application to INTOR*

In the case of INTOR we aim to control not only the shape of the plasma boundary but also the position of the stagnation point of the divertor. A study of this kind has been carried out in R.AYMAR et al., [2], [3], and we repeat here certain of the results of that work in order to show the difficulty of controlling the stagnation point. In fact, since this point X_1 is a point of zero field, the function ψ is quasi-constant in a neighborhood of this point and the part of Γ_d close to X_1 carries little weight in the functional J_5 given by (II.7). This is why in R.AYMAR et al., [2] a cost function J_6 is defined as follows :

(II.45) $$J_6 = J_5 + K_4 \sum_{i=1}^{m} I_i^2 + K_5 \int_V |\nabla\psi|^2 \, dS$$

where V is a neighbourhood of X_1.

The second term in J_6 is intended to prevent the appearance of too significant current dipoles, while the third term of J_6 aims to make X_1 the point of zero field in the configuration. The choice of coefficients K_4 and K_5 allows more or less importance to be given to controlling the plasma shape, zero field point or total electric energy of the system.

Fig. II.10 shows various configurations obtained with 4,5 or 6 control parameters. In the case of Fig.II.10i the accent is put on controlling the X−point and in this case we observe that the plasma Γ_p is strongly displaced towards the exterior relative to Γ_d. In the case of Fig.II.10ii the plasma shape is very close to Γ_d, but it is the stagnation point which is not at the desired position. Finally, in the case of Fig.II.10iii the boundary Γ_p is very close to Γ_d, the zero field point is very close to X_1, but at the price of substantial current dipoles (I_7 = 5500 kA, I_8= −8000 kA, I_9 = 10000 kA). As is further explained in R.AYMAR et al. [2], it is very difficult to obtain simultaneously the right plasma shape, correct position of divertor stagnation point and an energetically favourable configuration.

The study by R.AYMAR et al. [3] shows that a double divertor which is symmetric with respect to the equatorial plane would permit much better control of the plasma shape and the position of the stagnation point than a simple but asymmetric divertor.

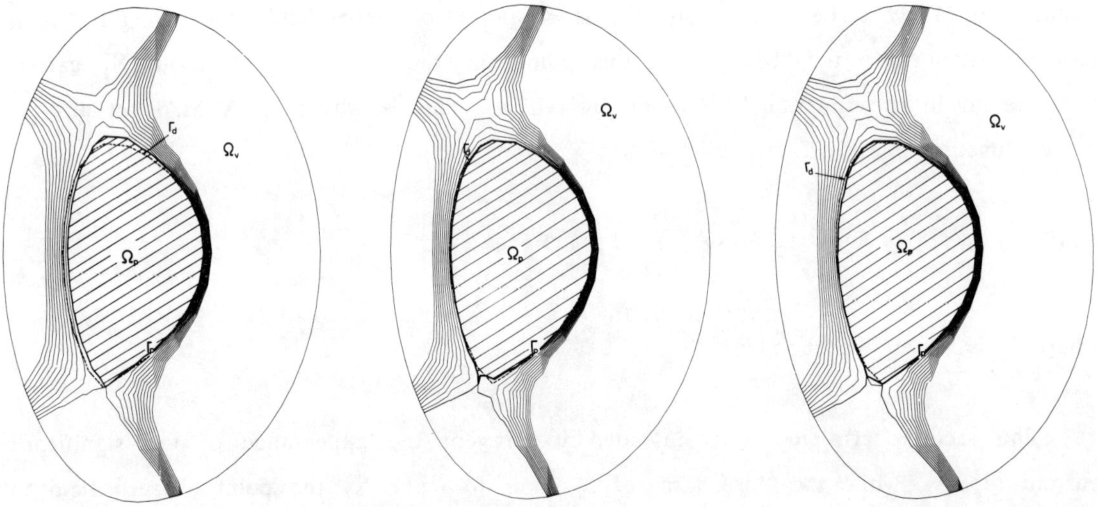

(i) 4 parameters (ii) 5 parameters (iii) 6 parameters

Figure II.10 : *Optimization of the equilibrium configuration in INTOR as a function of the number of control parameters.*

3. Existence and control of a solution to the equilibrium problem in a simple case

In this chapter we shall consider a simplified model for the equations of equilibrium in the case of a Tokamak without iron and where the plasma current density function is given by a linear law in ψ_N.

The problem of equilibrium in a perfect superconducting shell in the absence of inducing currents and of limiter has been studied in R. TEMAM [2], H. BERESTYCKI – H. BREZIS, J. P. PUEL. Here we shall first of all generalize some of these results to the case with inducing currents, and then use these results to give sufficient conditions for the existence of solutions to the problem with limiter.

Later, under certain hypotheses, we establish the existence of a solution and the optimality conditions of the control problem for the shape of the plasma formulated in Chapter II, the equations of state being those of the simplified model.

The results of this chapter are presented in J. BLUM – T. GALLOUËT – J. SIMON [1] and [2].

III.1 THE EQUATIONS FOR THE SIMPLIFIED MODEL

The following three assumptions will be made:

(i) We are dealing with the case of a Tokamak without iron, i.e. an air transformer: $\Omega_f = \emptyset$ so $\mu = \mu_0$ throughout Ω.

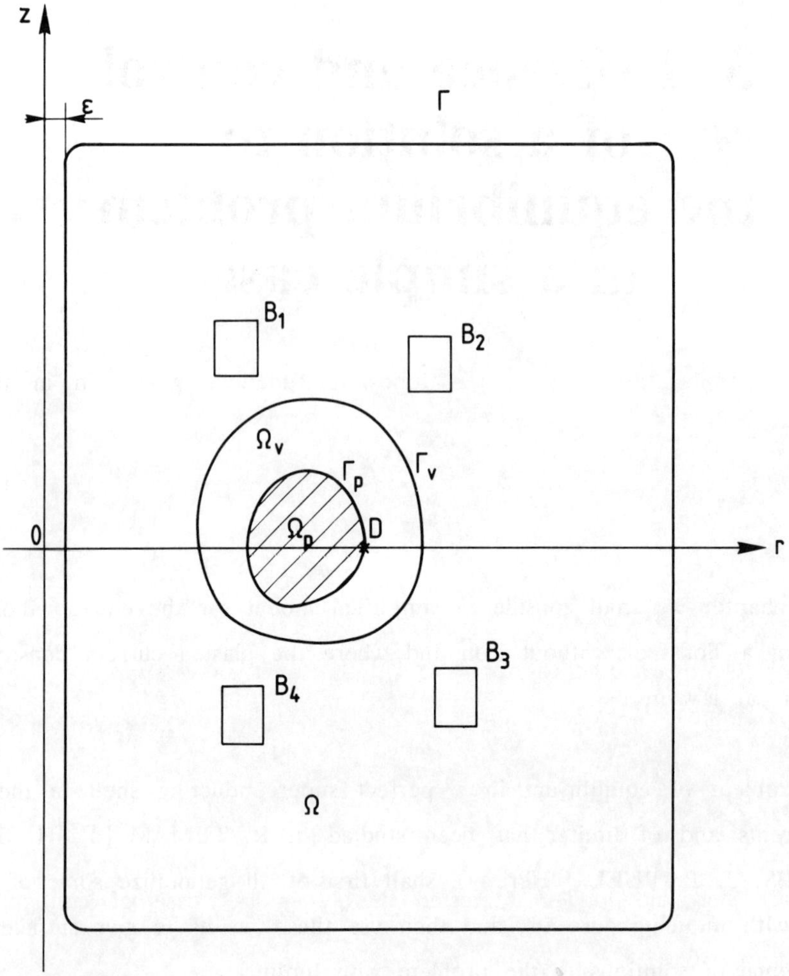

Figure III.1 : *Domain of study Ω in a meridian section of the torus.*

(ii) We suppose

$$\Omega \subset \{x = (r,z) \ , \ r \geqslant \varepsilon\} \quad \text{with } \varepsilon > 0 \ .$$

Remark III.1

This is always true in the case when the boundary Γ of Ω is a perfect shell. In the case when there is no shell the problem is theoretically posed in $R^+ \times R$. However, we restrict ourselves in fact to a bounded domain Ω in $R^+ \times R$ whose boundary Γ (supposed regular) is taken sufficiently far from the vacuum chamber and the coils that the poloidal flux ψ can be taken to be zero on Γ. In particular we suppose that $\psi = 0$ for $r = \varepsilon$ (ε being sufficiently small), while according to the definition (I.8) of ψ it is proportional to ε^2, and the boundary Γ of Ω will in general contain a segment of the straight line $r = \varepsilon$ (see Fig III.1). □

(iii) The plasma current density function can be written simply as

(III.1) $j_T(r,\psi) = \lambda(1-\psi_N) \quad \text{in } \Omega_p$

where we recall that ψ_N is the normalized flux given in (I.21).

Remark III.2

This corresponds to choosing the following functions g and h in the expression (I.21) for j_T :

$$g(\psi_N) = 1 - \psi_N$$

$$h(r) = 1 \ .$$

The function h(r) given in (I.21) may be taken equal to 1 in the case of Tokamaks with high aspect ratio where r is close to R_o throughout the plasma. The above choice of $g(\psi_N)$ corresponds to the so-called "linear" case (see Remark I.3) since j_T is linear in ψ_N. □

We are here no longer interested merely in the case of configurations having symmetry relative to the equatorial plane as in Problem (P_I) but in the case of arbitrary axisymmetric configurations.

The equations (I.30) for the pair (ψ,λ) for the problem (P_I) thus become

(III.2)
$$\begin{cases} L\psi = j_B & \text{in } \Omega_a = \Omega - \bar{\Omega}_p \\ L\psi = \lambda(1-\psi_N) & \text{in } \Omega_p \\ I_p = \lambda \int_{\Omega_p} (1 - \psi_N)\,dS & \\ \psi = 0 & \text{on } \Gamma \\ \begin{cases} (\frac{\partial\psi}{\partial n})_e = (\frac{\partial\psi}{\partial n})_i \\ \psi_e = \psi_i \end{cases} & \text{on } \Gamma_p \end{cases}$$

with

$$\Gamma_p = \{M \in \Omega_V | \psi(M) = \sup_D \psi\}$$

$$\Omega_p = \{M \in \Omega_V | \psi(M) > \sup_D \psi\}$$

and where j_B is given by (I.20).

We suppose here for simplicity that the current j_{CV} in the vacuum chamber is zero. The support for j_B is thus the union of the coils B_i.

The operator L defined by (I.13) is linear here because of the hypothesis (i).

We carry out the following change of variables:

$$\begin{cases} \psi' = \psi - \psi_p \\ \lambda' = \dfrac{\lambda}{\psi_a - \psi_p} \end{cases}$$

with

$$\psi_p = \sup_D \psi \quad \text{and} \quad \psi_a = \sup_{\Omega_p} \psi .$$

The equations (III.2) for the pair (ψ',λ') then become:

$$\text{(III.3)} \quad \begin{cases} L\psi' = j_B & \text{in } \Omega_a \\ L\psi' = \lambda'\psi' & \text{in } \Omega_p \\ I_p = \lambda' \displaystyle\int_{\Omega_p} \psi' \, dS & \\ \psi' = \text{constant on } \Gamma & \\ \begin{cases} \left(\dfrac{\partial\psi'}{\partial n}\right)_e = \left(\dfrac{\partial\psi'}{\partial n}\right)_i \\ \psi'_e = \psi'_i \end{cases} & \text{on } \Gamma_p \\ \sup_D \psi' = 0 & \end{cases}$$

with

$$\Gamma_p = \{M \in \Omega_V | \psi'(M) = 0\}$$

$$\Omega_p = \{M \in \Omega_V | \psi'(M) > 0\} \ .$$

The system (III.3) may also be written

$$\text{(III.4)} \quad \begin{cases} L\psi' = \lambda' \psi'^+ 1_{\Omega_V} + j_B \text{ in } \Omega \\ I_p = \lambda' \int_{\Omega_V} \psi'^+ \, dS \\ \psi' = \text{constant on } \Gamma \end{cases}$$

and

$$\text{(III.5)} \qquad \sup_D \psi' = 0$$

with $\psi'^+ = \sup(\psi',0)$ and where 1_{Ω_V} denotes the characteristic function for the vacuum region Ω_V.

The plasma is thus defined as being the zone of positivity of ψ' in the interior of Ω_V. There may be a second zone of positivity of ψ' outside Ω_V: in this zone there is no plasma and the term $\lambda'\psi'^+ 1_{\Omega_V}$ is zero.

<u>*Remark III.3*</u>

For a pair (ψ',λ') satisfying (III.4) and (III.5) the boundary Γ_p of the plasma Ω_p defined by $\Omega_p = \{M \in \Omega_V | \psi'(M) > 0\}$ may have nonempty intersection with the boundary Γ_V of Ω_V (see Fig III.2). In this case Γ_p is no longer the flux line $\psi' = 0$ in Ω_V and this solution is not interesting from a physical point of view. This shows that the limiter D is not correctly situated since in this case it does not prevent the plasma from touching the vacuum vessel. □

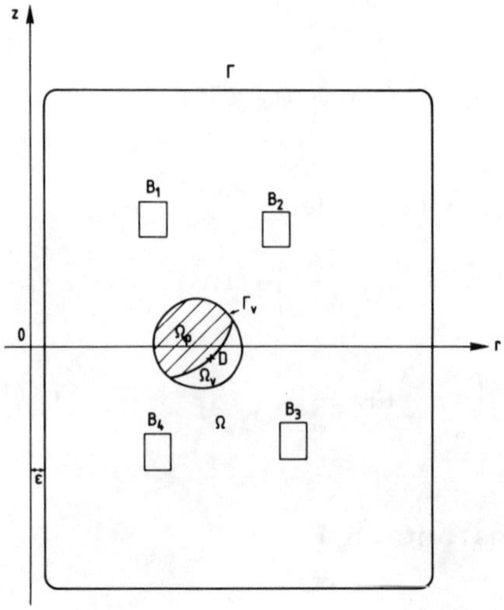

Figure III.2 : Example of a non-physical solution of equations (III.4) – (III.5).

Problem (P_{III}) consists of the following: given $j_B \in L^2(\Omega)$ and $I_p \in R^+$, find a pair $(\psi',\lambda') \in C^0(\bar{\Omega}) \times R^+$ that is a solution to equations (III.4) and (III.5).

In order to solve this problem we have to solve an ancillary problem, namely determination of the existence and properties of a solution ψ' to (III.4) when λ' is fixed.

III.2 AN ANCILLARY PROBLEM

III.2.1 An Existence Result

Let $H_c^2(\Omega) = \{v \in H^2(\Omega)$ such that v is constant on $\Gamma\}$

where $H^2(\Omega) = \{v \in L^2(\Omega)$ such that

$$(\frac{\partial v}{\partial r}, \frac{\partial v}{\partial z}, \frac{\partial^2 v}{\partial r^2}, \frac{\partial^2 v}{\partial r\, \partial z}, \frac{\partial^2 v}{\partial z^2}) \in (L^2(\Omega))^5\} .$$

and

$$|v|^2_{H^2(\Omega)} = |v|^2_{L^2(\Omega)} + \left|\frac{\partial v}{\partial r}\right|^2_{L^2(\Omega)} + \left|\frac{\partial v}{\partial z}\right|^2_{L^2(\Omega)} + \left|\frac{\partial^2 v}{\partial r^2}\right|^2_{L^2(\Omega)} + \left|\frac{\partial^2 v}{\partial r \partial z}\right|^2_{L^2(\Omega)} + \left|\frac{\partial^2 v}{\partial z^2}\right|^2_{L^2(\Omega)}$$

Recall that by the theorem of Sobolev (cf. J. L. LIONS [1], R.A. ADAMS)

$$H^2(\Omega) \subset C^0(\bar{\Omega}) \ .$$

Let

$$H^1_o(\Omega) = \{v \in H^1(\Omega) \quad \text{such that} \quad v = 0 \text{ on } \Gamma\}$$

where

$$H^1(\Omega) = \{v \in L^2(\Omega) \quad \text{such that} \ (\frac{\partial v}{\partial r}, \frac{\partial v}{\partial z}) \in (L^2(\Omega))^2\} \ .$$

Given $\theta > 0$, $I_p > 0$ and $j_B \in L^2(\Omega)$ we are interested in solutions v to

$$\text{(III.6)} \qquad \begin{cases} v \in H^2_c(\Omega) \\ Lv = \theta\, v^+ 1_{\Omega_v} + j_B \quad \text{in } \Omega \\ I_p = \theta \displaystyle\int_{\Omega_v} v^+ \, dS \ . \end{cases}$$

Many papers have been devoted to the case $j_B = 0$, $\Omega_v = \Omega$. In particular, existence results have been proved in R. TEMAM [2] and H. BERESTYCKI – H. BREZIS by functional minimization methods, and also in H. BERESTYCKI – H. BREZIS using the topological degree defined in J. LERAY – J. SCHAUDER.

Here, as in H. BERESTYCKI – H. BREZIS, we shall use the method of topological degree since it allows us to show not only the existence of a solution for each value of the parameters θ, I_p and of j_B, but also the existence of a connected set of solutions as θ varies. More precisely, we shall prove in Section III.2.6 the following theorem.

Theorem III.1

For each $I_p > 0$ and each $j_B \in L^2(\Omega)$ there exists a set $\mathscr{C}$ of pairs (v, θ) which satisfy (III.6), such that $\mathscr{C}$ is connected in $H^2_c(\Omega) \times R^+$ and such that θ runs through all of $R^+ - \{0\}$ when (v, θ) runs through $\mathscr{C}$. □

To prove this result we shall use the existence and uniqueness of a solution v to (III.6) when θ is small enough, and *a priori* estimates on the solutions to (III.6). We begin by establishing these properties.

III.2.2 *Uniqueness for Small* θ

Let $\kappa_1, \kappa_2, \ldots$ be the sequence of eigenvalues of the problem:

$$Lv = \kappa \, 1_{\Omega_v} v \quad , \quad v \in V$$

where

$$V = \{ v \in H^2_c(\Omega) \mid \int_{\Omega} L\, v = 0 \} .$$

Lemma III.1

If $0 < \theta < \kappa_2$ the equation (III.6) admits at most one solution. □

Remark III.4

Let $\lambda_1, \lambda_2, \ldots$ be the ascending sequence of eigenvalues of the problem:

$$Lv = \lambda \, 1_{\Omega_v} v \quad , \quad v \in H^1_0(\Omega) .$$

We have $\kappa_1 = 0$ and $0 < \lambda_1 < \kappa_2 \leqslant \lambda_2$ (cf. T. GALLOUET, for example).

Recall that in the case when $L = -\Delta$, $j_B = 0$, $\Omega = \Omega_v$, lemma III.1 is still true with λ_2 in place of κ_2 (cf R. TEMAM [2] and J.P PUEL). □

Proof of Lemma III.1

This result was already established in H. BERESTYCKI – H. BREZIS. We prove it here in a different way. Let v_1, v_2 be two distinct solutions of (III.6). Their difference $w = v_1 - v_2$ satisfies

(III.7) $$Lw = \theta \rho 1_{\Omega_V} w \quad , \quad w \in V$$

where

$$\rho(M) = \frac{v_2^+(M) - v_1^+(M)}{v_2(M) - v_1(M)} \quad \text{if} \quad v_2(M) - v_1(M) \neq 0$$

$$\rho(M) = 0 \quad \text{if} \quad v_2(M) - v_1(M) = 0 .$$

Let θ^{ρ}_1, θ^{ρ}_2,... be the ascending sequence of eigenvalues of (III.7). These eigenvalues decrease as ρ increases. Since $0 \leqslant \rho \leqslant 1$ we have $\theta^{\rho}_2 \geqslant \kappa_2$. Since $\theta < \kappa_2$ we must have $\theta = \theta^{\rho}_1 = 0$ which is absurd. □

III.2.3 A priori Estimate

Lemma III.2

Every solution v of (III.6) satisfies

$$|v|_{H^2(\Omega)} \leqslant C(\theta^2 + \frac{1}{\theta})(I_p + \frac{1}{I_p})(1 + |j_B|_{L^2(\Omega)})^2$$

where C depends only on Ω and Ω_V. □

Because of the hypothesis (ii) of Section III.1, the operator L is uniformly elliptic with coefficients regular in $\bar{\Omega}$, and is an isomorphism of $H^1_0(\Omega) \cap H^2(\Omega)$ onto $L^2(\Omega)$. Let K denote its inverse.

For every $s \leqslant 0$ there exists $\nu_s \in R$ (depending only on Ω) such that

(III.8) $$|Kv|_{H^{s+2}(\Omega)} \leqslant \nu_s |v|_{H^s(\Omega)} \quad , \quad \forall v \in L^2(\Omega) .$$

Proof of Lemma III.2

Let v be a solution of (III.6). Let v_Γ denote its (constant) value on Γ, and $P(v) = \{M \in \Omega_V | v(M) > 0\}$.

We have

$$v = v_\Gamma + \theta\, K(v^+ 1_{\Omega_v}) + K(j_B) .$$

(i) upper bound on $v - v_\Gamma$ in $L^2(\Omega)$

We have $|v^+ 1_{\Omega_v}|_{L^1(\Omega)} = I_p/\theta$ and so the Sobolev theorem implies that :

$$|v^+ 1_{\Omega_v}|_{H^{-2}(\Omega)} \leqslant C_1\, I_p/\theta .$$

Hence with (III.8) :

$$|v - v_\Gamma|_{L^2(\Omega)} \leqslant C_2(I_p + |j_B|_{L^2(\Omega)}) . \qquad \text{(III.9)}$$

(ii) upper bound on $|v_\Gamma|$:

If $v_\Gamma \leqslant 0$ we have

$$2|v_\Gamma|v^+ \leqslant (v^+ + |v_\Gamma|)^2 = |v - v_\Gamma|^2 \quad \text{in} \quad P(v)$$

and integrating over P(v) we deduce

$$2|v_\Gamma| I_p/\theta \leqslant |v - v_\Gamma|^2_{L^2(\Omega)} .$$

On the other hand, if $v_\Gamma \geqslant 0$ we have

$$v_\Gamma \leqslant |v - v_\Gamma| + v^+ \quad \text{in } \Omega$$

and on integrating over Ω_v we deduce

$$v_\Gamma |\Omega_v| \leqslant |\Omega|^{\frac{1}{2}} |v - v_\Gamma|_{L^2(\Omega)} + I_p/\theta .$$

where $|\Omega_v|$ denotes the measure of Ω_v. By regrouping the two cases and using the majorization of $v - v_\Gamma$ we obtain

$$|v_\Gamma| \leqslant C_3(\theta + \frac{1}{\theta})(I_p + \frac{1}{I_p})(1 + |j_B|_{L^2(\Omega)})^2 .$$

(iii) upper bound on v in $H^2(\Omega)$

From (III.8) we have

$$|v|_{H^2(\Omega)} \leqslant |v_\Gamma| \, |\Omega|^{\frac{1}{2}} + \theta \, \nu_0 |v|_{L^2(\Omega)} + \nu_0 |j_B|_{L^2(\Omega)}$$

and thus the lemma results from the estimates for $|v - v_\Gamma|$ and for $|v_\Gamma|$. □

III.2.4 Transformation of the Equation (III.6)

According to the Lax–Milgram theorem, for each $w \in L^2(\Omega)$ there exists a unique $q \in H^1_c(\Omega)$ which is a solution to :

$$\int_\Omega \left(\frac{1}{\mu_0 r} \nabla q \cdot \nabla u + qu\right) dS = \int_\Omega w \, u \, dS \; , \; \forall u \in H^1_c(\Omega)$$

where

$$H^1_c(\Omega) = \{u \in H^1(\Omega) \mid u \text{ is constant on } \Gamma\} \; .$$

We can verify that q is the unique solution in $H^2_c(\Omega)$ to :

$$\begin{cases} Lq + q = w \\ \int_\Omega Lq \, dS = 0 \end{cases}$$

We thus define a continuous linear operator $\mathscr{L}$ from $L^2(\Omega)$ into $H^2_c(\Omega)$ by $\mathscr{L}w = q$.

The Lax–Milgram theorem likewise gives the existence (and uniqueness) of the solution $\varphi \in H^1_c(\Omega)$ to :

$$\int_\Omega \left(\frac{1}{\mu_0 r} \nabla\varphi \cdot \nabla u + \varphi u\right) dS = -\left(I_p + \int_\Omega j_B \, dS\right) u_\Gamma, \; \forall u \in H^1_c(\Omega) .$$

In fact φ is a solution in $H^2_c(\Omega)$ to:

$$\begin{cases} L\varphi + \varphi = 0 \\ \int_\Omega L\varphi \, dS = I_p + \int_\Omega j_B \, dS \; . \end{cases}$$

We define a map H from $H_c^2(\Omega)$ into itself by:

$$H(v) = \varphi + \mathscr{L}(\theta\, v^+ 1_{\Omega_v} + j_B + v) .$$

Lemma III.3

Problem (III.6) is equivalent to:

$$\text{(III.10)} \qquad v \in H_c^2(\Omega) \quad , \quad v - H(v) = 0 . \qquad \square$$

Proof

The equation $v - H(v) = 0$ is equivalent to

$$\begin{cases} L(v - \varphi) + v - \varphi = \theta\, v^+ 1_{\Omega_v} + j_B + v \\ \int_\Omega L(v - \varphi)\, dS = 0 \end{cases}$$

and so (III.10) is equivalent to (III.6). $\square$

The operator H is compact from $H_c^2(\Omega)$ to itself and hence by J. LERAY – J. SCHAUDER we can define the index for solutions of (III.10). By using the results of J. LERAY – J. SCHAUDER we shall show in Section III.2.5 that if θ is sufficiently small there exists a unique solution with nonzero index, and in Section III.2.6 we shall deduce the existence of a connected set $\mathscr{C}$ of pairs (v, θ) in which θ runs through all of $R^+ - \{0\}$.

III.2.5 Existence of a Solution for Small θ, and Calculation of its Index

Lemma III.4

There exists $\alpha > 0$, depending only on Ω, Ω_v, I_p and j_B such that when $0 < \theta \leqslant \alpha$ the problem (III.10) has a unique solution v, and the index of v is -1. $\square$

We shall prove this by transforming the equation continuously with respect to a parameter t into an equation having a unique solution φ_0 whose index we are able to calculate.

Proof

Let α and θ be given such that

$$0 < \theta \leqslant \alpha < \kappa_2 .$$

(i) Transformation of (III.10)

There exists a unique real number b such that $\varphi_0 = \varphi + b$ satisfies

(III.11) $$\theta \int_{\Omega_v} \varphi_0^+ \, dS = I_p .$$

Define a map G from $H_c^2(\Omega) \times [0,1]$ into $H_c^2(\Omega)$ by

$$G(v,t) = t\,\varphi + (1-t)\varphi_0 + \mathscr{L}[\theta\, v^+ 1_{\Omega_v} + v + t\, j_B - (1-t)(\theta\, \varphi_0^+ 1_{\Omega_v} + \varphi_0)]$$

where $\mathscr{L}$ was defined in Section II.2.4.

The equation

(III.12) $$w \in H_c^2(\Omega) \quad , \quad w - G(w,t) = 0$$

is equivalent to

$$\begin{cases} L(w - t\varphi - (1-t)\varphi_0) + (w - t\varphi - (1-t)\varphi_0) \\ \quad = \theta\, w^+ 1_{\Omega_v} + w + t\, j_B - (1-t)(\theta\, \varphi_0^+ 1_{\Omega_v} + \varphi_0) \\ \displaystyle\int_\Omega L(w - t\varphi - (1-t)\varphi_0)\, dS = 0 . \end{cases}$$

Since φ_0 has been chosen so that $\int_\Omega L\,\varphi_0 = \int_\Omega L\,\varphi = I_p + \int_\Omega j_B dS$, it follows that (III.12) is equivalent to

(III.13) $$\begin{cases} w \in H_c^2(\Omega) \\ Lw = \theta\, w^+ 1_{\Omega_v} + t\, j_B + (1-t)(L\varphi_0 - \theta\varphi_0^+ 1_{\Omega_v}) \\ \theta \displaystyle\int_{\Omega_v} w^+\, dS = I_p . \end{cases}$$

For $t = 1$ we recover (III.6) which is equivalent to (III.10).

For all t, Lemma III.1 shows there exists at most one solution w. Moreover, the operator $v \to G(v,t)$ is compact from $H_c^2(\Omega)$ into itself and therefore we can define the index of the solution w when it exists.

(ii) The case $t = 0$. Existence of a solution and calculation of its index

For $t = 0$, we have that φ_0 is the solution to (III.13) and hence to (III.12). We shall compute its index relative to this equation.

The condition (III.11) shows that we can choose α sufficiently small so that when $\theta \leqslant \alpha$ we have $\varphi_0 > 0$ in $\bar{\Omega}$. Then w is positive in a neighborhood of φ_0 and therefore

$$G(w,0) = \varphi_0 + \theta \mathscr{L} [(w-\varphi_0) 1_{\Omega_v}] + \mathscr{L}(w-\varphi_0) .$$

It follows that $G(., 0)$ is Fréchet differentiable at the point φ_0, and has the operator $v \to \mathscr{L}(\theta 1_{\Omega_v} v + v)$ as its derivative.

According to J. LERAY – J. SCHAUDER ("Conclusion", p.56), the index of φ_0 for the equation (III.12) is equal to the index of 0 for the equation

$$z - \mathscr{L}(\theta 1_{\Omega_v} z + z) = 0$$

in $H_c^2(\Omega)$, when this second index exists. Now this equation is equivalent to

$$\begin{cases} Lz = \theta 1_{\Omega_v} z \\ \displaystyle\int_\Omega Lz \, dS = 0 \end{cases}$$

and therefore its eigenvalues are the κ_i (cf. III.2.2). Since $0 = \kappa_1 < \theta < \kappa_2$ and since κ_2 is a simple eigenvalue, this index is equal to -1. This establishes that the index of φ_0 for (III.12) is equal to -1.

(iii) Existence of a solution for $t = 1$

The function $G : H_c^2(\Omega) \times [0,1] \to H_c^2(\Omega)$ is continuous, the maps $t \to G(v,t)$ are equicontinuous, and for each t the operator $v \to G(v,t)$ is compact.

Moreover, Lemma III.2 shows that the possible solution w to (III.13) (and hence to (III.12)) is bounded independently of t.

By J. LERAY – J. SCHAUDER there exists a solution to (III.12) for $t = 1$ and its index is equal to the index of φ_0 relative to (III.12) for $t = 0$. Since $G(., 1) = H$ this proves Lemma III.4. □

III.2.6 *Proof of Theorem III.1*

Let $\underline{\theta}$ and $\overline{\theta}$ be two real numbers such that : $0 < \underline{\theta} \leqslant \alpha \leqslant \overline{\theta}$ where α is given by Lemma III.4. As the map H defined in Section III.2.4 depends on θ, we define a map F from $H_c^2(\Omega) \times [\underline{\theta},\overline{\theta}]$ to $H_c^2(\Omega)$ by $F(v,\theta) = H(v)$.

Lemma III.3 shows that equation (III.6) is equivalent to

$$\text{(III.14)} \qquad v \in H_c^2(\Omega) \quad , \quad v - F(v,\theta) = 0 \ .$$

Lemma III.2 shows that the solutions of (III.14) are bounded (in $H_c^2(\Omega)$) independently of θ. Furthermore, F is continuous, the maps $\theta \to F(v,\theta)$ are equicontinuous when v remains in a bounded region of $H_c^2(\Omega)$, and for each θ the map $v \to F(v,\theta)$ is compact. Finally, for $\theta = \alpha$ the total index of the solutions to (III.14) is equal to -1 by Lemma III.4.

The theorem 1 of J. LERAY – J. SCHAUDER then shows that there exists a set $\mathscr{C}_{\underline{\theta},\overline{\theta}}$ of pairs (v,θ) satisfying (III.14) such that $\mathscr{C}_{\underline{\theta},\overline{\theta}}$ is connected in $H_c^2(\Omega) \times R^+$ and such that θ runs through $[\underline{\theta},\overline{\theta}]$ as (v,θ) runs through $\mathscr{C}_{\underline{\theta},\overline{\theta}}$.

For every positive integer n we obtain a connected set $\mathscr{C}_n$ by taking $\underline{\theta} = \alpha/n$ and $\overline{\theta} = n\alpha$. Since each $\mathscr{C}_n$ contains the point (v,α) where v is the unique solution to (III.10) corresponding to α, it follows that the union $\mathscr{C}$ of the $\mathscr{C}_n$ is connected, and in it θ takes all strictly positive values. This proves Theorem III.1. □

III.2.7 *Lower Bound on the Solutions for Small* θ

Lemma III.5

If $\theta \leqslant I_p \,/\, d(I_p + |j_B|_{L^2(\Omega)})$ every solution v of (III.6) satisfies

$$v \geqslant \frac{I_p}{\theta|\Omega|} - d'(I_p + |j_B|_{L^2(\Omega)}) \quad \text{in } \bar{\Omega}$$

where the constants d and d' depend only on Ω. □

<u>Proof</u>

Recall that $P(v) = \{M \in \Omega_v \text{ such that } v(M) > 0\}$. We have

$$\int_{P(v)} (v-v_\Gamma)\,dS \leqslant \int_\Omega |v-v_\Gamma|\,dS \leqslant |\Omega|^{\frac{1}{2}}|v-v_\Gamma|_{L^2(\Omega)}$$

and (III.9) gives

$$I_p/\theta - v_\Gamma\,|P(v)| \leqslant d(I_p + |j_B|_{L^2(\Omega)}) .$$

If $I_p/\theta \geqslant d(I_p + |j_B|_{L^2(\Omega)})$ we have $v_\Gamma \geqslant 0$ and so majorizing $|P(v)|$ by $|\Omega|$ gives

$$v_\Gamma \geqslant \frac{I_p}{\theta|\Omega|} - \frac{d}{|\Omega|}(I_p + |j_B|_{L^2(\Omega)}) .$$

With the notation of Section III.2.3 we have

$$v = v_\Gamma + \theta\, K(v^+ 1_{\Omega_v}) + K(j_B) .$$

The maximum principle shows that $K(v^+ 1_{\Omega_v}) \geqslant 0$ and (III.8) implies that

$$|K(j_B)|_{L^\infty(\Omega)} \leqslant d_1\,|j_B|_{L^2(\Omega)} ,$$

hence

$$v \geqslant \frac{I_p}{\theta|\Omega|} - \frac{d}{|\Omega|}(I_p + |j_B|_{L^2(\Omega)}) - d_1\,|j_B|_{L^2(\Omega)} . \quad □$$

III.2.8 <u>*An Estimate for P(v) when θ is Large*</u>

Suppose $\text{supp}(j_B) \cap \Omega_v = \emptyset$; we then have the following lemma :

Lemma III.6

Let $\varepsilon_0 > 0$; then there exists $\theta_0 > 0$ such that for $\theta > \theta_0$ and v a solution of (III.6) we have $B \not\subset P(v)$ for every $M_0 \in \Omega$ and every ball B centered at M_0 with radius ε_0. □

Proof Suppose there is a ball B of radius ε_0 contained in P(v). Then

$$\begin{cases} Lv = \theta\, v & \text{in } B \\ v \geqslant 0 & \text{on } \partial B \end{cases}.$$

Let $\lambda_1(B)$ be the first eigenvalue of

$$L\varphi = \lambda\varphi \quad , \quad \varphi \in H_0^1(B)$$

and φ_1 the corresponding eigenvector.

Then

$$\theta \int_B v\,\varphi_1\, dS = \int_B Lv\cdot\varphi_1\, dS$$

$$= \int_B v\cdot L\varphi_1\, dS + \int_{\partial B} \frac{v}{\mu_0 r}\frac{\partial\varphi_1}{\partial n}\, d\Gamma$$

$$= \lambda_1 \int_B v\,\varphi_1\, dS + \int_{\partial B} \frac{v}{\mu_0 r}\frac{\partial\varphi_1}{\partial n}\, d\Gamma ,$$

hence

$$(\theta-\lambda_1)\int_B v\,\varphi_1\, dS = \int_{\partial B} \frac{v}{\mu_0 r}\frac{\partial\varphi_1}{\partial n}\, d\Gamma \leqslant 0$$

and therefore

$$\theta \leqslant \lambda_1 < \frac{C}{\varepsilon_0^2}$$

where C is a constant depending only on Ω. If $\theta_0 = C\varepsilon_0^{-2}$ then for $\theta > \theta_0$ this is absurd; thus $B \not\subset P(v)$. □

III.2.9 Variational Solutions

For $v \in H^1_c(\Omega)$ consider the functional

$$E(v) = \int_\Omega \left(\frac{1}{\mu_0 r} |\vec{\text{grad}}\, v|^2 - \theta(v^+)^2 1_{\Omega_v} - 2j_B v\right)dS + 2\left(I_p + \int_\Omega j_B dS\right)v_\Gamma .$$

Using the method presented in R. TEMAM [2] we show that E is bounded below on the set

$$K_0 = \{v \in H^1_c(\Omega) \ , \int_{\Omega_v} v^+ \, dS = I_p/\theta\} \ .$$

Furthermore, E attains its minimum on K_0 at a point v which is a solution to (III.6).

For each $\theta > 0$ we thus obtain at least one solution, called a variational solution.

The behaviour of these solutions when $\theta \to +\infty$ has been studied in T. GALLOUET and L.A. CAFFARELLI – A. FRIEDMAN in the absence of inducing current j_B and with $\Omega_v = \Omega$. The results obtained have been extended to the case with currents for the operator $L = -\Delta$ by J. SIMON [2], [3]. It can be shown (cf. KACEMI) that for every variational solution v of (III.6) we have

$$\text{(III.15)} \qquad v_\Gamma \leqslant -\frac{I_p r_0}{4\pi} \text{Log}\ \theta + C$$

where r_0 is the maximum of r on Ω_v and where C depends only on Ω, Ω_v, I_p and j_B. On the other hand, it can be shown that as $\theta \to \infty$ the set P(v) becomes concentrated on the part S of Γ_v where r is maximal, defined by $S = \{M \in \bar{\Omega}_v$ such that $r(M) \geqslant r(M')$, $\forall M' \in \Omega_v\}$.

More precisely, there exist

(i) constants $C > 0$ and C' such that diam $P(v) \leqslant C'/\theta^C$

(ii) a function $g(\theta)$ with $g(\theta) \to 0$ as $\theta \to \infty$ such that

(III.16) $d(M,S) \leqslant g(\theta) \quad , \quad \forall M \in P(v)$

where d denotes the distance from M to S.

III.3 EXISTENCE OF SOLUTIONS TO THE PROBLEM WITH LIMITER

The currents $I_1 \ldots I_k$ in the coils and the plasma current I_p are given. The aim then is to find a solution (ψ',λ') to the equations (III.4) and (III.5).

The function j_B belongs to $L^2(\Omega)$; its support consists of the union of the coils B_i which are subsets of Ω that are mutually disjoint and have empty intersection with Ω_v.

The limiter D is a closed subset of Ω_v.

Theorem (III.1) and Lemma (III.1) show that for every θ such that $0 < \theta < \kappa_2$ there exists a unique solution $v = v_\theta$ to equation (III.6).

Recall that κ_2 is defined in III.2.2.

Theorem III.2

Suppose that

(III.17) there exists γ with $0 < \gamma < \kappa_2$ such that $v_\gamma \leqslant 0$ in D.

Then there exists a pair $(\psi',\lambda') \in H_c^2(\Omega) \times R^+$ which is a solution to (III.4) and (III.5).

Remark III.5

Each γ defines an admissible zone for the limiter D: in fact if $D \subset \Omega_v - P(v_\gamma)$ then there exists a solution to (III.4) and (III.5). These zones depend on j_B since v_γ does. They are nonempty for $\gamma > \lambda'_1$ where λ'_1 is the first eigenvalue of the Dirichlet problem associated to the operator L in Ω_v. □

Proof of Theorem III.2

By assumption we have

$$\sup_{M \in D} v_\gamma(M) \leqslant 0 .$$

Moreover, Lemma III.5 shows that there exists β, $0 < \beta \leqslant \gamma$, such that $\sup v_\beta(M) \geqslant 0$.

It follows easily from Theorem III.1 that $\{(v_\theta, \theta) \mid \beta \leqslant \theta \leqslant \gamma\}$ is connected in $H_c^2(\Omega) \times R^+$ and therefore in $C^0(\overline{\Omega}) \times R^+$. Hence $\{\sup_{M \in D} v_\theta(M) \mid \beta \leqslant \theta \leqslant \gamma\}$ is connected in R. Thus there exists λ' between β and γ such that $\sup_{M \in D} v_{\lambda'}(M) = 0$. The pair $(v_{\lambda'}, \lambda')$ is a solution to (III.4) and (III.5). □

III.3.1 The Case of Point Limiters

Suppose that D is reduced to one point d.

Theorem III.3

There is a dense subset ω of Ω_V such that if $d \in \omega$ there exists a pair $(\psi', \lambda') \in H_c^2(\Omega) \times R^+$ that is a solution to (III.4) and (III.5). □

Remark III.6 *The Circular Cylindrical Case*

If the machine is cylindrical and not toroidal, the operator L is replaced by $(-\Delta)$. Moreover, if Ω is a ball centered at P_0 and $\Omega_V = \Omega$, $j_B = 0$, then there exists a solution to (III.4) – (III.5) if and only if $P_0 \notin D$. The set ω in Theorem III.3 can then be taken to be $\Omega - \{P_0\}$. □

Remark III.7

Theorem III.3 extends to the case of arbitrary limiters. In fact if $\tilde{D}$ is a closed subset of Ω_V then there exists a closed subset D of Ω_V, as close to D (in the sense of Hausdorff metric) as may be desired, such that the system (III.4) – (III.5) admits a solution (cf. J. BLUM – T. GALLOUËT – J. SIMON [1]). □

Proof of Theorem III.3

Let d_0 be a point of Ω_v and ε_0 a strictly positive scalar such that $\varepsilon_0 < \text{dist}(d_0,\Omega_v^c)$, where Ω_v^c is the complement of Ω_v in Ω. Let $\mathscr{C}$ be the connected set defined by Theorem III.1. Then by Lemma III.6 there exists a point $d \in B(d_0,\varepsilon_0) \cap \Omega_v$ and a scalar θ_0 such that if $(v,\theta_0) \in \mathscr{C}$ then $v(d) < 0$.

By Lemma III.5 there is a scalar δ between 0 and $\inf(\theta_0,\kappa_2)$ such that $v_\delta(d) \geqslant 0$. By Theorem III.1 there is a connected branch $\mathscr{C}'$ of solutions (u,θ) to (III.6) such that θ covers the interval $[\delta,\theta_0]$. Therefore there exists λ' between δ and θ_0 and $(\psi',\lambda') \in \mathscr{C}'$ such that $\psi'(d) = 0$. The pair (ψ',λ') is then a solution to (III.4) and (III.5). □

III.3.2 *Open Problems*

Let $\mathscr{C}^*$ be the connected component of the set of all solutions to (III.6) that contains the pairs (v,θ) for $\theta < \kappa_2$. By the same proof as for Theorem III.2 we can show that there exists a solution to (III.4) and (III.5) if the following hypothesis is satisfied :

(III.18) there exists a pair $(v,\theta) \in \mathscr{C}^*$ such that $v \leqslant 0$ on D.

This condition is hard to verify in practice because we do not know if a given solution (v,θ) to (III.6) belongs to $\mathscr{C}^*$ unless we suppose $\theta < \kappa_2$ as in Theorem III.2.

Since D is a closed subset of Ω_v it follows from (III.16) that for θ large enough every variational solution satisfies $v \leqslant 0$ on D. To verify condition (III.18) it therefore suffices to show that $\mathscr{C}^*$ contains variational solutions for θ large. This is the case if one of the following hypotheses is satisfied:

(i) the set of pairs (v,θ) satisfying (III.6) is connected in $C^0(\bar{\Omega})\times R^+$;

(ii) the set of pairs (v,θ) such that v is a variational solution to (III.6) is connected in $C^0(\bar{\Omega}) \times R^+$;

(iii) for θ large enough the system (III.6) has a unique solution.

The truth or otherwise of these three hypotheses are open problems.

Remark III.8

The latter hypothesis (iii) is not satisfied for all geometries and all currents. In fact the counter example for $(-\Delta)$ given in D.G.SCHAEFFER can be adapted to the operator L. □

III.4 CONTROL OF THE PLASMA SHAPE

III.4.1 Statement of the Problem

We aim to determine the currents I_i in the coils B_i that will make the region Ω_p occupied by the plasma approach "as well as possible" a given open set Ω_d contained in Ω_v and with regular boundary $\Gamma_d = \partial\Omega_d$.

As we are supposing here that the current in the vacuum vessel is zero, we recall that

$$j_B = \frac{I_i}{S_i} \quad \text{in } B_i \ , \quad i \in \{1,\dots,k\}$$

$$j_B = 0 \quad \text{elsewhere} \ .$$

We no longer assume as in Chapter II that the sum of the currents in the coils is fixed, nor that the currents in certain coils are fixed.

We minimize the functional

$$J(\psi') = \frac{1}{2}\int_{\Gamma_d} \psi'^2 \, d\Gamma \tag{III.19}$$

on the set Ψ of solutions to (III.4) – (III.5) when the I_i vary, i.e. on

$$\Psi = \{\psi' \in C^0(\bar{\Omega}) \mid \exists\ \lambda' > 0 \ , \ I_i,\dots,I_k \ \text{ such that}$$

$$\text{(III.4) and (III.5) are satisfied}\} \ .$$

We therefore seek a solution ψ_0' to

(III.20) $$\psi'_o \in \Psi\ ,\ J(\psi'_o) = \inf_{\psi' \in \Psi} J(\psi')\ .$$

<u>*Remark III.9*</u>

Suppose there exists $\psi'_o \in \Psi$ such that $P(\psi'_o) = \Omega_d$. Then ψ'_o is zero on Γ_d and so $J(\psi'_o) = 0$ and ψ'_o minimises J. Conversely, if J is zero, ψ' is zero on Γ_d and Γ_p then contains Γ_d without, however, necessarily being identical with Γ_d. □

It is reasonable to assume that Γ_d is in contact with the limiter, since every plasma must satisfy this condition, but we shall not exploit this in what follows. Instead we shall use the following hypothesis:

(III.21) $$\Omega - \bigcup_{i=1}^{k} B_i \quad \text{is connected.}$$

III.4.2 <u>*An a Priori Estimate*</u>

<u>Lemma III.7</u>

Every solution $\psi' \in C^0(\overline{\Omega})$ of (III.4) satisfies

$$\sum_i |I_i| + |\psi'_\Gamma| \leqslant e(I_p + \sqrt{J(\psi')})$$

where the constant e depends only on Ω, Γ_d and the B_i. □

<u>Proof</u>

(i) With the notation of Section III.2.3 we have

$$\psi' = \psi'_\Gamma + K(\lambda'\ \psi'^{+}\ 1_{\Omega_V}) + K(j_B)\ .$$

We have

$$|\lambda'\ \psi'^{+}\ 1_{\Omega_V}|_{L^1(\Omega)} = I_p\ .$$

Since $L^1(\Omega) \subset H^{-5/4}(\Omega)$ (cf. R.A. ADAMS) it follows from (III.8) that

$$|K(\lambda' \psi' + 1_{\Omega_V})|_{H^{\frac{3}{2}}(\Omega)} \leqslant e_1 I_p ,$$

hence

$$|K(\lambda' \psi' + 1_{\Omega_V})|_{L^2(\Gamma_d)} \leqslant e_2 I_p .$$

Since $|\psi'|_{L^2(\Gamma_d)} = \sqrt{2J(\psi')}$, we have

$$|K(j_B) + \psi'_\Gamma|_{L^2(\Gamma_d)} \leqslant e_2 I_p + \sqrt{2J(\psi')} .$$

Let $\bar{u}$ be the mean value of a function u that is integrable over Γ_d, that is

$$\bar{u} = \frac{1}{|\Gamma_d|} \int_{\Gamma_d} u \, d\Gamma .$$

We have

$$|\bar{u}|_{L^2(\Gamma_d)} \leqslant |u|_{L^2(\Gamma_d)} .$$

Let $g = K(j_B) - \overline{K(j_B)}$. Then

$$|g|_{L^2(\Gamma_d)} \leqslant 2|K(j_B) + \psi'_\Gamma|_{L^2(\Gamma_d)} \leqslant 2(e_2 I_p + \sqrt{2J(\psi')}) .$$

(ii) We now show there exists $e_3 > 0$ such that

$$|g|_{L^2(\Gamma_d)} \geqslant e_3 \sum_{i=1}^{k} |I_i| .$$

Suppose $|g|_{L^2(\Gamma_d)} = 0$, i.e. $g = 0$ on Γ_d. We have $Lg = j_B$ which vanishes in $\Omega - \bigcup_i B_i$. Since Γ_d is disjoint from the B_i it follows that $g = 0$ in Ω_d and therefore in $\Omega - \bigcup_i B_i$ which we supposed was connected. For each i we then have

$$I_i = \int_{B_i} j_B \, dS = \int_{B_i} Lg \, dS = - \int_{\partial B_i} \frac{1}{\mu_0 r} \frac{\partial g}{\partial n} \, d\Gamma = 0 .$$

The map $(I_1,\ldots,I_k) \rightarrow |g|_{L_2(\Gamma_d)}$ is therefore a norm on R^k, which establishes the existence of the constant e_3.

(iii) It remains to bound ψ'_Γ. We have

$$|\psi'_\Gamma| \leqslant |\overline{K(j_B) + \psi'_\Gamma}| + |\overline{K(j_B)}|$$

and

$$|\overline{K(j_B) + \psi'_\Gamma}| \leqslant |\Gamma_d|^{-\frac{1}{2}} \, |K(j_B) + \psi'_\Gamma|_{L^2(\Gamma_d)}$$

$$|\overline{K(j_B)}| = |\sum_i \frac{I_i}{S_i} \overline{K(1_{B_i})}| \leqslant e_4 \sum_i |I_i| \; .$$

Hence

$$|\psi'_\Gamma| \leqslant (|\Gamma_d|^{-\frac{1}{2}} + \frac{2e_4}{e_3}) |K(j_B)+\psi'_\Gamma|_{L^2(\Gamma_d)} \leqslant e_5(I_p + \sqrt{J(\psi')}) \; . \qquad \square$$

III.4.3 Existence of an Optimal Control

Suppose that λ' remains bounded when j_B and ψ'_Γ are bounded. More precisely, we make the following hypothesis :

(III.22) For every $a > 0$ and every $\psi' \in \Psi$ such that

$$\sum_i |I_i| + |\psi'_\Gamma| \leqslant a \;, \quad \text{we have} \quad \lambda' \leqslant A \;,$$

where A depends only on a, Ω, I_p and the B_i.

Remark III.10

This hypothesis (III.22) is satisfied (thanks to (III.15)) if the equation (III.6) has only variational solutions, thus in particular if for every θ and for every j_B there exists a unique solution v to (III.6) (cf. Remark III.8). □

Theorem III.4

Suppose the hypotheses (III.21) and (III.22) are satisfied. Then there exists a solution ψ'_0 to (III.20). □

Proof

Let $\{\psi'^n\}$ be a sequence minimizing J in Ψ, that is such that :

$$J(\psi'^n) \to \inf_{\psi' \in \Psi} J(\psi') \quad \text{as } n \to \infty .$$

Let λ'^n and $I_1^n,\ldots,I_k^n$ be the coefficient of proportionality and the currents relative to ψ'^n. Lemma III.7 shows that the I_i^n and $\psi_\Gamma'^n$ are bounded independently of n. Then j_B^n is bounded in $L^2(\Omega)$ and Lemma III.5 shows that there exists $b > 0$, independent of n, such that $\lambda'^n \geqslant b$ (otherwise the ψ'^n would not be bounded).

The hypothesis (III.22) shows furthermore that $\lambda'^n \leqslant B$. Finally, Lemma III.2 shows that the ψ'^n are bounded in $H^2(\Omega)$.

As the inclusion of $H^2(\Omega)$ into $C^0(\bar{\Omega})$ is compact, we may extract a subsequence, also denoted by ψ'^n, such that as $n \to \infty$

$$\psi'^n \to \psi'_o \quad \text{in} \quad C^0(\bar{\Omega})$$

$$\lambda'^n \to \lambda'_o \quad , \quad I_i^n \to I_i^o \quad .$$

Passing to the limit in equations (III.4), (III.5) applied to ψ'^n shows that $\psi'_o \in \Psi$. Moreover $J(\psi'^n) \to J(\psi'_o)$ and therefore ψ'_o satisfies (III.20). □

Remark III.11 *Control of variational solutions*

Let Ψ_{var} be the subset of Ψ consisting of variational solutions. There exists $\psi'_o \in \Psi_{var}$ such that

$$J(\psi'_o) = \inf_{\psi' \in \Psi_{var}} J(\psi') .$$

In fact if ψ'_n is a minimizing sequence in Ψ_{var} then λ'_n is bounded, according to the estimate (III.15). We conclude the argument as in Theorem III.4, observing that every limit of variational solutions is a variational solution. □

III.4.4 Differentiating a Solution with Respect to Currents in the Coils.

Given a solution (ψ',λ') to (III.4) – (III.5) relative to $j_B = \sum_i (I_i/S_i) 1_{B_i}$ we aim to find, in a neighborhood of (ψ',λ'), a solution depending regularly on the currents. To do this it suffices to vary the currents $I_1,\ldots,I_k$ independently. Consider the variation with respect to I_1.

Suppose (and this may be generalized, as we shall see later) that

(III.23) D is reduced to one point d.

Given $t \in R$ we look for a solution to (III.4) – (III.5) relative to $j_t = j_B + t/S_1 . 1_{B_1}$, that is a solution to:

$$
\text{(III.24)} \quad \begin{cases} \psi'_t \in H^2_c(\Omega) \quad , \quad \lambda'_t > 0 \\ L\psi'_t - \lambda'_t \, \psi'^{+}_t \, 1_{\Omega_v} = j_t \\ \int_\Omega L\psi'_t \, dS = I_p + \int_\Omega j_t \, dS \\ \psi'_t(d) = 0 \end{cases}
$$

such that $t \to (\psi'_t, \lambda'_t)$ is differentiable.

Let us begin by differentiating this equation formally in order to obtain the equations satisfied by the derivative $\tilde{\psi}' = \partial\psi'_t/\partial t$, $\tilde{\lambda}' = \partial\lambda'_t/\partial t$ at $t = 0$. We obtain

$$
\text{(III.25)} \quad \begin{cases} \tilde{\psi}'_t \in H^2_c(\Omega) \quad , \quad \tilde{\lambda}' \in R \\ L\tilde{\psi}' - \lambda' \, 1_{\Omega_p} \, \tilde{\psi}' - \tilde{\lambda}' \, \psi'^{+} \, 1_{\Omega_v} = 1_{B_1}/S_1 \\ \int_\Omega L\tilde{\psi}' \, dS = 1 \\ \tilde{\psi}'(d) = 0 \; . \end{cases}
$$

We shall justify this result by the Implicit Function Theorem. For this we introduce the following eigenvalue problem:

(III.26)
$$\begin{cases} \varphi \in V = \{v \in H^2_c(\Omega) \ , \ \int_\Omega L\, v \, dS = 0\} \\ L\,\varphi = \theta\, 1_{\Omega_p}\, \varphi \ . \end{cases}$$

If λ' is not an eigenvalue of (III.26) we define φ_1 uniquely by

(III.27)
$$\begin{cases} \varphi_1 \in V \\ L\,\varphi_1 - \lambda' 1_{\Omega_p} \varphi_1 = \psi'^+ 1_{\Omega_v} \ . \end{cases}$$

Theorem III.5

Given a solution $(\psi',\lambda') \in H^2_c(\Omega) \times R^+$ to (III.4) – (III.5), suppose that

(III.28) $\nabla\psi'(M) \neq 0$ for all $M \in \bar{\Omega}_v$ such that $\psi'(M) = 0$.

(III.29) λ' is not an eigenvalue of (III.26) and $\varphi_1(d) \neq 0$.

Then for all sufficiently small t there exists a unique solution to (III.24) such that $t \to (\psi'_t,\lambda'_t)$ is continuously differentiable with values in $H^2_c(\Omega) \times R$ and such that $(\psi'_0,\lambda'_0) = (\psi',\lambda')$. The derivative for $t = 0$ is the unique solution $(\tilde{\psi}',\tilde{\lambda}')$ of (III.25).□

Remark III.12 (on the hypotheses):

(i) It follows from (III.24) that $\psi' \in C^1(\bar{\Omega})$ and so (III.28) makes sense.

(ii) Suppose $\lambda' < \kappa_2$ where κ_2 is defined in Section III.2.2. Then λ' is not an eigenvalue of (III.26). In fact, as we saw in the proof of Lemma III.1, the first eigenvalue of (III.26) is zero and the second is greater than κ_2.

(iii) The hypothesis (III.28) is identical with Hypothesis H3 of Section I.3.

Remark III.13

Analogous results on differentiability have been obtained by A. DERVIEUX for the problem of equilibrium without currents in the coils and without limiter. □

Proof of Theorem III.5

Given $p > 2$, let

$$X = \{u \in W_c^{2,p}(\Omega) \text{ such that } u(d) = 0\}$$

where

$$W_c^{2,p}(\Omega) = \{u \in L^p(\Omega) \text{ such that}$$

$$\left(\frac{\partial u}{\partial r}, \frac{\partial u}{\partial z}, \frac{\partial^2 u}{\partial r^2}, \frac{\partial^2 u}{\partial r\, \partial z}, \frac{\partial^2 u}{\partial z^2}\right) \in [L^p(\Omega)]^5 \text{ and } u = \text{constant on } \Gamma\} .$$

The space X is equipped with the $W^{2,p}(\Omega)$ norm which makes it into a Banach space contained in $C^1(\bar{\Omega})$ (cf. R.A. ADAMS).

Define $F: R \times X \times R \to L^p(\Omega) \times R$ by

$$F(t;v,b) = (Lv - bv^+ 1_{\Omega_v} - j_t\ ,\ \int_\Omega Lv\, dS - I_p - \int_\Omega j_t\, dS) .$$

We are thus looking for a solution to $F(t;\psi_t',\lambda_t') = 0$ in a neighborhood of the solution (ψ',λ') that corresponds to $t = 0$. The Implicit Function Theorem gives the results stated, provided we verify the following properties:

(III.30) there exists a neighborhood $\mathscr{V}$ of ψ' in X such that

$$F \in C^1(R \times \mathscr{V} \times R,\ L^p(\Omega) \times R)$$

(III.31) the derivative $A = \partial F/\partial(v,b)\ (0;\psi',\lambda')$ is an isomorphism from $X \times R$ onto $L^p(\Omega) \times R$.

(i) Proof of (III.30)

The only delicate point arises from the nonlinear term $v^+ 1_{\Omega v}$. The map $v \to v^+ 1_{\Omega v}$ is differentiable from $W^{2,p}(\Omega)$ into $L^p(\Omega)$ at every point w such that $\{M \in \Omega_v,\ w(M) = 0\}$ has measure zero, and its derivative is the map $G(w) : v \to 1_{P_w} \cdot v$ where $P_w = \{M \in \Omega_v,\ w(M) > 0\}$ (cf. for example T. GALLOUËT). Therefore to establish (III.30) it suffices to verify that

(III.32) for every $w \in \mathscr{V}$ the set $\{M \in \Omega_v,\ w(M) = 0\}$ has measure zero

(III.33) $G \in C^0(\mathscr{V}, \mathscr{L}(W^{2,p}(\Omega), L^p(\Omega)))$.

Now hypothesis (III.28) implies the existence of a neighbourhood of ψ in $C^1(\bar{\Omega})$, and hence a neighborhood $\mathscr{V}$ in X, such that for every $w \in \mathscr{V}$ we have

$$\nabla w(M) \neq 0 \text{ for all } M \in \bar{\Omega}_v \text{ such that } w(M) = 0.$$

This implies (III.32) since from G. STAMPACCHIA we have

$$1_{w=0} \nabla w = 0 \quad \text{a.e. in } \Omega .$$

Moreover, when $w_n \to w$ in $\mathscr{V}$ we have $1_{P_{w_n}} \to 1_{P_w}$ in $L^1(\Omega)$ because $\{M \in \Omega_v,\ w(M) = 0\}$ has measure zero. Therefore for all $v \in W^{2,p}(\Omega)$ we have

$$|(1_{P_{w_n}} - 1_{P_w})v|_{L^p(\Omega)} \leqslant |1_{P_{w_n}} - 1_{P_w}|^{1/p}_{L^1(\Omega)} \, |v|_{L^\infty(\Omega)}$$

which implies (III.33).

(ii) Proof of (III.31)

The operator $A \in \mathscr{L}(X \times R;\ L^p(\Omega) \times R)$ is defined by

$$A(\varphi,\xi) = (L\varphi - \lambda'\, 1_{\Omega_p}\, \varphi - \xi\psi'^{+}\, 1_{\Omega_v} \,,\ \int_\Omega L\varphi \, dS) .$$

We therefore have to show that for every $(g, a) \in L^p(\Omega) \times R$ there exists a unique solution (φ,ξ) to

$$\text{(III.34)} \quad \begin{cases} \varphi \in W^{2,p}_c(\Omega) \quad , \quad \xi \in R \\ L\varphi - \lambda'\, 1_{\Omega_p}\, \varphi - \xi\, \psi'^{+}\, 1_{\Omega_v} = g \\ \int_\Omega L\varphi \, dS = a \\ \varphi(d) = 0 . \end{cases}$$

Since λ' is not an eigenvalue of (III.26) there exists a unique function φ_2 such that

$$\begin{cases} \varphi_2 \in H^2_c(\Omega) \\ L\varphi_2 - \lambda' \, 1_{\Omega_p} \varphi_2 = g \\ \int_\Omega L\varphi_2 \, dS = a \, . \end{cases}$$

It is clear that φ_1 and φ_2 belong to $W^{2,P}_c(\Omega)$; therefore (φ,ξ) is a solution to (III.34) if and only if

$$\text{(III.35)} \qquad \begin{cases} \varphi = \xi \, \varphi_1 + \varphi_2 \\ \varphi(d) = 0 \, . \end{cases}$$

Since we have assumed $\varphi_1(d) \neq 0$, it follows that (III.35) (and therefore (III.34)) has a unique solution (φ,ξ), which proves (III.31). Theorem III.5 is thus proved. □

Remark III.14 (on the limiter)

In Theorem III.5 we supposed that D is reduced to one point. This hypothesis is not necessary. In fact if D consists of a finite number of points and if we assume that the maximum of ψ on D is attained at a unique point, it is easy to see that Theorem III.5 is still true.

More generally, suppose that D is a closed subset of Ω with regular boundary and that the maximum of ψ on D is attained at a unique point d. It is then possible to prove a theorem analogous to Theorem III.5. With additional hypotheses, in particular that the maximum of ψ on D is nondegenerate or that the "curvatures" of the boundaries of Ω_p and of D at the point d are different, it can be shown that ψ'_t attains its maximum at a point d_t which depends regularly on t. The derivative of $t \to \psi'_t$ at $t = 0$ is still given by (III.25).

These hypotheses are identical to Hypothesis H1 of Section I.3 and the differentiability of $\sup_D \psi$ with respect to ψ is given by Lemma I.2. □

III.4.5 Optimality Conditions

In this section we continue to assume that D is reduced to one point d (although this hypothesis could be generalized as we have seen in Remark III.14).

Let ψ'_o be an optimal control, that is a solution to (III.20). There are thus $\lambda'_o > 0$ and unique currents $I^o{}_1,\ldots,I^o{}_k$ such that (ψ'_o,λ'_o) satisfies (III.4) and (III.5).

Denote by δ_d the Dirac measure concentrated at d, and let $\delta_{\Gamma d}$ be the measure supported on Γ_d and defined by

$$(\delta_{\Gamma_d}, v) = \int_{\Gamma_d} v \, d\Gamma \quad , \quad \forall\, v \in C^o(\bar{\Omega}) \; .$$

Let $\Omega_p^o = \{M \in \Omega_v, \; \psi'_o(M) > 0\}$.

Theorem III.6

Suppose that (ψ'_o,λ'_o) satisfies the hypotheses (III.28) and (III.29). Then there exists a unique solution χ' to

$$\text{(III.36)} \quad \begin{cases} \chi' \in W^{1,p'}_c(\Omega) \quad \text{where} \quad 1 < p' < 2 \quad , \\ L\chi' - \lambda'_o \, 1_{\Omega_p^o} \chi' = \psi'_o \, \delta_{\Gamma_d} - (\lambda'_o \int_{\Omega_p^o} \chi' \, dS + \int_{\Gamma_d} \psi'_o \, d\Gamma) \, \delta_d \\ \int_{\Omega_v} \chi' \, \psi'^{+}_o dS = 0 \end{cases}$$

and we have the following necessary first order optimality condition:

$$\text{(III.37)} \qquad \frac{1}{S_i} \int_{B_i} \chi' \, dS = \chi'_\Gamma \quad , \quad \forall\, i \in \{1,\ldots,k\} \; . \qquad \square$$

Proof of Theorem III.6

(i) Differentiation of the cost Theorem III.5 allows us to define uniquely for small enough t a solution (ψ'_t,λ'_t) to (III.4) – (III.5) relative to $j_t = j_o + (t/S_1)1_{B_1}$.

Since ψ_o' is optimal we have

$$\int_{\Gamma_d} \psi_t'^{\,2}\, d\Gamma \geqslant \int_{\Gamma_d} \psi_o'^{\,2}\, d\Gamma \quad , \quad \forall\, t .$$

Since $t \to \psi_t'$ is differentiable with values in $H_c^2(\Omega)$ it follows that

$$\int_{\Gamma_d} \psi_o' \, \tilde{\psi}' \, d\Gamma = 0 \qquad \text{(III.38)}$$

where $(\tilde{\psi}',\tilde{\lambda}')$ is the unique solution to (III.25) (relative to ψ_o', λ_o').

We shall transform this condition (III.38) by introducing an appropriate adjoint state.

(ii) Definition of the adjoint state χ'

As we saw in part (ii) of the proof of Theorem III.5, the operator A which is defined by :

$$A(\varphi,\xi) = (L\varphi - \lambda_o' \, 1_{\Omega_p^o} \varphi - \xi \, \psi_o'^{+} \, 1_{\Omega_V} \, , \, \int_\Omega L\varphi \, dS)$$

is continuous linear from $X \times R$ into $L^p(\Omega) \times R$ and invertible, where $X = \{v \in W_c^{2,p}(\Omega),\ v(d) = 0\}$.

The adjoint operator A^* of A is therefore continuous linear from $L^{p'}(\Omega) \times R$ to $X' \times R$ and is invertible, with $p' = p/(p-1)$.

Since X embeds continuously in $C^o(\bar{\Omega})$, the measure $\psi_o' \, \delta_{\Gamma_d}$ is an element of X'.

Define (χ',ν) uniquely by

$$A^*(\chi',\nu) = (\psi_o' \, \delta_{\Gamma_d}, 0)$$

i.e. by

$$\text{(III.39)} \quad \begin{cases} \chi' \in L^{p'}(\Omega) \quad , \quad \nu \in R \\ \displaystyle\int_\Omega \chi' (L\varphi - \lambda_o' \, 1_{\Omega_p^o} \varphi - \xi \, \psi_o'^{+} \, 1_{\Omega_V}) dS \\ \displaystyle + \nu \int_\Omega L\varphi \, dS = \int_{\Gamma_d} \psi_o' \, \varphi \, d\Gamma \quad , \quad \forall\, \varphi \in X \quad , \quad \xi \in R . \end{cases}$$

As u runs through $W_c^{2,p}(\Omega)$, so u − u(d) runs through X and therefore (III.39) is equivalent for $\chi' \in L^{p'}(\Omega)$, $\nu \in R$ to

$$\text{(III.40)}\quad \begin{cases} \int_\Omega \chi'(Lu - \lambda_o' 1_{\Omega_p^o} u)\,dS + \nu \int_\Omega Lu\,dS = \int_{\Gamma_d} \psi_o' u\,d\Gamma \\ - u(d)\,[\lambda_o' \int_{\Omega_p^o} \chi'\,dS + \int_{\Gamma_d} \psi_o'\,d\Gamma]\ , \quad \forall u \in W_c^{2,p}(\Omega) \\ \int_{\Omega_v} \chi' \psi_o'^{+}\,dS = 0\ . \end{cases}$$

(iii) <u>Characterization (III.36) of χ'</u>

Let

$$\zeta = \lambda_o' 1_{\Omega_p^o} \chi' + \psi_o' \delta_{\Gamma_d} - (\lambda_o' \int_{\Omega_p^o} \chi'\,dS + \int_{\Gamma_d} \psi_o'\,d\Gamma)\,\delta_d\ .$$

We have $\zeta \in W^{-1,p'}(\Omega)$ since the measures belong to this space.

Consider the solution η to

$$\text{(III.41)}\qquad \eta \in W_o^{1,p'}(\Omega)\quad ,\quad L\eta = \zeta$$

and suppose for a moment that η is the unique solution to

$$\text{(III.42)}\qquad \eta \in L^{p'}(\Omega)\quad ,\quad \int_\Omega \eta\, Lu\,dS = (\zeta, u)\ ,$$

$$\forall u \in W^{2,p}(\Omega) \cap W_o^{1,p}(\Omega)\ .$$

Since $\chi' + \nu$ satisfies (III.42) we have $\chi' + \nu = \eta$ which establishes (III.36). Conversely, if χ' satisfies (III.36) we have $\chi' - \chi_\Gamma' = \eta$, so for $\nu = -\chi_\Gamma'$ the pair (χ', ν) satisfies (III.40) and thus (III.36) has a unique solution.

(iv) <u>Equivalence of (III.41) and (III.42)</u>

First of all we show that η satisfies (III.42). We decompose u as $u = u_1 + u_2$ where u_1 is zero on a neighborhood of Γ and u_2 is zero on a neighborhood of $\Gamma_d \cup \{d\}$.

If $u_1 \in \mathcal{D}(\Omega)$ we have

$$\int_\Omega \eta \, Lu_1 \, dS = (L\eta, u_1)_{\mathcal{D}'(\Omega) \times \mathcal{D}(\Omega)} = (\zeta, u_1)$$

and by density we have the result for $u_1 \in W^{2,p}(\Omega)$ on a neighborhood of Γ. On the other hand, since $\zeta = \lambda_o' 1_{\Omega_p^o} \chi'$ on the support S of u_2 we have

$$\eta \in W^{2,p'}(S) \quad \text{and}$$

$$\int_\Omega \eta \, Lu_2 \, dS = \int_S \eta \, Lu_2 \, dS = \int_S L\eta \, u_2 \, dS = (\zeta, u_2) \, .$$

By adding these equalities we obtain (III.42).

It remains to verify that (III.42) has a single solution. In fact the difference w of two solutions satisfies $w \in L^{p'}(\Omega)$ and

$$\int_\Omega w \, Lu \, dS = 0 \quad , \quad \forall u \in Y$$

$$\text{where} \quad Y = W^{2,p}(\Omega) \cap W_o^{1,p}(\Omega) \, .$$

As L is an isomorphism of Y onto $L^p(\Omega)$, its adjoint L^* is an isomorphism of $L^{p'}(\Omega)$ onto Y'. We have

$$(L^* w, u)_{Y' \times Y} = 0 \quad , \quad \forall u \in Y$$

and so $L^*w = 0$ and $w = 0$.

(v) Necessary condition for optimality

From the definition of (χ', ν) we have

$$A^*(\chi', \nu) \cdot (\tilde{\psi}', \tilde{\lambda}') = (\psi_o' \, \delta_{\Gamma_d}, 0) \cdot (\tilde{\psi}', \tilde{\lambda}')$$

$$= \int_{\Gamma_d} \psi_o' \, \tilde{\psi}' \, d\Gamma \, .$$

On the other hand, from the definition (III.25) of $(\tilde{\psi}',\tilde{\lambda}')$ we have

$$(\chi',\nu)\cdot A(\tilde{\psi}',\tilde{\lambda}') = (\chi',\nu)\cdot(1_{B_1}/S_1,\ 1)$$

$$= \frac{1}{S_1}\int_{B_1} \chi'\ dS + \nu\ .$$

The optimality condition (III.38) shows that

$$\frac{1}{S_1}\int_{B_1} \chi'\ dS = -\nu = \chi'_{\Gamma}\ .$$

Naturally we could replace B_1 by any of the coils B_i, and hence obtain the optimality condition (III.37). □

Remark III.15

If, as in Chapter II, we had imposed the condition that the sum $\sum_i I_i$ of the currents in the coils should be fixed, the optimality condition would be replaced by

$$\frac{1}{S_i}\int_{B_i} \chi'_i\ dS = \frac{1}{S_1}\int_{B_1} \chi'_1\ dS\ , \quad \forall\ i \in \{2,\ldots,k\}\ .$$ □

4. Study of equilibrium solution branches and application to the stability of horizontal displacements

The object of this chapter is to study the equilibrium solution branches when the distribution of currents among the various coils is varied. More precisely, let $I_1,\ldots,I_k$ be the currents in the k poloidal coils. Suppose $I_2,\ldots,I_{k-1}$ are fixed, as also is the sum I_S of I_1 and I_k. Put

$$\text{(IV.1)} \qquad \begin{cases} I_1 = \Lambda I_S \\ I_k = (1-\Lambda) I_S \ . \end{cases}$$

The parameter Λ thus represents the proportion of the current in the innermost coil relative to the sum of the currents in the coils at the extremes. The difference between two configurations corresponding to distinct Λ values thus resides in a current dipole situated in the coils B_1 and B_k; this dipole creates a vertical field in the plasma region which modifies the radial position Δ_H of the latter. The aim of the problem is to describe the equilibrium solution branches for the plasma when the parameter Λ varies, and in particular to study the function $\Delta_H(\Lambda)$ representing the radial position of the plasma as a function of Λ.

Section IV.1 treats the computation of these equilibrium solution branches by a continuation method. In Section IV.2 we present a new method of determining these branches using the theory of optimal control. Section IV.3 deals with loss of stability of horizontal displacements of the plasma at limit points of these branches.

IV.1 DETERMINATION OF EQUILIBRIUM SOLUTION BRANCHES BY A CONTINUATION METHOD : THE PROBLEM (P_Λ).

We formulate the problem (P_Λ) of computing equilibrium solutions as a function of Λ using the discrete finite element formulation of Section I.4.1. We are given the currents $I_2,\ldots,I_{k-1}$ as well as the sum I_S of the currents in the coils B_1 and B_k and the total plasma current I_p; we are also given the functions $h(r)$, $g(\psi_N)$ and $\bar{\mu}(B_p^2)$. The current density is assumed to be zero in the vacuum vessel.

The problem (P_Λ) consists of determining a family of equilibria $u(\Lambda) = (\psi(\Lambda), \lambda(\Lambda)) \in V_\ell \times R$ with $\Lambda \in R$ such that :

$$\text{(IV.2)}\quad \begin{cases} a_\mu(\psi,\varphi) - \sum_{i=2}^{k-1} \frac{I_i}{S_i}\int_{B_i} \varphi\, dS - \frac{I_S}{S_k}\int_{B_k} \varphi\, dS - \lambda\int_{\Omega_p} h(r)g(\psi_N)\varphi dS \\ \quad - \Lambda I_S\left(\frac{1}{S_1}\int_{B_1} \varphi\, dS - \frac{1}{S_k}\int_{B_k} \varphi\, dS\right) = 0, \ \forall\varphi \in V_\ell \\ I_p - \lambda\int_{\Omega_p} h(r)g(\psi_N) dS = 0 \end{cases}$$

with $\Omega_p = \{M \in \Omega_v | \psi(M) > \sup_D \psi\}$, $\mu = \bar{\mu}(\frac{1}{r^2}\nabla^2 \psi)$.

IV.1.1 Review of Notions of Regular Solutions, Limit Points and of Continuation Methods :

Let us recall some definitions concerning solution branches. Let G be a map from $R \times H$ to H where H is a given Hilbert space. The solution branches are the set of solutions to

$$\text{(IV.3)}\qquad G(\Lambda,u) = 0$$

as Λ varies. In our problem u denotes the pair (ψ,λ) and G is the system of equations (IV.2). The space H is then a finite-dimensional space. Let D_uG, $D_\Lambda G$ be the Fréchet derivatives of G with respect to u and to Λ respectively. Let (Λ_0,u_0) be a solution to (IV.3), and denote by $(D_uG)_0$ and $(D_\Lambda G)_0$ the derivatives D_uG and $D_\Lambda G$ at the point (Λ_0,u_0).

A solution (Λ_0, u_0) of (IV.3) is called <u>regular</u> if $(D_uG)_0$ is nonsingular, i.e. is an isomorphism of H. There is then a pair $(\dot{\Lambda}_0, \dot{u}_0) \in R \times H$ such that

$$\text{(IV.4)} \quad \begin{cases} (D_uG)_o \, \dot{u}_o + (D_\Lambda G)_o \dot{\Lambda}_o = 0 \\ |\dot{u}_o|^2 + |\dot{\Lambda}_0|^2 = 1 \, . \end{cases}$$

A solution (Λ_0, u_0) of (IV.3) is said to be a <u>limit point</u> (or <u>turning point</u>) of the branch if :

$$\text{(IV.5)} \quad \begin{cases} \text{i)} \quad (D_uG)_0 \text{ is singular} \\ \text{ii)} \quad \dim \operatorname{Ker} (D_uG)_0 = \operatorname{codim} \operatorname{Im}(D_uG)_0 = 1 \\ \text{iii)} \quad (D_\Lambda G)_0 \notin \operatorname{Im}(D_uG)_0 \end{cases}$$

where Ker and Im denote the kernel and image, respectively. The proof of convergence of approximation of solution branches that have regular points or limit points in a finite-dimensional space is given, under suitable hypotheses, in F. BREZZI – J. RAPPAZ – P.A. RAVIART. To obtain numerical solutions of the discretized problem we shall use the continuation methods defined in H.B. KELLER [1]. In particular, in the neighborhood of the limit points where the Jacobian of the system becomes singular, the continuation methods consist of adopting as parameter no longer Λ but the arc length s of the curve $|u|(\Lambda)$ where $|u|$ is a norm of u in H. The parameter s is defined by an equation called a <u>continuation equation</u>:

$$\text{(IV.6)} \qquad W(u, \Lambda, s) = 0$$

where W is a map of $H \times R \times R$ into R. The continuation method consists of finding the solution pair $(u(s), \Lambda(s))$ for (IV.3) and (IV.6). The Jacobian $\mathscr{A}$ of this new system can be written :

$$\text{(IV.7)} \qquad \mathscr{A} = \begin{bmatrix} D_uG & D_\Lambda G \\ D_uW & D_\Lambda W \end{bmatrix} .$$

We recall in respect of this the fundamental lemma of H.B. KELLER [1].

<u>Lemma IV.1</u> :

Let $\mathscr{A}$ be the Jacobian defined by (IV.7). Then

i) if D_uG is nonsingular then $\mathscr{A}$ is nonsingular if and only if

$$D_\Lambda W - (D_u W)(D_u G)^{-1}(D_\Lambda G) \neq 0$$

ii) if D_uG is singular and if

$$\dim \mathrm{Ker}(D_u G) = \mathrm{codim}\ \mathrm{Im}(D_u G) = 1$$

then $\mathscr{A}$ is nonsingular if and only if

$$D_\Lambda G \notin \mathrm{Im}(D_u G)$$

and

$$D_u W \notin \mathrm{Im}(D_u G)^*$$

where $(D_uG)^*$ is the adjoint of D_uG. □

In the continuation methods the function W is chosen in such a way that $\mathscr{A}$ is nonsingular at a limit point.

Next we consider the various methods used for following the solution branches $u(\Lambda)$ of (IV.2).

IV.1.2 The Euler–Newton method :

Let $u_0 = u(\Lambda_0) = (\psi_0,\lambda_0) = (\psi(\Lambda_0),\lambda(\Lambda_0))$; knowing a solution u_0 to the equilibrium problem corresponding to Λ_0, we propose to calculate $u(\Lambda)$ where Λ is "close" to Λ_0.

The predictor for Euler's method is written

$$u^1 = u_0 + \frac{du}{d\Lambda}(\Lambda_o)(\Lambda-\Lambda_o) \quad . \tag{IV.8}$$

We calculate $u^1 = (\psi^1,\lambda^1) \in V_\ell \times R$ by the two equations of (I.89) where we put

$$\begin{cases} I_1 = \Lambda I_S \\ I_k = (1-\Lambda) I_S \end{cases} \tag{IV.9}$$

and where the Jacobian is calculated at $u_0 = (\psi_0, \lambda_0)$.

Starting with this predictor u^1 we continue the Newton iterations (I.89) with I_1 and I_k given always by (IV.9), and in the limit we obtain $u(\Lambda) = (\psi(\Lambda), \lambda(\Lambda))$.

The Euler–Newton method can thus be written :

. initialization : $u^0 = u(\Lambda_0)$;

. starting with u^n we calculate $u^{n+1} = (\psi^{n+1}, \lambda^{n+1}) \in V_\ell \times R$ such that :

(IV.10)

$$\begin{cases} b_{u^n}(\psi^{n+1},\varphi) - \lambda^{n+1} c_{u^n}(\varphi) = \sum_{i=1}^{k} \frac{I_i}{S_i} \int_{B_i} \varphi \, dS + a'_{u^n}(\psi^n, \varphi), \quad \forall \varphi \in V_\ell \\ d_{u^n}(\psi^{n+1}, 1) + \lambda^{n+1} c_{u^n}(1) = I_p \end{cases}$$

where I_1 and I_k are given by (IV.9) and a'_u, b_u, c_u, d_u are defined by (I.48)–(I.51).

. $u^\infty = u(\Lambda)$.

We fix a step $\delta\Lambda = \Lambda - \Lambda_0$ and can thus describe the solution branch as Λ varies, provided all the time that the hypotheses H'_1, H'_2, H'_3 of Section I.4.4.1 are satisfied and the matrix of the system (IV.10) is invertible. This last condition is no longer satisfied at a singular point, and we then have recourse to a continuation method.

IV.1.3 *The Continuation Method* :

Since the parametrization by Λ no longer permits the curve to be followed in the neighborhood of limit points, the idea is then to use another parameter – namely, the arc length of the curve representing u as a function of Λ. Let Δ_H be the radial position of the plasma corresponding to u (see Section I.5.2) and denote by s the path length of the curve $\Delta_H(\Lambda)$. Note that it is not the norm of u, as in H.B. KELLER [1], but the radial position Δ_H of the plasma that is chosen as ordinate for the curve.

IV.1.3.1 *Definition of Pseudo–Arc–Length by the Continuation Equation* :

The arc length s of the curve $\Delta_H(\Lambda)$ is defined by :

$$(\frac{d\Delta_H}{ds})^2 + (\frac{d\Lambda}{ds})^2 = 1 . \quad \text{(IV.11)}$$

As this definition is rather impractical from the numerical point of view, we define a pseudo–arc–length (cf. H.B. KELLER [1]) by

$$\dot{\Delta}_H(s_0)[\Delta_H(s)-\Delta_H(s_0)]+\dot{\Lambda}(s_0)[\Lambda(s)-\Lambda(s_0)] = s-s_0, \quad s_0 \leqslant s \leqslant s_1 \quad \text{(IV.12)}$$

where $\dot{\Delta}_H(s_0)$ and $\dot{\Lambda}(s_0)$ represent the derivatives of Δ_H and Λ with respect to s at s_0, calculated from the discrete equivalent of the derived system (I.52).

From the definition (I.98) of Δ_H we can write :

$$\Delta_H(s)-\Delta_H(s_0) = \frac{1}{2} [R_e(s)-R_e(s_0) + R_i(s)-R_i(s_0)] .$$

Assuming $s-s_0$ to be small we can, by using formula (I.67) for the displacement of Γ_p, make the following approximations :

$$R_e(s) - R_e(s_0) \simeq \frac{\psi(M_0) - \psi(M_e)}{\frac{\partial\psi_0}{\partial n}(M_e)}$$

$$R_i(s) - R_i(s_0) \simeq \frac{\psi(M_0) - \psi(M_i)}{\frac{\partial\psi_0}{\partial n}(M_i)}$$

where M_0, M_i, M_e are respectively the point of contact between Γ_p and D, and the inner and outer points of intersection of Γ_p with the r–axis for the equilibrium corresponding to s_0, and where ψ_0 and ψ denote the flux associated to s_0 and s respectively.

The continuation equation (IV.12) can thus be written :

$$\frac{1}{2}\dot{\Delta}_H(s_0)\left[\frac{\psi(M_0)-\psi(M_e)}{\frac{\partial\psi_0}{\partial n}(M_e)} + \frac{\psi(M_0)-\psi(M_i)}{\frac{\partial\psi_0}{\partial n}(M_i)}\right] + \dot{\Lambda}(s_0)[\Lambda(s) - \Lambda(s_0)] = s-s_o \quad \text{(IV.13)}$$

We define $\dot{\Delta}_H(s_0)$ and $\dot{\Lambda}(s_0)$ from the derivative vector $(\dot{\psi}_0, \dot{\lambda}_0) \in V_\ell \times R$ obtained from the discrete equivalent of the linearized system (I.52) as follows :

(IV.14)

$$\begin{cases} b_{u_0}(\dot{\psi}_0, \varphi) - \dot{\lambda}_0 c_{u_0}(\varphi) = \dot{\Lambda}(s_0) I_S \left[\frac{1}{S_1} \int_{B_1} \varphi \, dS - \frac{1}{S_k} \int_{B_k} \varphi \, dS \right], \; \forall \varphi \in V_\ell \\ d_{u_0}(\dot{\psi}_0, 1) + \dot{\lambda}_0 c_{u_0}(1) = 0 \\ \dot{\Delta}_H(s_0) = \frac{1}{2} \left[\dfrac{\dot{\psi}_0(M_0) - \dot{\psi}_0(M_e)}{\frac{\partial \psi_0}{\partial n}(M_e)} + \dfrac{\dot{\psi}_0(M_0) - \dot{\psi}_0(M_i)}{\frac{\partial \psi_0}{\partial n}(M_i)} \right] \\ [\dot{\Delta}_H(s_0)]^2 + [\dot{\Lambda}(s_0)]^2 = 1 \\ \dot{\Delta}_H(s_0)\dot{\Delta}_H(s_{-1}) + \dot{\Lambda}(s_0)\dot{\Lambda}(s_{-1}) > 0 \end{cases}$$

where u_0 denotes $u(s_0)$ and where s_{-1} denotes the arc length from the equilibrium previously obtained to the one corresponding to s_0.

IV.1.3.2 *The System of Equilibrium Equations Augmented by the Continuation Equation:*

Once $\dot{\Delta}_H(s_0)$ and $\dot{\Lambda}(s_0)$ have been calculated by (IV.14), in order to compute the equilibrium corresponding to $s = s_1$ we have to solve the system of the equilibrium equations (IV.2) and the continuation equation (IV.13) with $s = s_1$, the unknowns being $\Lambda(s_1) \in R$ and $u(s_1) = (\psi(s_1), \lambda(s_1)) \in V_\ell \times R$.

The Jacobian of this system with respect to (u,Λ) is

$$K' = \begin{bmatrix} K & K_2 \\ K_2' & \dot{\Lambda}(s_0) \end{bmatrix}$$

where K is the Jacobian of the system (IV.2) with respect to $u = (\psi,\lambda)$, the last row of K' corresponds to equation (IV.13) and the vector K_2 has components

$$\text{(IV.15)} \quad \begin{cases} k_i = I_S\left[\frac{1}{S_k}\int_{B_k}\varphi_i dS - \frac{1}{S_1}\int_{B_1}\varphi_i dS\right], \quad 1 \leqslant i \leqslant N_\ell \\ k_{N_\ell+1} = 0 \end{cases}$$

where $\varphi_i (1 \leqslant i \leqslant N_\ell)$ are the basis functions of V_ℓ (see Section I.4.1).

As in H.B. KELLER [1] we can prove the following proposition :

<u>Proposition IV.1</u> :

The Jacobian K' of the system of equations (IV.2) – (IV.13) with respect to (u,Λ) is nonsingular at every regular point and at every limit point. □

<u>Proof</u> :

Let $u(s_0) = (\psi(s_0),\lambda(s_0))$ be a limit point; this implies that K is singular and $K_2 \notin \text{Im } K$. From this we deduce that $\dot{\Lambda}(s_0) = 0$ in (IV.14) and therefore that $\dot{u}(s_0) = (\dot{\psi}_0,\dot{\lambda}_0) \in \text{Ker } K$ and that $[\dot{\Delta}_H(s_0)]^2 = 1$. The row vector K'_2 then does not belong to $(\text{Ker } K)^{\perp}$, so not to $\text{Im } K^*$. We are thus in a situation where option (ii) of Lemma IV.1 can be applied. The Jacobian K' is nonsingular.

If now $u(s_0)$ is a regular point then $\dot{\Lambda}(s_0) \neq 0$ and $([\dot{\Delta}_H(s_0)]^2 + [\dot{\Lambda}(s_0)]^2)/\dot{\Lambda}(s_0) \neq 0$. We can thus apply option (i) of Lemma IV.1, which implies that K' is nonsingular. □

The introduction of the continuation equation (IV.13) has thus allowed us to obtain a nonsingular Jacobian while that for the system (IV.2) alone was singular at the limit point. We can then solve the system of equations (IV.2), (IV.13) by the Euler–Newton method.

IV.1.4 Practical Application

Starting from a solution $u(\Lambda_0)$ we vary Λ and use the Euler–Newton method defined in Section IV.1.2; to do this we solve the linear system with matrix K^n by the factorization method defined in Section I.4.4.1. As we saw earlier, this method no longer converges when we come close to a singular point.

To determine the moment at which it becomes necessary to pass from the Euler–Newton method to the continuation method we use the factorization formula (I.90) which allows us to obtain the determinant of K from the determinant of K_0. We saw that in practice K_0 is a positive definite matrix, so its determinant does not vanish; it suffices to follow the behaviour of the other factors involved in det K, and when the product P of these factors becomes lower in absolute value than some pre-determined size we abandon the Euler–Newton method and pass to the continuation method of Section IV.1.3. We return to the Euler–Newton method when P, having changed sign, becomes once again greater in magnitude than the prescribed value.

In the continuation method we first have to compute $\dot{\Delta}_H(s_0)$and $\dot{\Lambda}(s_0)$; this necessitates solving a system with matrix K. The latter was factorized during the last iteration in the computation of $u(s_0)$, which makes the computation very inexpensive in time. Solution of the linear systems associated to the matrix K' is carried out using formula (I.90) for factorization. In comparison with the Euler–Newton method this therefore represents the supplementary solution of a system with matrix K that has been previously factorized, which causes only a very slight increase in calculation time by iteration. To describe a branch close to a singular point it suffices to fix a sufficiently small step-size δs, but if we want to determine the singular point exactly we use a method of dichotomy that enables us to find the value of s which annihilates the determinant of K, and to determine the associated pair $u(s) = (\psi(s), \lambda(s))$.

To determine whether the singular point is a limit point we calculate the eigenvectors $\bar{u}_0 = (\bar{\psi}_0, \bar{\lambda}_0)$ of K and $u_0^* = (\psi_0^*, \lambda_0^*)$ of K^* associated to the eigenvalue 0 at the singular point. For this we have recourse to the method of inverse iterations which enables us to obtain these eigenvectors right from the first iteration since K and K^* are quasi-singular. The condition (iii) of (IV.5) can then be written :

(IV.16) $$\frac{1}{S_1}\int_{B_1}\psi_o^*\,dS \neq \frac{1}{S_k}\int_{B_k}\psi_o^*\,dS\ .$$

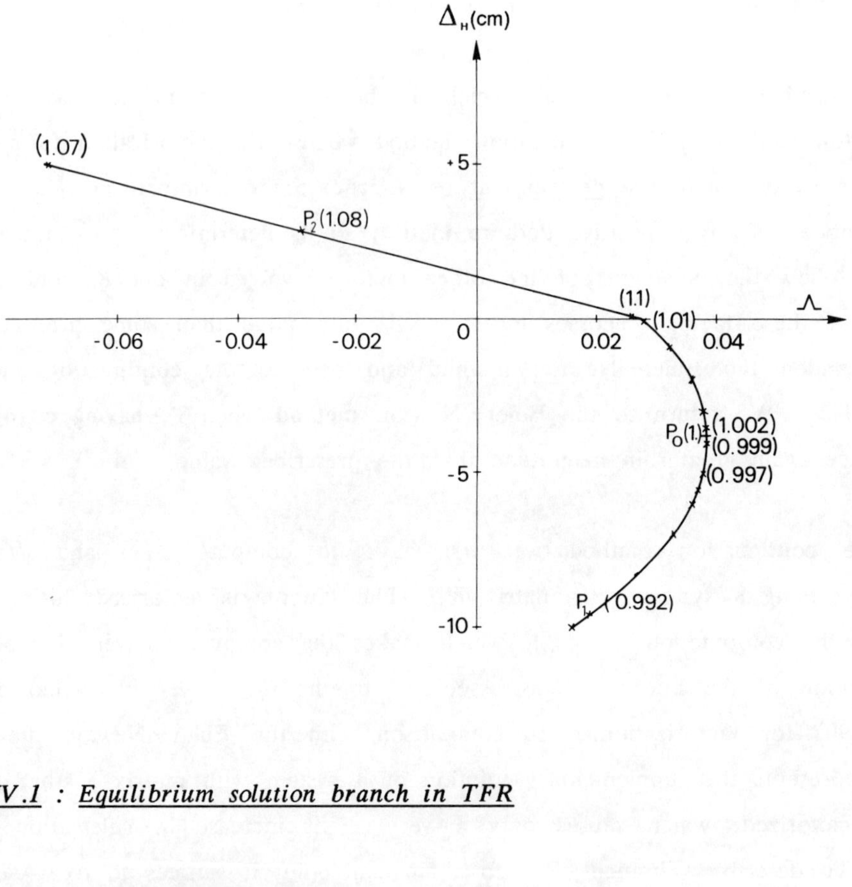

Figure IV.1 : Equilibrium solution branch in TFR

IV.1.5 Numerical Results

IV.1.5.1 Application to TFR

There are just two types of poloidal coils in TFR, the inner coil (with current I_1) and the outer coil (with current I_2). The parameter Λ represents the ratio of current in the inner coil to the total current I_S. We plan to follow the equilibrium solution branch corresponding to the following data :

$$I_p = 100\ \text{kA},\ I_S = -102\ \text{kA}$$

$$\beta = 1,\quad g(\psi_N) = (1-\psi_N^{1.5})^3\ .$$

s	0	1	2	3
Λ	0.0378	0.0385	0.0380	0.0355
Δ_H	-5.	-4.	-3.02	-2.08
P	0.36×10^{-3}	0.08×10^{-3}	-0.19×10^{-3}	-0.47×10^{-3}
Number of iterations		5	5	6
Computation time (CDC7600)		23s	23s	28s

Table IV.1

To represent this branch we trace in Fig. IV.1 the curve relating the radial position Δ_H of the plasma to the parameter Λ. We start from the configuration No.1 of Section I.5.3 corresponding to $\Lambda = 0.0196$, for which $\Delta_H = -9.63$ cm and which is represented by the point P_1 on Fig. IV.1. We use the Euler–Newton method as far as $\Lambda = 0.0378$ ($\Delta_H \simeq -5$ cm). The product P of the factors involved in det K' then becomes less than 10^{-3} (recall that det $K' = P \times \det K$ where K is positive definite), and we pass to the continuation method with which we obtain the results given in Table IV.1 for $\delta s = 1$.

Observe that the computation time for one iteration is almost identical to that for Newton's method, which proves that the continuation equation hardly increases at all the computation time for the algorithm. The product P becomes once again greater than 10^{-3} in absolute value, and we return to the Euler–Newton method which enables us to trace the branch as far as $\Lambda = 0.028$ ($\Delta_H = -0.1$ cm). As Λ continues to decrease, the algorithm converges to equilibria already obtained earlier and corresponding to values of Δ_H near -8 cm. This is due to the fact that in passing through $\Delta_H = 0$ the derived system no longer exists because the hypothesis H_1' is no longer satisfied; the plasma touches the limiter simultaneously at the two points J_o and K_o (cf. Fig. I.9). We then begin again from configuration No.2 of Section I.5.2, represented by the point P_2 on Fig. IV.1. By the Euler–Newton method we obtain the part of the curve corresponding to $\Delta_H > 0$. We confirm by verifying condition (IV.16) that the point P_o is indeed a limit point : it corresponds to $\Lambda = 0.0385$, $\Delta_H = -3.8$ cm. It is clear that for $\Lambda > \Lambda(P_o)$ there is no equilibrium solution corresponding to the above data, while for $\Lambda < \Lambda(P_o)$ there are two possible equilibria.

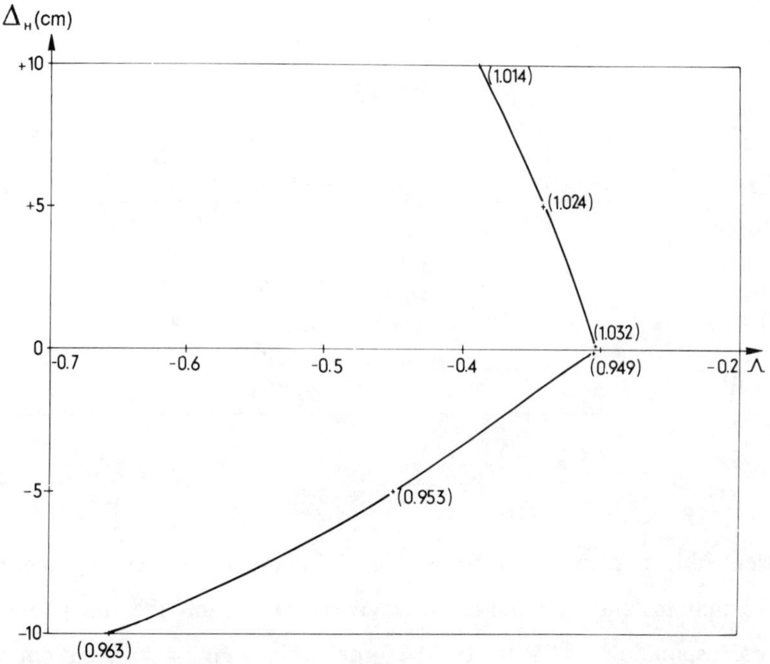

Figure IV.2 : Equilibrium solution branch in TFR without iron.

In Fig. IV.1 the spectral radius $\rho(M_u)$ corresponding to the Picard algorithm AP2 of Section I.4.2.2 is also shown in brackets; note that this algorithm converges very slowly, since $\rho(M_u)$ is very close to 1 for equilibria such that $\Delta_H < \Delta_H(P_0)$ and diverges for equilibria with $\Delta_H > \Delta_H(P_0)$. The jump in $\rho(M_u)$ at $\Delta_H = 0$ is explained by the fact that there is one derivative operator for $\Delta_H \to 0^+$ and another for $\Delta_H \to 0^-$.

To study the influence of the iron on these solution branches we simulated the equilibria in a Tokamak having the characteristics of TFR but without iron. For the same data as before, the two semi–branches $\Delta_H > 0$ and $\Delta_H < 0$ were drawn separately and the diagram of Fig. IV.2 was obtained. Note the absence of a turning point; the point corresponding to $\Delta_H = 0$ is a cusp point, and we have

$$\text{(IV.17)} \qquad \lim_{\Delta_H \to 0^-} \rho(M_u) < 1 < \lim_{\Delta_H \to 0^+} \rho(M_u) .$$

The Picard algorithm AP2 thus converges as long as the plasma rests on the inner point of the limiter, but diverges when the plasma touches the outer point. This phenomenon has already been demonstrated for air transformer Tokamaks in W. FENEBERG – K. LACKNER and G. CENACCHI – R. GALVAO – A. TARONI.

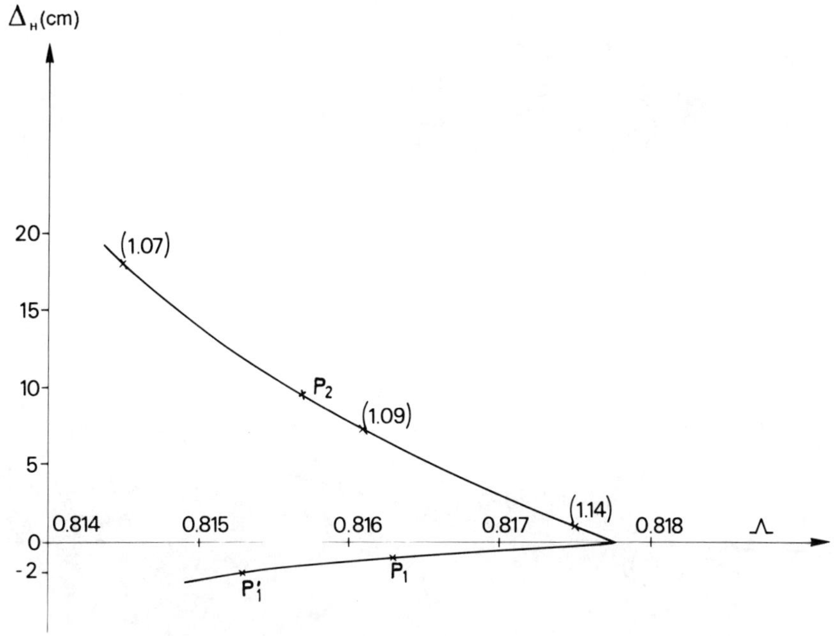

Figure IV.3 : *Branch of equilibrium configurations in JET (elliptical plasmas)*

IV.1.5.2 *Application to JET* :

As we saw in Fig. I.17 the poloidal system of JET in the upper half-plane consists of four coils. The currents I_2, I_3 in the coils B_2, B_3 are kept fixed, and Λ represents the ratio $I_1/(I_1+I_4)$.

First let us study those plasmas having an elliptical D shape and corresponding to the following data :

I_p = 1.9 MA, I_S = -7.524 MA, I_2 = 0, I_3 = 0.105 MA,

β = 1.5, $g(\psi_N) = 1-\psi_N$.

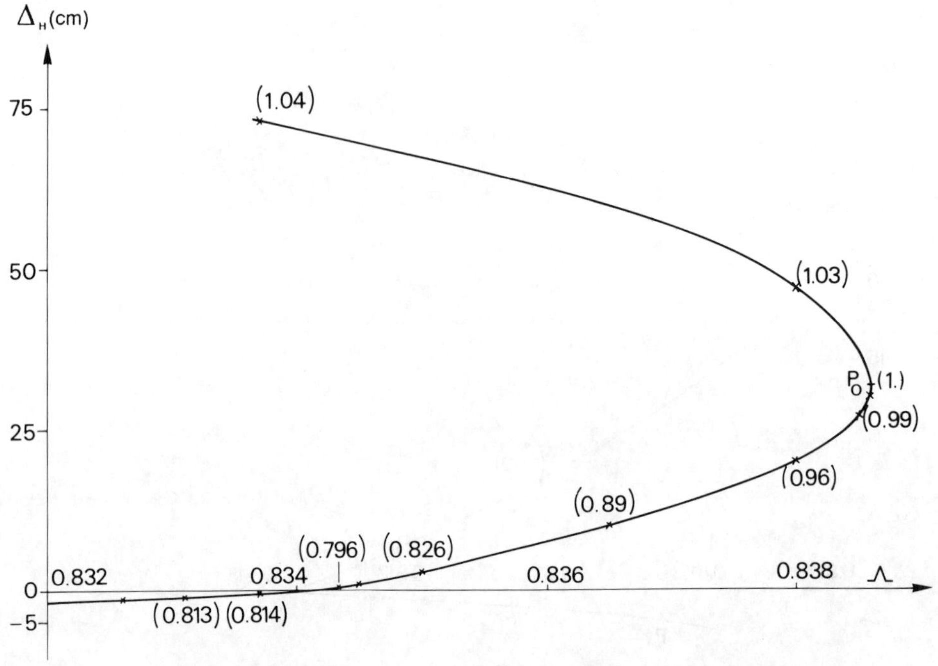

Figure IV.4 : Branch of equilibrium configurations in JET (circular plasmas)

We obtain the diagram of Fig. IV.3 by the Euler– Newton method. Observe the absence of a limit point, but a cusp point at $\Delta_H = 0$. The spectral radius $\rho(M_u)$ could not be obtained by the power method for $\Delta_H < 0$, probably because the largest eigenvalue is not real.

Next we are interested in the case of equilibria with circular section corresponding to the following data :

$$I_p = 1.3 \text{ MA}, \; I_S = -4.978 \text{ MA}, \; I_2 = 0, \; I_3 = -0.394 \text{ MA}$$

$$\beta = 2, \; g(\psi_N) = 1-\psi_N \; .$$

Using the Euler– Newton method and the continuation method close to the limit point we obtained the diagram of Fig. IV.4, where we observe, in contrast to TFR, that the limit point P_0 corresponds to a positive value of Δ_H. As the values of $\rho(M_u)$ in brackets indicate, the Picard algorithm converges for equilibria corresponding to $\Delta_H <$ 30 cm.

JET is a Tokamak with iron for which the diagrams are nevertheless very different from those of TFR. The existence and position of the limit point thus depends largely on the Tokamak being studied, on the degree of saturation of the iron and on the type of equilibrium considered (circular or elliptic plasma). We shall see in Section IV.3 that the existence of these limit points corresponds to loss of stability of horizontal displacements of the plasma.

IV.2 APPLICATION OF THE METHOD OF CONTROL OF RADIAL POSITION TO THE STUDY OF EQUILIBRIUM SOLUTION BRANCHES

For the control of the radial position of the plasma we used in Chapter II only two coils in the optimisation, namely the innermost coil B_1 and the outermost coil B_k. These two coils create a vertical field in the plasma region which enables the radial position Δ_H of the plasma to be controlled (cf. Section II.4.1). The control vector Λ then has a single component $\Lambda = I_1/(I_1+I_k)$ and is identical to the parameter Λ defined by (IV.1). We seek to minimize the functional J_1 defined by (II.3), where being given the points I and E amounts to fixing a radial position Δ_H given by (II.43).

We shall use the solution of this problem in order to describe the equilibrium solution branches and to compare this method with the continuation method used earlier.

IV.2.1 Study of the Equilibrium Solution Branches

In Section IV.1 the solution branches were represented by the curve $\Delta_H(\Lambda)$. Observe that in Figs. IV.1 – IV.4 on each of the branches there is a unique solution for fixed Δ_H. The idea here is to use this property. To describe a branch we vary Δ_H by modifying the points I and E in the functional J_1, and for each pair (I,E) the control problem is solved in order to obtain the parameter Λ and the corresponding pair (ψ,λ). We therefore have to solve a sequence of control problems (P_{II}). The solution algorithm is that of Section II.3, where each external iteration consists of a single internal gradient iteration since it is a problem with just one parameter. Each external iteration thus requires the solution of two linear systems whose matrix is that of the system (II.29) just as in the continuation method. The main difficulty is the existence of turning points as in Figs. IV.1 and IV.4. We shall show for the discrete problem that at the turning points P_0 the matrix for the optimality system for the minimization of the functional J_1 is regular, while the Jacobian of the state equations is singular. Specifically, we have the following proposition.

Proposition IV.2 :

The matrix for the linearized discrete equivalent of the optimality system (II.2), (II.14), (II.27) for the functional J_1 is regular at the limit point P_0 for $(\psi,\lambda,\chi,\nu,\Lambda) \in V_\ell \times R \times V_\ell \times R \times R$.

Proof :

We shall write down the Jacobian matrix for the optimality system for the discrete problem corresponding to equations (II.2), (II.14), (II.27) at $((\psi,\lambda), (\chi,\nu), \Lambda) \in (V_\ell \times R) \times (V_\ell \times R) \times R$. Recall that K is the Jacobian matrix for equations (II.2) at $(\psi,\lambda) \in V_\ell \times R$, and K_2 is the vector defined by (IV.15). The matrix for the linearized optimality system for $((\psi,\lambda), (\chi,\nu), \Lambda) \in (V_\ell \times R) \times (V_\ell \times R) \times R$ is thus

$$K'' = \begin{bmatrix} K & 0 & K_2 \\ K_3 & K^* & 0 \\ 0 & K_2^* & 0 \end{bmatrix}$$

where K_3 is the matrix corresponding to the term $[\psi(E)-\psi(I)][\varphi(E)-\varphi(I)]$.

Let K_4 be the following $(2N_\ell+2) \times (2N_\ell+2)$ matrix :

$$K_4 = \begin{bmatrix} K & 0 \\ K_3 & K^* \end{bmatrix}$$

and K_5, K_6 the following $(2N_\ell+2)$ vectors :

$$K_5 = \begin{bmatrix} K_2 \\ 0 \end{bmatrix} \qquad K_6 = \begin{bmatrix} 0 \\ K_2 \end{bmatrix}.$$

Then K'' may be written :

$$K'' = \begin{bmatrix} K_4 & K_5 \\ K_6^* & 0 \end{bmatrix}$$

At limit point P_0 we have by (IV.5) :

$$\text{(IV.18)} \qquad \begin{cases} \det K = 0 \\ \dim(\operatorname{Ker} K) = 1 = \operatorname{codim}(\operatorname{Im} K) \\ K_2 \notin \operatorname{Im} K \,. \end{cases}$$

We deduce from (IV.18) that K_4 is singular and that

$$\dim(\operatorname{Ker} K_4) = \operatorname{codim}(\operatorname{Im} K_4) = 1 \,.$$

By Lemma IV.1, the matrix K'' is nonsingular if and only if $K_5 \notin \operatorname{Im}K_4$ and $K_6^* \notin \operatorname{Im}K_4^*$. Now, if K_5 belonged to $\operatorname{Im}K_4$ or if K_6^* belonged to $\operatorname{Im}K_4^*$ then K_2 would belong to $\operatorname{Im}K$, which contradicts (IV.18). We thus deduce that K'' is nonsingular. □

This explains the fact that the control method works perfectly well in the neighborhood of the limit points. Since each iteration requires the solution of two linear systems with matrix K for the control method just as for the continuation method presented in Section IV.1, the computation time needed to describe a solution branch is identical by the two methods. The difference is that in the continuation method the arc length coordinate s along the curve $\Delta_H(\Lambda)$ is varied, while in the control method it is Δ_H that is varied, which is more physical. Furthermore, in using the control method it is possible to pass from a point of the branch to another very distant point without describing the whole branch, for the control agorithm is sufficiently robust to permit this.

IV.2.2 Application to TFR

As in Section IV.1 we represent the equilibrium branches by the diagram $\Delta_H(\Lambda)$. The four curves of Figure IV.5 correspond to the common data $I_p = 100$ kA, $I_S = -102$ kA; they differ by the values of β (0.25 or 1) or by the peaked quality of the function $g(\psi_N)$:

. flat current : $g(\psi_N) = 1$ $(\ell_i = 0.5)$

or

. peaked current : $g(\psi_N) = (1-\psi_N{}^{1.5})^3$ $(\ell_i = 1.5)$.

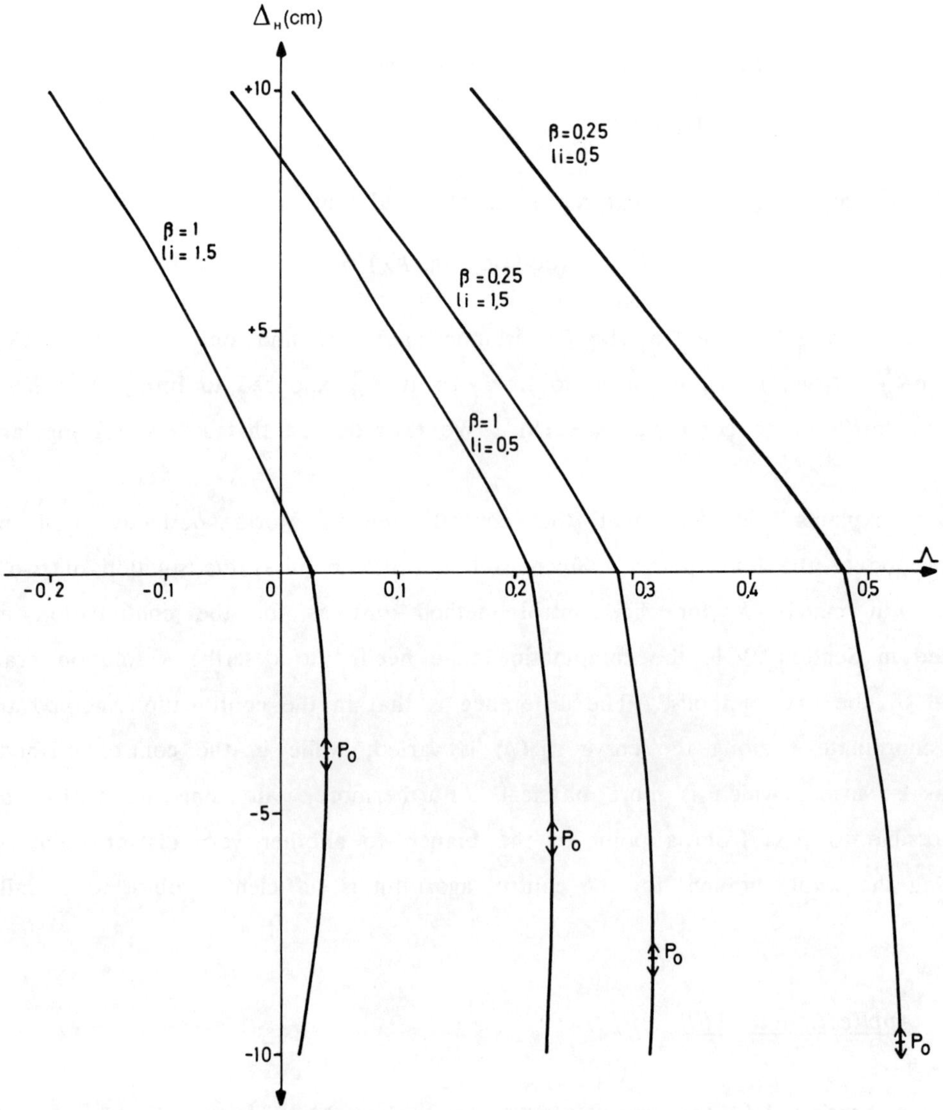

Figure IV.5 : Four equilibrium solution branches in TFR corresponding to various values of β and ℓ_i

These four curves were obtained by varying I and E in such a way that Δ_H takes all values from -10 cm to $+$ 10 cm with equidistant 1 cm steps. For each value of Δ_H, and therefore for each pair (I,E) corresponding to it, the algorithm for minimizing J_1 was initialized at the solution corresponding to the previous Δ_H, and one single iteration of "steepest descent" is necessary to make each external iteration converge since it is a one-parameter problem; the corresponding Λ value was thus obtained. Note that in Fig. IV.5 each of the four curves contains a limit point P_0; the position of this point depends on the values of β and ℓ_i.

IV.2.3 Generalization to the Study of Certain Solution Branches

The method used above for describing the equilibrium solution branches for the plasma in a Tokamak can be used in a very general way to describe certain solution branches having one or more limit points.

Consider a solution to an equilibrium problem governed by (IV.3) with $\Lambda \in R$, $u \in H$ where H is a Hilbert space that we identify with its dual, and where G is a continuously differentiable map from $R \times H$ into H. We make the following assumption:

(IV.19) for $|u|$ fixed there exists a unique solution u satisfying (IV.3), where $|u|$ denotes the norm of u in H.

We shall seek a solution to (IV.3) such that $|u| = k$, and vary k so as to describe the whole branch. The problem is formulated as follows :

. find the pair $(\Lambda_0,u_0) \in R \times H$ such that $G(\Lambda_0,u_0) = 0$ and

$$J(\Lambda_0,u_0) = \inf_{\substack{\Lambda\in R,\, u\in H \\ G(\Lambda,u)=0}} J(\Lambda,u) \tag{IV.20}$$

with $J(\Lambda,u) = 1/2[|u|^2 - k^2]^2$.

The optimality system for problem (IV.20) is the following

$$\begin{cases} G(\Lambda_0,u_0) = 0 \\ (D_uG)^*_0\ \chi' = [|u|^2-k^2]u \\ (D_\Lambda G)^*_0\ \chi' = 0 \end{cases} \tag{IV.21}$$

for $(u,\chi',\Lambda) \in H \times H \times R$ where χ' denotes the adjoint state.

The Jacobian K'' of this system in $(u,\chi',\Lambda) \in H \times H \times R$ is

$$K'' = \begin{bmatrix} D_uG & 0 & D_\Lambda G \\ K_3 & (D_uG)^* & 0 \\ 0 & (D_\Lambda G)^* & 0 \end{bmatrix}.$$

If we are at a limit point of the branch then

$$\dim(\mathrm{Ker}\, D_u G) = \mathrm{codim}\,(\mathrm{Im}\, D_u G) = 1$$
$$D_\Lambda G \notin \mathrm{Im}\, D_u G \;.$$

As before, observe that the submatrix

$$K_4 = \begin{bmatrix} D_u G & 0 \\ K_3 & (D_u G)^* \end{bmatrix}$$

is singular, but that $\begin{bmatrix} D_\Lambda G \\ 0 \end{bmatrix} \notin \mathrm{Im}\, K_4$ and $(0, (D_\Lambda G)^*) \notin \mathrm{Im}\, K_4^*$.

Thus K'' is nonsingular.

For the numerical solution of (IV.20) we adopt a sequential quadratic algorithm of the type presented in Chapter II for the problem (P_{II}), where each external iteration required the solution of two linear systems with matrix $D_u G$, i.e. the same number of operations as each Newton iteration in the continuation method. We vary k in order to describe the solution branch, while in the continuation method we vary the arc length s of the curve. The ability to fix k arbitrarily is an advantage in certain cases, because it is then possible to obtain the physically interesting solutions on the branch without having to describe it entirely.

This method therefore seems more robust than the continuation method, but it is less general since it assumes that (IV.19) is satisfied.

Remark IV.1 :

In the case of Problem (P_{II}), since the minimum of J_1 has to be zero, rather than write out the complete optimality system we could have adjoined to the equations of state (IV.2) the equation

(IV.22) $$\psi(E) = \psi(I)$$

in order to determine Λ and (ψ, λ). In the general case (IV.3) we could likewise have adjoined the equation

(IV.23) $$|u| = k \;.$$

Equations (IV.22) and (IV.23) thus play the rôle of continuation equation (of the type of equation (IV.13) or (IV.6)), with the parameters to be varied being the pair (I, E) and k respectively. □

IV.3 STABILITY OF HORIZONTAL DISPLACEMENTS OF THE PLASMA

IV.3.1 The Equations of the Evolution Problem $P_\Lambda(t)$:

We are concerned here with the equations of "quasi– static" evolution of the equilibrium, namely the equations (I.30) where we take account solely of the phenomena of diffusion of current in the vessel chamber Ω_{cv} (see Fig. I.4).

With the notation of Fig. I.3 the integral form of Faraday's equation is :

$$\int_C E \cdot d\ell = -\frac{d}{dt}\int_D B \cdot \underline{dS} \; . \qquad \text{(IV.24)}$$

Using the assumption of axial symmetry, equation (IV.24) may be written

$$rE_T = -\dot{\psi} \qquad \text{(IV.25)}$$

where $\dot{\psi}$ denotes the time derivative of ψ at a fixed point $x = (r,z)$, and E_T is the toroidal component of the electric field.

If σ_V denotes the conductivity of the vacuum vessel, Ohm's law can be written

$$j_{cv} = \sigma_V \, E_T = -\frac{\sigma_V}{r}\dot{\psi} \; . \qquad \text{(IV.26)}$$

By (I.20) the equation for ψ in Ω_{cv} is then

$$\frac{\sigma_V}{r}\dot{\psi} + L\psi = 0 \quad \text{in} \quad \Omega_{cv} \; . \qquad \text{(IV.27)}$$

The transmission conditions for ψ at the interface between the vacuum vessel and the air are those of continuity of ψ and its normal derivative.

The problem $(P_\Lambda(t))$ therefore has for its equations the system (I.30) where j_B is given by (I.20), with I_1 and I_k being calculated from Λ by (IV.1) and with j_{cv} given by (IV.26).

The weak formulation of this problem in discrete form can then be written : find $u(t) = (\psi(t), \lambda(t)) \in L^2([0,T] \times V_\ell) \times L^2[0,T]$ such that

$$(IV.28) \quad \begin{cases} \dfrac{d}{dt} \displaystyle\int_{\Omega_{cv}} \frac{\sigma_v}{r} \psi\varphi \, dS + a_\mu(\psi,\varphi) = \sum_{i=2}^{k} \frac{I_i}{S_i} \int_{B_i} \varphi \, dS + \frac{I_S}{S_k} \int_{B_k} \varphi \, dS \\ + \Lambda I_S \Big(\dfrac{1}{S_1} \displaystyle\int_{B_1} \varphi \, dS - \frac{1}{S_k} \int_{B_k} \varphi \, dS\Big) + \lambda \int_{\Omega_p} h(r) g(\psi_N) \varphi \, dS, \\ \qquad \forall \varphi \in V_\ell \\ I_p = \lambda \displaystyle\int_{\Omega_p} h(r) g(\psi_N) \, dS \end{cases}$$

with $\Omega_p = \{M \in \Omega_v | \psi(M) > \sup_D \psi\}$.

IV.3.2 Physical Interpretation of the Evolution Problem ($P_\Lambda(t)$)

In order to interpret this evolution system physically we have to consider the time scales of the various phenomena. The one related to the diffusion equation (IV.27) is the penetration time τ_v of the magnetic field into the vacuum vessel and is given by

$$\tau_v = 2\pi\sigma_v \, d_v \, b_o$$

where d_v is the thickness of the vessel and b_o is its mean minor radius.

In the plasma, the equilibrium equation (I.17) is satisfied at every instant because if we suppose the MHD activity to be stabilized the plasma is in equilibrium at all times. However, the functions $p(\psi)$ and $f(\psi)$ in (I.17) are governed as time goes on by the diffusion equations that will be studied in Chapter VI. Suppose here that the time constant τ_v for the vessel is small compared with the diffusion time τ_p for the plasma; this is equivalent to saying that on the time-scale τ_v the functions $p(\psi)$ and $f(\psi)$ remain constant : they are given by the formulae (I.24) where the function $g(\psi_N)$ remains unchanged during the time interval considered. Moreover, we suppose the coil currents I_i and the total plasma current I_p likewise remain constant, which implies the shell effect of the coils is neglected (either because they are far from the plasma as in TFR, or because their time constant is large compared with τ_v).

We work on the time-scale τ_V. The evolution problem $P_\Lambda(t)$ then represents the evolution of a plasma whose internal characteristics are fixed and whose movements induce eddy currents j_{CV} in the vacuum vessel.

The assumption $\tau_V \ll \tau_p$ is reasonable in many actual Tokamaks where the plasma is surrounded by a thin vacuum vessel : in TFR, for example, $\tau_V = 300\ \mu s$ while τ_p is of the order of 10 ms. In experimental reality this model corresponds to the evolution of the plasma, after a sudden perturbation of the magnetic field that causes a displacement of the free boundary Γ_p of the plasma, stabilized or not by the currents induced in the vessel. In Chapter VI we shall study a complete model for the evolution of the equilibrium, taking account of diffusion phenomena within the plasma.

IV.3.3 Study of the Stability of $P_\Lambda(t)$ in the Neighborhood of the Limit Point P_o

The linear stability in the sense of Lyapunov for the evolution problem $P_\Lambda(t)$ can be studied by considering solutions to the system (IV.28) that can be written in the form

$$U(t) = u + \epsilon e^{-\theta t}\bar{u}$$

where u is a solution to the stationary problem (P_Λ) defined by (IV.2). Expanding about $\varepsilon = 0$ and retaining only the first order terms of (IV.28) we obtain for $\bar{u} = (\bar{\psi},\bar{\lambda}) \in V_\ell \times R$:

$$\text{(IV.29)}\qquad \begin{cases} b_u(\bar{\psi},\varphi) - c_u(\varphi)\bar{\lambda} = \theta \displaystyle\int_{\Omega_{CV}} \frac{\sigma_V}{r}\bar{\psi}\varphi\, dS, \quad \forall\varphi \in V_\ell \\ d_u(\bar{\psi},1) + c_u(1)\bar{\lambda} = 0 . \end{cases}$$

Let A be the operator that to $(\psi_1,\lambda_1) \in V_\ell \times R$ assigns $(\psi_2,\lambda_2) \in V_\ell \times R$ such that

$$\psi_2(M) = \frac{\sigma_V}{r}\psi_1(M) \quad \text{if } M \in \Omega_{CV}$$

$$\psi_2(M) = 0 \qquad \text{if } M \notin \Omega_{CV}$$

$$\lambda_2 = 0 .$$

The parameter θ defined by (IV.29) is an A–eigenvalue of the linearized operator with matrix K. If Re $\theta > 0$ for all A–eigenvalues of K then u is linearly stable; if at least one A–eigenvalue of K has Re$\theta < 0$ then u is linearly unstable.

We observed in Figs. IV.1 and IV.4 the existence of limit points P_0 on the branch $u(\Lambda)$. We shall now show that there is loss of stability for the problem $P_\Lambda(t)$ at these limit points.

Recall that the system defined by (IV.2) can be written

$$G(\Lambda,u) = 0 .$$

Let (Λ_0,u_0) be the pair corresponding to the limit point P_0. Eigenvectors $\bar{u}_0 = (\bar{\psi}_0,\bar{\lambda}_0) \in V_\ell \times R$ and $u_0^* = (\psi_0^*,\lambda_0^*) \in V_\ell \times R$ of D_uG and $(D_uG)^*$ respectively associated to the eigenvalue 0 are defined by :

$$\text{(IV.30)} \quad \begin{cases} b_{u_0}(\bar{\psi}_0,\varphi) - c_{u_0}(\varphi)\bar{\lambda}_0 = 0, \ \forall\varphi \in V_\ell \\ d_{u_0}(\bar{\psi}_0,1) + c_{u_o}(1)\bar{\lambda}_o = 0 \\ |\bar{\psi}_o|^2_{V_\ell} + |\bar{\lambda}_0|^2 = 1 \end{cases} \qquad \begin{cases} b_{u_0}(\varphi,\psi_0^*-\lambda_0^*) = 0 , \ \forall\varphi\in V_\ell \\ c_{u_o}(\psi_o^*-\lambda_o^*) = 0 \\ \langle\bar{\psi}_o,\psi^*_o\rangle_{V_\ell} + \bar{\lambda}_0\,\lambda^*_0 = 1 . \end{cases}$$

Define $V_1 = \text{Ker}\,(D_uG)_0 = R\bar{u}_0$

$$V_2 = \text{Im}\,(D_uG)_0 = \{v = (\psi,\lambda) \in V_\ell \times R | \langle\psi,\psi_0^*\rangle + \lambda\lambda_0^* = 0\} .$$

If the algebraic multiplicity of the zero eigenvalue is 1 we have

$$V_\ell \times R = V_1 \oplus V_2$$

and $(D_uG)_0$ is an isomorphism of V_2.

Now we proceed to parametrize the solutions to (IV.2) in the neighborhood of the limit point using the method of F. BREZZI – J. RAPPAZ – P.A. RAVIART or of M.G. CRANDALL – P.H. RABINOWITZ.

Proposition IV.3 :

If the hypotheses H_1', H_2' and H_3' of I.4.4.1, are satisfied at the limit point P_0 then there exists $s_0' > 0$ such that the solutions to (IV.2) in a neighborhood of P_0 form a branch $(\Lambda(s'), u(s'))$ with :

$$\Lambda(s') = \Lambda_0 + \Lambda'(s')$$

$$u(s') = u_0 + s'\bar{u}_0 + u'(s')$$

where $s' \in [-s_0', +s_0'] \to (\Lambda'(s'),u'(s')) \in R \times V_2$ is a continuously differentiable function in the neighborhood of $s' = 0$ and where

$$\dot{\Lambda}'(0) = (\frac{d\Lambda'}{ds'})(0) = 0$$

$$\dot{u}'(0) = (\frac{du'}{ds'})(0) = 0 .$$

Proof :

The hypotheses H_1', H_2', H_3' and the finite element method chosen imply (as was seen in Section I.4.4) that G is continuously differentiable in a neighborhood of P_0. Consider the function $G':R \times R \times V_2 \to R^{N\ell+1}$ defined by :

$$G'(s',\Lambda',u') = G(\Lambda_0+\Lambda',u_0+s'\bar{u}_0+u')-G(\Lambda_0,u_0) .$$

We have

- $G'(0,0,0) = 0$;
- $D_{\Lambda',u'}G'(0,0,0)$ is an isomorphism $R\times V_2 \to R^{N\ell+1}$ in view of the condition (iii) of (IV.5) and the fact that $(D_uG)_0$ is an isomorphism of V_2;
- $D_{s'}G'(0,0,0) = 0$ by definition of $\bar{u}_0$.

We deduce Proposition IV.3 immediately from this by the Implicit Function Theorem. □

According to M.G. CRANDALL – P.H. RABINOWITZ, since 0 is a simple A–eigenvalue of K there exist continuous functions $\theta(s') \in R$ and $\bar{u}(s') = (\bar{\psi}(s'),\bar{\lambda}(s'))$ $\in V_\ell \times R$ defined in the neighborhood of $s' = 0$ and such that :

$$
\text{(IV.31)}\quad
\begin{cases}
b_{u(s')}(\bar{\psi}(s'),\varphi) - c_{u(s')}(\varphi)\bar{\lambda}(s') = \theta(s')\displaystyle\int_{\Omega_{cv}} \frac{\sigma_v}{r}\,\bar{\psi}(s')\varphi\, dS, \\
\qquad\qquad\qquad\qquad\qquad\qquad\qquad\qquad \forall\varphi \in V_\ell \\
d_{u(s')}(\bar{\psi}(s'),1) + c_{u(s')}(1)\bar{\lambda}(s') = 0 \\
(\bar{u}(s') - \bar{u}_o) \in V_2 \\
\bar{u}(0) = \bar{u}_o, \quad \theta(0) = 0 .
\end{cases}
$$

<u>Proposition IV.4</u> :

With the assumptions of Proposition IV.3 and with $\theta(s')$, $\bar{u}(s')$ defined by (IV.31) where 0 is a simple A–eigenvalue, the quantities

$$
\theta(s')\int_{\Omega_{cv}} \frac{\sigma_v}{r}\,\bar{\psi}_o\,\psi_o^*\, dS \quad\text{and}\quad \dot{\Lambda}(s')I_S\Big[\frac{1}{S_1}\int_{B_1}\psi_o^*\, dS - \frac{1}{S_k}\int_{B_k}\psi_o^*\, dS\Big]
$$

have the same zeros in a neighborhood of $s' = 0$. If $\dot{\Lambda}(s') \neq 0$ then they have the same sign, and moreover

$$
\lim_{s'\to 0}\frac{\theta(s')}{\dot{\Lambda}(s')} = \frac{I_S\Big[\frac{1}{S_1}\displaystyle\int_{B_1}\psi_o^*\, dS - \frac{1}{S_k}\int_{B_k}\psi_o^*\, dS\Big]}{\displaystyle\int_{\Omega_{cv}}\frac{\sigma_v}{r}\,\bar{\psi}_o\,\psi_o^*\, dS} .
$$

□

<u>Proof</u> :

Differentiating equations (IV.2) with respect to s' we obtain

(IV.32)

$$
\begin{cases}
b_{u(s')}(\dot{\psi}(s'),\varphi) - c_{u(s')}(\varphi)\dot{\lambda}(s') = \dot{\Lambda}(s')I_S\Big[\frac{1}{S_1}\displaystyle\int_{B_1}\varphi\, dS \\
\qquad\qquad\qquad\qquad - \frac{1}{S_k}\displaystyle\int_{B_k}\varphi\, dS\Big], \quad \forall\varphi \in V_\ell \\
d_{u(s')}(\dot{\psi}(s'),1) + c_{u(s')}(1)\dot{\lambda}(s') = 0 .
\end{cases}
$$

By Proposition IV.3 we have $\dot{u}(s') - \bar{u}_0 \in V_2$.

By (IV.31) we have $\bar{u}(s') - \bar{u}_0 \in V_2$.

We deduce that $\dot{u}(s') - \bar{u}(s') \in V_2$ and we extract from (IV.31) and (IV.32) the following equation for $\dot{u}(s') - \bar{u}(s')$:

(IV.33)

$$\begin{cases} b_{u(s')}(\dot{\psi}(s')-\bar{\psi}(s'),\varphi) - c_{u(s')}(\varphi)[\dot{\lambda}(s')-\bar{\lambda}(s')] \\ = \dot{\Lambda}(s')I_S[\frac{1}{S_1}\int_{B_1}\varphi\, dS - \frac{1}{S_k}\int_{B_k}\varphi\, dS] - \theta(s')\int_{\Omega_{cv}}\frac{\sigma_v}{r}\bar{\psi}(s')\varphi\, dS \\ \qquad \forall\varphi \in V_\ell \\ d_{u(s')}(\dot{\psi}(s') - \bar{\psi}(s'),1) + c_{u(s')}(1)[\dot{\lambda}(s')-\bar{\lambda}(s')] = 0 \end{cases}$$

Since $(D_uG)_0$ is an isomorphism of V_2 there exists a neighborhood of $s' = 0$ in which D_uG is also an isomorphism of V_2 and we deduce

(IV.34) $$|\dot{u}(s')-\bar{u}(s')| \leqslant C[|\dot{\Lambda}(s')| + |\theta(s')|]$$

where $|\dot{u}(s') - \bar{u}(s')|^2_{V_\ell\times R} = |\dot{\psi}(s')-\bar{\psi}(s')|^2_{V_\ell} + |\dot{\lambda}(s')-\bar{\lambda}(s')|^2$.

We now compute from (IV.33) the following quantities :

(IV.35)

$$\dot{\Lambda}(s')I_S[\frac{1}{S_1}\int_{B_1}\varphi\, dS - \frac{1}{S_k}\int_{B_k}\varphi\, dS] - \theta(s')\int_{\Omega_{cv}}\frac{\sigma_v}{r}\bar{\psi}_0\varphi\, dS$$

$$= b_{u_0}(\dot{\psi}(s')-\bar{\psi}(s'),\varphi) - c_{u_0}(\varphi)[\dot{\lambda}(s')-\bar{\lambda}(s')]$$

$$+ (b_{u(s')}-b_{u_0})(\dot{\psi}(s')-\bar{\psi}(s'),\varphi)-(c_{u(s')}-c_{u_0})(\varphi)[\dot{\lambda}(s')-\bar{\lambda}(s')]$$

$$- \theta(s')\int_{\Omega_{cv}}\frac{\sigma_v}{r}(\bar{\psi}_0-\bar{\psi}(s'))\varphi\, dS \ , \ \forall\varphi \in V_\ell \ .$$

and

$$d_{u_0}(\dot{\psi}(s')-\bar{\psi}(s'),1)+c_{u_0}(1)[\dot{\lambda}(s')-\bar{\lambda}(s')]$$

$$(IV.36) \qquad + (d_{u(s')}-d_{u_0})(\dot{\psi}(s')-\bar{\psi}(s'),1)$$

$$+ (c_{u(s')}-c_{u_0})(1)[\dot{\lambda}(s')-\bar{\lambda}(s')] = 0 \; .$$

Putting $\varphi = \psi_0^*$ in (IV.35) and multiplying (IV.36) by λ_0^* and adding these two equations we obtain using (IV.30) :

$$B = \dot{\Lambda}(s') I_S [\frac{1}{S_1}\int_{B_1} \psi_0^* \, dS - \frac{1}{S_k}\int_{B_k} \psi_0^* \, dS] - \theta(s')\int_{\Omega_{cv}} \frac{\sigma_v}{r} \bar{\psi}_0 \psi_0^* \, dS$$

$$= (b_{u(s')}-b_{u_0})(\dot{\psi}(s')-\bar{\psi}(s'),\psi_0^*)-(c_{u(s')}-c_{u_0})(\psi_0^*)[\dot{\lambda}(s')-\bar{\lambda}(s')]$$

$$+ (d_{u(s')}-d_{u_0})(\dot{\psi}(s')-\bar{\psi}(s'),1)\lambda_0^*+(c_{u(s')}-c_{u_0})(1)[\dot{\lambda}(s')-\bar{\lambda}(s')]\lambda_0^*$$

$$- \theta(s')\int_{\Omega_{cv}} \frac{\sigma_v}{r}(\bar{\psi}_0-\bar{\psi}(s'))\psi_0^* \, dS \; .$$

By (IV.34) and the fact that D_uG is continuously differentiable in a neighborhood of P_0 we have

$$B \leqslant o(1)[|\dot{\Lambda}(s')|+|\theta(s')|] \quad \text{when } s' \to 0.$$

We immediately deduce Proposition IV.4. □

Note in figures (IV.1) and (IV.4) the change of sign of $\dot{\Lambda}(s')$ at the limit point P_0. From Proposition IV.4 we deduce the change of sign of $\theta(s')$, which implies a loss of stability for the evolution problem $P_\Lambda(t)$ at the point P_0.

Remark IV.2 :

In Propositions IV.3 and IV.4 we assumed that hypotheses of continuous differentiability of the state equations were satisfied. If we suppose that the state function is twice differentiable, we can calculate the derivative of the eigenvalue $\theta(s')$ with respect to s' at 0 as in H.B. KELLER [2] or F. MIGNOT – J.P. PUEL. We obtain

(IV.37) $$\dot{\theta}(0) = \frac{\langle u_o^*, D^2_{uu}G(\bar{u}_o,\bar{u}_o)\rangle}{\int_{\Omega_{cv}} \frac{\sigma_v}{r}\,\bar{\psi}_o\psi_o^*\,dS}\ .$$

The eigenvalue $\theta(s)$ changes sign if the right hand side of (IV.37) is nonzero. This is the condition of nondegeneracy of the limit point (cf. F. BREZZI – J. RAPPAZ – P.A. RAVIART) which is evidently satisfied in the diagrams IV.1 and IV.4. This is therefore another way of discerning the loss of stability at the point P_o. □

Remark IV.3 :

Another way of proving the change of sign of an eigenvalue at P_o is to use the theory of topological degree, which enables us to show as in D.H. SATTINGER that the determinant of the Jacobian of the system changes sign at P_o if the eigenvalue 0 there is simple. We saw numerically in Section IV.1.5 that det K changed sign at P_o, which confirms the change of sign of an eigenvalue at this point. □

IV.3.4 *Study of the Global Stability of Solutions of the Evolution Problem* $P_\Lambda(t)$.

In Section IV.3.3 we showed that an eigenvalue changes sign at the limit point P_o, if it exists. However, the solution to the equilibrium problem is stable only if all the other eigenvalues θ defined by (IV.29) are in the positive half-plane. Since the spectrum of a 1500 × 1500 matrix is not something which it is easy to compute numerically, we discretized the evolution problem $P_\Lambda(t)$ by a finite difference method in time with implicit scheme, a P_1 finite element method in space and a Newton iteration at each time interval. If (ψ_j,λ_j) denotes the values of ψ_j and λ_j at the instant $j\Delta t$ then the system which calculates $(\psi_{j+1},\lambda_{j+1})$ from (ψ_j,λ_j) is

(IV.38) $$\begin{cases} \int_{\Omega_{cv}} \frac{\sigma_v}{r}\,\psi_{j+1}\,\varphi\,dS + \Delta t\,[b_{u_j}(\psi_{j+1},\varphi) - c_{u_j}(\varphi)\lambda_{j+1}] \\ = \int_{\Omega_{cv}} \frac{\sigma_v}{r}\,\psi_j\,\varphi\,dS + \Delta t\,[\sum_{i=1}^{k} \frac{I_i}{S_i}\int_{B_i}\varphi\,dS + a'_{u_j}(\psi_j,\varphi)]\,,\forall\varphi\in V_\ell \\ d_{u_j}(\psi_{j+1},1) + c_{u_j}(1)\lambda_{j+1} = I_p \\ \quad \text{with } I_1 = \Lambda I_S,\ I_k = (I-\Lambda)I_S. \end{cases}$$

Consider the example of the equilibrium solution branches in JET where the plasma has elliptical shape of the type shown in Fig. IV.3. We begin with the equilibrium corresponding to the point P_1 ($\Delta_H = -1.35$ cm) and perturb the value of Λ so that the plasma is no longer in equilibrium. It is allowed to evolve according to the equations for the problem $P_\Lambda(t)$, discretized according to (IV.38). In Figs. IV.6 the time evolution of the radial position Δ_H of the plasma and its height b are shown. Observe that there is convergence towards a new equilibrium position represented by the point P_1' in Fig. IV.3 and corresponding to the perturbed value of Λ.

Next we consider the equilibrium corresponding to the point P_2 in Fig. IV.3. We perturb Λ slightly and represent the evolution of Δ_H and of b in Fig. IV.7. Observe that first of all the plasma evolves in an unstable way, but then when it comes into contact with the inner point of the limiter ($\Delta_H < 0$) it begins to converge towards an interior position of equilibrium corresponding to the perturbed value of Λ.

By a simple continuity argument we can thus say that the solution branch corresponding to $\Delta_H < 0$ is stable, while the branch corresponding to $\Delta_H > 0$ is unstable.

We have carried out the same type of study for branches having limit points, as in Figs. IV.1 and IV.4. We noted that the equilibria such that $\Delta_H < \Delta_H(P_0)$ are stable while those corresponding to $\Delta_H > \Delta_H(P_0)$ are unstable. Observe that in Fig. IV.5 the zone of stability is the larger as β_p and ℓ_i are the greater.

We noticed the following equivalences in the collection of diagrams IV.1 to IV.4 :

. the branch is stable for $P_\Lambda(t)$

$\Leftrightarrow$ the Picard algorithm AP_2 converges ($\rho(M_u) < 1$)

$\Leftrightarrow$ Δ_H is an increasing function of Λ

. the branch is unstable for $P_\Lambda(t)$

$\Leftrightarrow$ the Picard algorithm AP_2 diverges ($\rho(M_u) > 1$)

$\Leftrightarrow$ Δ_H is a decreasing function of Λ.

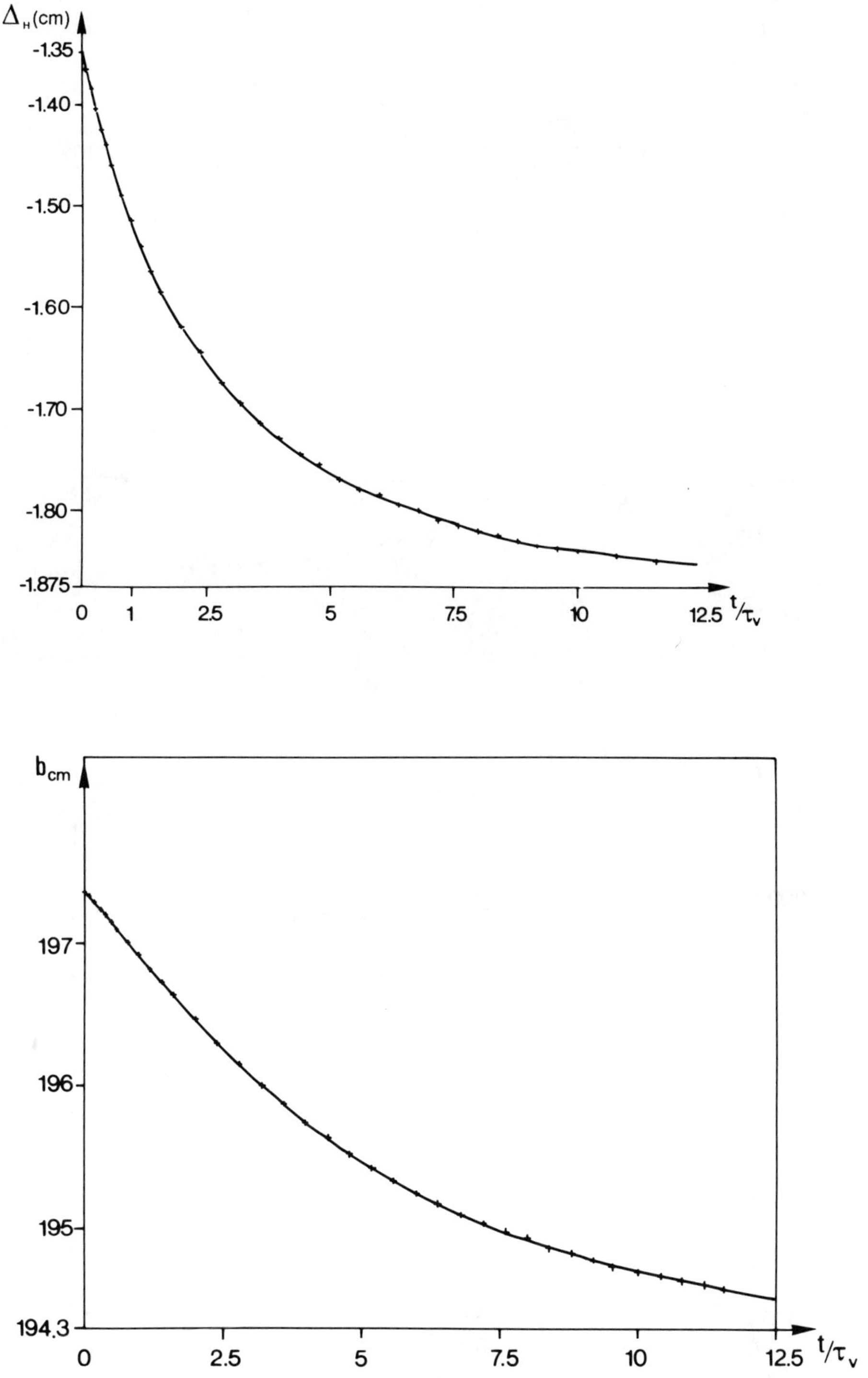

Figure IV.6 : Evolutions of the radial position Δ_H and the height b of the plasma after a perturbation in JET (initial plasma displaced towards the interior)

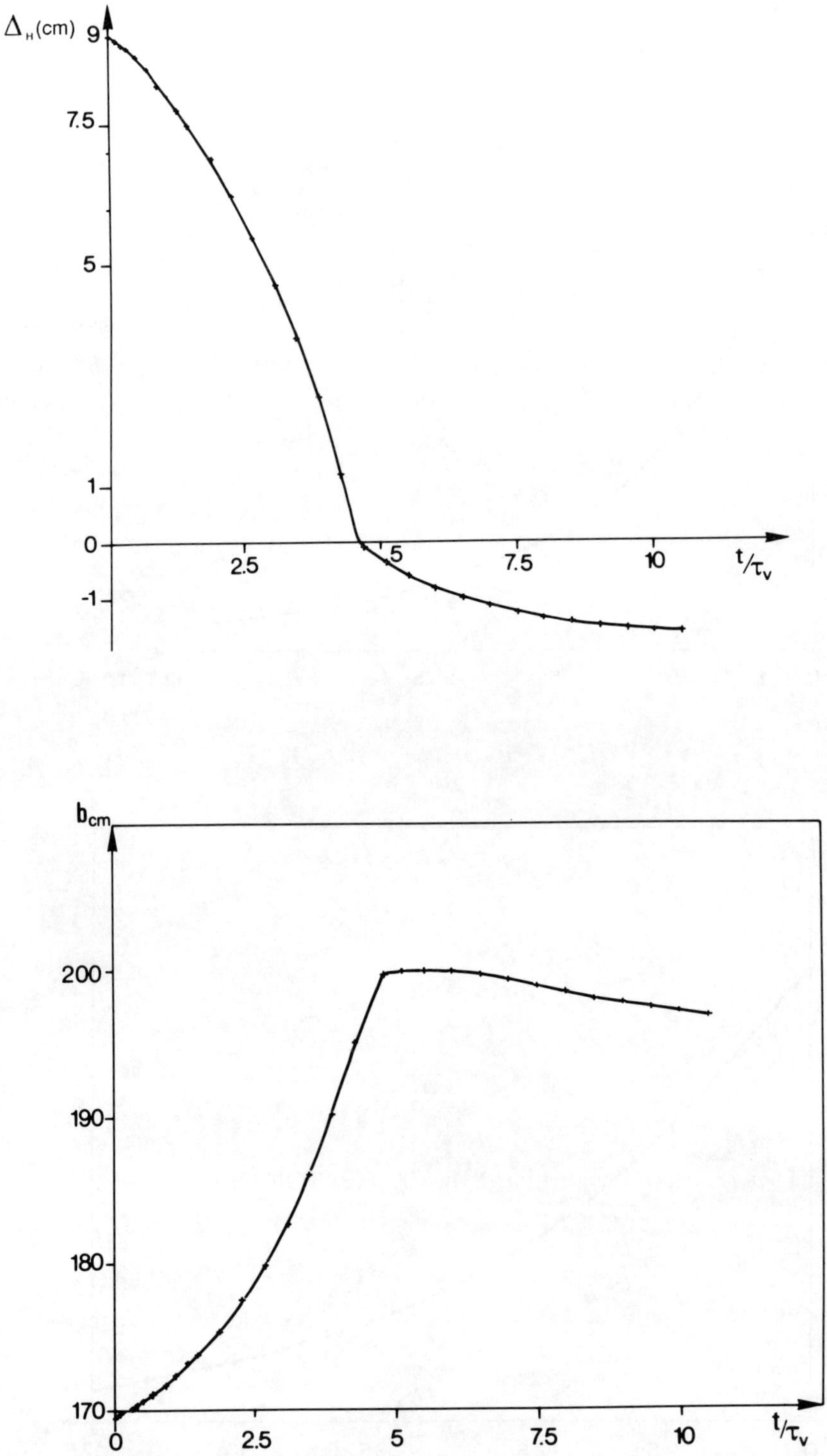

Figure IV.7 : Evolutions of the radial position Δ_H and the height b of the plasma after a perturbation in JET (initial plasma displaced towards the exterior)

In Chapter VII we shall return to studying the stability of horizontal displacements of the plasma in TFR by a different method, and will make comparisons with experimental observations.

***Remark IV.4* :**

In W. FENEBERG – K. LACKNER the convergence of the algorithms for the equilibrium problem with free boundary in a Tokamak without iron is studied and the following "rule" is stated : "The Picard algorithm converges when the plasma touches the inner point of the limiter ($\Delta_H < 0$) and diverges when it touches the outer point ($\Delta_H > 0$)". Figure IV.2, corresponding to the case of TFR where the iron is removed, seems to confirm this rule. In G. CENACCHI – R. GALVAO – A. TARONI it has been demonstrated that this rule is related to the stability of horizontal displacements of the plasma when it is in contact with a limiter. We should make it clear that this "rule" is in general no longer valid in Tokamaks with iron. Figures IV.1 and IV.4 show clearly that the passage from convergence to divergence of the Picard algorithm and the loss of stability of horizontal displacements of the plasma take place at the limit point P_0 and not at $\Delta_H = 0$. The "rule" of W. FENEBERG – K. LACKNER and G. CENACCHI – R. GALVAO – A. TARONI is valid only when the condition (IV.17) is satisfied, which is generally the case in Tokamaks with air transformer but no longer in Tokamaks with iron. □

***Remark IV.5* :**

Limit points have been demonstrated by J. HELTON – T.S. WANG in studying equilibrium solution branches for the machine DOUBLET : for the same currents in the coils it is possible to have three equilibrium solutions, one corresponding to an elliptical plasma, the second to a "doublet" and the third to a "droplet" (cf. Remark I.8). □

***Remark IV.6*:**

Bifurcation points have elsewhere been demonstrated by D.G. SCHAEFFER and M. SERMANGE in the case of a shell in the form of a hourglass; one branch corresponds to solutions that are symmetric with respect to the equatorial plane, and the other to asymmetric solutions. □

5. Identification of the plasma boundary and plasma current density from magnetic measurements

In a Tokamak it is essential to know the shape and position of the plasma boundary at every moment of the discharge. It is the first link in a chain of diagnostics that enable the internal characteristics of the plasma (densities, temperatures, etc...) to be interpreted.

The plasma boundary is identified from measurements of the poloidal flux ψ in the vacuum vessel, and of the tangential component of the poloidal field B_P, equal to $1/r\ \partial\psi/\partial n$ where $\partial\psi/\partial n$ is the derivative of ψ in the direction normal to the vessel. Mathematically we are in the situation of a Cauchy problem for an elliptic equation, i.e. an "ill-posed" problem in the sense of Hadamard. In this chapter we shall use two different methods to solve this plasma boundary identification problem.

In Section V.1 we consider the equation $L\psi = 0$ for the poloidal flux ψ in the vacuum region, i.e. between the wall of the vacuum vessel and the plasma boundary. It is therefore one of the boundaries of the domain that is unknown. We suppose that there exists an "elliptic" extension of ψ (i.e. in the sense of $L\psi = 0$) as far as a fixed boundary Γ_0' lying within the plasma. In this way we are faced with solving a Cauchy problem for $L\psi = 0$ in a fixed domain, with the plasma boundary appearing simply as the line of flux inside this domain that is tangent to the limiter D. Optimal control techniques with regularization are employed to solve this Cauchy problem and obtain a stable formulation. If we solve the optimality system by a finite element method, the poloidal flux ψ at each point of the vacuum region is written as a linear combination of measurements of flux and field tangent to the vacuum vessel. This method then enables the plasma boundary to be determined in ultra-rapid fashion, and is a promising tool for on-line control of the plasma shape.

In Section V.2 we solve the equation for ψ not only in the vacuum vessel but also in the interior of the plasma. To do this we assume *a priori* the following law for the toroidal component of the plasma current density:

$$j_T(r,\psi) = \lambda[\beta\ r/R_o + (1-\beta)R_o/r](1-\psi_N)^\gamma\ .$$

We shall make use of the fact that we measure superabundant data (Dirichlet and Neumann conditions) on the boundary of the domain in order to identify the parameters β and γ in the expression for $j_T(r,\psi)$. For this we use the measurement of ψ in the vacuum vessel as a boundary condition for the equation for ψ, and we minimize the quadratic distance between the measured tangent field and the computed one, the optimization parameters being β and γ. This optimal control problem is solved by the same sequential quadratic optimization algorithm as in Chapter II. The plasma boundary is a free boundary, defined by its contact with the limiter, and this constitutes an additional nonlinearity in the problem. The numerical solution of this optimization on CRAY 1 requires a CPU time of the order of one second, while the method developed in Section V.1 needs computation time of the order of a millisecond. However, the method developed in this section gives additional information on the plasma current density. Moreover, it enables the flux lines within the plasma to be identified, and this is essential for the analysis of measurements of density and temperature.

In Section V.3 the two methods presented in the preceding sections are applied to the cases of the Tokamaks TORE Supra and JET. The measurements of flux and field in the vacuum chambers of these Tokamaks being discrete measurements, we first of all show how the methods of solution presented in Sections V.1 and V.2 for continuous measurements can be adapted to the case of discrete measurements. In the case of TORE Supra we study by the rapid method of Section V.1 the influence of perturbations of the magnetic measurements on the determination of the plasma boundary and the role of the regularization parameter. We show how this method can be used in the on-line control of the plasma shape. In the case of JET these methods were used first of all to optimize the position and number of magnetic measurements that enable the plasma boundary to be identified. The method of Section V.2 is used on JET to identify the magnetic configuration at each moment of the discharge. For plasmas of elliptic section it also allows separate identification of the coefficients β_P (poloidal beta) and ℓ_i (internal inductance of the plasma per unit length). On the other hand, for plasmas of circular section with small β_P only the coefficient $\beta_P + \ell_i/2$ can be identified. This program is the first in a chain of interpretation codes used to analyse the discharges in

JET; the parametric identification of the density of the plasma current and of the lines of flux within the plasma are subsequently used to analyse the transport in a discharge.

V.1 RAPID IDENTIFICATION OF THE PLASMA BOUNDARY BY SOLVING THE CAUCHY PROBLEM IN A VACUUM

V.1.1 Formulation of the Problem

Let Γ_V denote the section of the inner wall of the vacuum vessel, which we assume to be of class C^2. The poloidal flux ψ is measured on Γ_V, and this function is denoted f_V. By (I.11) the component of the poloidal magnetic field B_P tangent to Γ_V is $1/r\ \partial\psi/\partial n$ where $\partial\psi/\partial n$ denotes the derivative of ψ normal to Γ_V; this component is measured and denoted by g_V.

The vacuum region surrounding the plasma is denoted by $\Omega_X = \Omega_V - \bar{\Omega}_p$ (see Fig. V.1.) By (I.20) and (I.28) the problem is formulated as follows in Ω_X :

$$\text{(V.1)} \quad \begin{cases} \psi = f_V & \text{on} \quad \Gamma_V \\ \dfrac{1}{r}\dfrac{\partial\psi}{\partial n} = g_V & \text{on} \quad \Gamma_V \\ \psi = \sup\limits_{D} \psi & \text{on} \quad \Gamma_p \\ L\psi = 0 & \text{in} \quad \Omega_X \end{cases}$$

In the formulation (V.1) the domain Ω_X is unknown, since Γ_p is an unknown for the problem. Moreover, the problem is "ill-posed" in the sense of Hadamard (cf. R. COURANT – D.HILBERT) since there are two Cauchy conditions on the outer boundary Γ_V.

To determine the free boundary Γ_P we could use techniques of optimal design (cf. O. PIRONNEAU and references). However, this would imply working in a domain that varies as the iterations proceed, and thus with an adaptive mesh. This seems incompatible with the requirements of speed demanded by this problem.

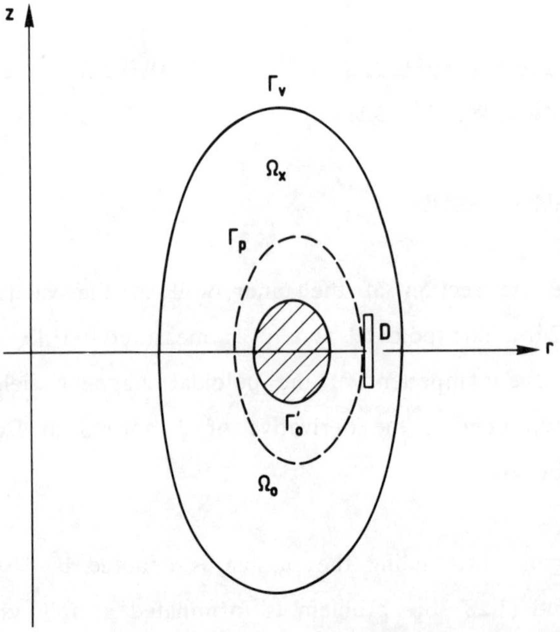

Figure V.1 : *Domain of resolution for the rapid method of determining the plasma boundary.*

We shall modify the problem (V.1) as follows:

i) Assume that the function ψ can be extended in the sense of $L\psi = 0$ as far as boundary Γ'_0 fixed within the plasma and of class C^2 (see Fig. V.1.). By the Cauchy– Kowalewska Theorem (cf. R. COURANT–D. HILBERT) this extension exists in a neighborhood of Γ_p; hence if Γ'_0 is not "too far" from Γ_p we may legitimately suppose that this extension exists. The problem then becomes one on a fixed domain Ω_0 contained between Γ_V and Γ'_0.

ii) Retain the Dirichlet condition on Γ_V and relax the Neumann condition by expressing it in the form of an optimization problem. More precisely, define $\psi(v)$ by :

$$\text{(V.2)} \quad \begin{cases} \psi = f_V & \text{on} \quad \Gamma_V \\ \dfrac{1}{r}\dfrac{\partial \psi}{\partial n} = v & \text{on} \quad \Gamma'_0 \\ L\psi = 0 & \text{in} \quad \Omega_0 \end{cases}$$

where v is a function defined on Γ'_0.

We seek u such that :

(V.3) $$J(u) = \inf J(v)$$

with

$$J(v) = \frac{1}{2} \int_{\Gamma_V} \left(\frac{1}{r}\frac{\partial\psi}{\partial n} - g_V\right)^2 d\sigma$$

where ψ is related to v by (V.2)

iii) The solution to the problem (V.3) is still unstable with respect to the data f_V and g_V. We regularize the problem by defining a functional J_ε by

(V.4) $$J_\varepsilon(v) = J(v) + \frac{\varepsilon}{2} \int_{\Gamma'_o} v^2 \, d\sigma$$

and we then seek u_ε such that

(V.5) $$J_\varepsilon(u_\varepsilon) = \inf J_\varepsilon(v)$$

where ψ is related to v by the system (V.2).

The solution ψ_ε related to u_ε by (V.2) satisfies the Dirichlet condition on Γ_V precisely, and the Neumann condition approximately. Later we shall show that ψ_ε is stable with respect to f_V and g_V for fixed ε. The plasma boundary is then simply the equipotential of ψ_ε that is tangent to the limiter D:

(V.6) $$\Gamma_p^\varepsilon = \{M \in \Omega_o \text{ such that } \psi_\varepsilon(M) = \sup_D \psi_\varepsilon\} .$$

Clearly, the restriction of ψ_ε to the region between Γ_p^ε and Γ'_o has no physical meaning, and has nothing to do with the real function ψ in the interior of the plasma.

__*Remark V.1.*__

Instead of including a Dirichlet boundary condition for ψ on Γ_V in (V.2), we could have used a Neumann boundary condition corresponding to the measurement g_V and introduced $(\psi - f_V)^2$ into the cost-function (V.3). Our choice was motivated by the fact that the boundary Γ_p of the plasma is more sensitive to measurement errors in f_V than in g_V, especially when Γ_p is close to Γ_V. This is clearly evident when we take the Taylor expansion of ψ to first order in the direction normal to Γ_V:

$$\psi(M) = f_V + d(M) \times g_V \ , \ \forall M \in \Gamma_p$$

where d is the distance from M to Γ_V along this normal. If d is small the first term dominates and the sensitivity to error in g_V is less than that in f_V. □

Remark V.2.

A similar formulation of the problem is given in J. BLUM – J. LE FOLL – B. THOORIS [3], but there the control vector v corresponds to a Dirichlet condition on Γ_o':

(V.7) $\psi = v$ on Γ_o'

and not to a Neumann condition as in (V.2). The advantage of (V.2) will appear in the next subsection in the form of a physical interpretation, while the equality (V.7), which likewise corresponds to an extension of ψ up to Γ_o', has no simple physical interpretation. □

V.1.2 *Physical Interpretation of the Method of Solution*

If Γ_o' were identical with Γ_p the vector v defined in (V.2) would represent the poloidal field on Γ_p. By the virtual shell principle (cf. V.D. SHAFRANOV – L.E. ZAKHAROV) v would then, up to a multiplicative factor of $1/\mu_o$, denote that surface current density on Γ_p for which the magnetic field created outside the plasma by this current sheet is identical to the field created by the real current density spread throughout the plasma.

In the case when Γ_o' is a fixed contour the vector v denotes a surface current density on Γ_o' such that, if ψ is equal to f_V on Γ_V then the field tangent to Γ_V is as nearly as possible equal to g_V, in the sense of least squares.

We thus recover a formulation close to that of W. FENEBERG – K. LACKNER – P. MARTIN where the surface current density is decomposed into a Fourier series and where a small number of modes is used in order to avoid oscillations connected with the unstable nature of the problem.

Remark V.3.

In D.K. LEE – Y.K. PENG a multipolar decomposition of the plasma current density together with knowledge of the currents in the poloidal field coils allows ψ to be written as a combination of Legendre functions whose coefficients are identified from field or flux measurements on Γ_V.

In D.W. SWAIN – G.H. NEILSON a simpler method consists of representing the plasma current by toroidal filaments carrying currents I_i' and localized at points $x_i = (r_i, z_i)$, so that the equation for ψ becomes

$$L\psi = \sum_i I_i' \, \delta(x-x_i) \ .$$

The currents I_i' are determined by minimizing the quadratic distance between the field measured on Γ_V and the computed field. According to W. FENEBERG – K. LACKNER – P. MARTIN the plasma boundary so obtained is more sensitive to the location of these filaments than to the choice of the contour Γ_0' on which a surface current distribution is considered. □

V.1.3 *Mathematical Study of the Problem* $(P^{\varepsilon}{}_V)$:

Let $f_V \in H^1(\Gamma_V)$ and $g_V \in L^2(\Gamma_V)$. Let $\mathcal{U} = L^2(\Gamma_0')$, the space of admissible controls v. Define $\psi(v)$ by the system (V.2), and call the following problem $(P^{\varepsilon}{}_V)$: find $u_\varepsilon \in \mathcal{U}$ such that

$$J_\varepsilon(u_\varepsilon) = \inf_{v \in \mathcal{U}} J_\varepsilon(v) \tag{V.8}$$

where J_ε is given by (V.4).

We have the theorem:

Theorem V.1

Problem $(P^{\varepsilon}{}_V)$ admits a unique solution $u_\varepsilon \in \mathcal{U}$ which is stable with respect to the data f_V and g_V. In fact for $f_V^1, f_V^2 \in H^1(\Gamma_V)$ and $g_V^1, g_V^2 \in L^2(\Gamma_V)$ we have:

$$|u_\varepsilon^1 - u_\varepsilon^2|_{L^2(\Gamma_0')} \leqslant \frac{1}{\sqrt{\varepsilon}} \left[|g_V^1 - g_V^2|_{L^2(\Gamma_V)} + C|f_V^1 - f_V^2|_{H^1(\Gamma_V)} \right] \tag{V.9}$$

Moreover, if there exists $u \in \mathcal{U}$ such that

$$\frac{1}{r}\frac{\partial\psi(u)}{\partial n} = g_V ,$$

then

$$u_\varepsilon \rightarrow u \quad \text{in} \quad \mathcal{U} \quad \text{when} \quad \varepsilon \rightarrow 0 . \qquad \square$$

We follow the same technique of proof as P. COLLI FRANZONE *et al*. for the solution of an inverse problem in electrocardiology.

Consider the map which associates to $v \in \mathcal{U}$ the trace of $1/r\ \partial\psi(v)/\partial n$ on Γ_V, where $\psi(v)$ is defined by (V.2). Then we have the following lemma:

Lemma V.1.

The map $v \rightarrow 1/r\ \partial\psi/\partial n|\Gamma_V$ is a continuous affine injection of $L^2(\Gamma_0')$ into $L^2(\Gamma_V)$. $\square$

Proof

First consider $\psi(0)$ defined from (V.2) by

$$\text{(V.10)} \qquad \begin{cases} \psi(0) = f_V & \text{on} \quad \Gamma_V \\ \dfrac{1}{r}\dfrac{\partial\psi(0)}{\partial n} = 0 & \text{on} \quad \Gamma_0' \\ L\psi(0) = 0 & \text{in} \quad \Omega_0 \end{cases}$$

and let

$$\text{(V.11)} \qquad B = \frac{1}{r}\frac{\partial\psi(0)}{\partial n}\bigg|_{\Gamma_V} .$$

If $f_V \in H^{1/2}(\Gamma_V)$ then by J.L. LIONS – E. MAGENES $\psi(O) \in H^1(\Omega_o)$ and $B \in H^{-1/2}(\Gamma_V)$. The map associating B to f_V is continuous linear from $H^{1/2}(\Gamma_V)$ to $H^{-1/2}(\Gamma_V)$. Likewise if $f_V \in H^{3/2}(\Gamma_V)$ then $\psi(0) \in H^2(\Omega_o)$, $B \in H^{1/2}(\Gamma_V)$ and the map associating B to f_V is continuous linear from $H^{3/2}(\Gamma_V)$ to $H^{1/2}(\Gamma_V)$. By linear interpolation (cf. J.L. LIONS – E. MAGENES) it is clear that the map which associates $B \in L^2(\Gamma_V)$ to $f_V \in H^1(\Gamma_V)$ is continuous linear from $H^1(\Gamma_V)$ to $L^2(\Gamma_V)$.

Now consider the map which to $v \in L^2(\Gamma'_0)$ associates the function $\psi_1(v)$ defined by

(V.12) $$\begin{cases} \psi_1 = 0 & \text{on} \quad \Gamma_V \\ \dfrac{1}{r}\dfrac{\partial \psi_1}{\partial n} = v & \text{on} \quad \Gamma'_0 \\ L\psi_1 = 0 & \text{in} \quad \Omega_0 \end{cases}$$

and let

(V.13) $$Av = \frac{1}{r}\frac{\partial \psi_1}{\partial n}\Big|_{\Gamma_V}.$$

The map $v \to \psi_1$ is a continuous linear map from $H^{-1/2}(\Gamma'_0)$ into $H^1(\Omega_0)$ (respectively, from $H^{1/2}(\Gamma'_0)$ into $H^2(\Omega_0)$). By a trace theorem, the map which associates $1/r\ \partial\psi_1/\partial n|\Gamma_V$ to ψ_1 is continuous linear from $H^1(\Omega_0)$ into $H^{-1/2}(\Gamma_V)$ (respectively, from $H^2(\Omega_0)$ into $H^{1/2}(\Gamma_V)$). By linear interpolation it is clear that the map A is continuous linear from $L^2(\Gamma'_0)$ into $L^2(\Gamma_V)$.

Since $1/r\ \partial\psi/\partial n|\Gamma_V = Av + B$, we deduce that the map $v \to 1/r\ \partial\psi/\partial n|\Gamma_V$ is continuous affine from $L^2(\Gamma'_0)$ into $L^2(\Gamma_V)$.

The injectivity of this map follows from the uniqueness of the solution to the Cauchy problem for an elliptic equation (cf. R. COURANT – D. HILBERT.) □

<u>Proof of Theorem V.1</u> :

Using Lemma V.1 and the coercivity of J_ε, from J.L. LIONS [2] it is clear that Problem (P^ε_V) admits a unique solution $u_\varepsilon \in \mathcal{U}$.

Let B_1 and B_2 be the vectors in $L^2(\Gamma_V)$ defined by (V.10) and (V.11) and corresponding to f^1_V and f^2_V respectively. The optimality condition for (P^ε_V) is written

(V.14) $$\int_{\Gamma_V} (A\, u^1_\varepsilon + B_1 - g^1_V) A\, v\, d\sigma + \varepsilon \int_{\Gamma'_0} u^1_\varepsilon\, v\, d\sigma = 0, \ \forall\, v \in \mathcal{U}$$

and

(V.15) $$\int_{\Gamma_V} (A\, u^2_\varepsilon + B_2 - g^2_V) A\, v\, d\sigma + \varepsilon \int_{\Gamma'_0} u^2_\varepsilon\, v\, d\sigma = 0, \ \forall\, v \in \mathcal{U}$$

respectively, where A is defined by (V.12) — (V.13).

Putting $v = u_\varepsilon^1 - u_\varepsilon^2$ in (V.14) and (V.15) and subtracting we obtain

$$\text{(V.16)} \qquad \int_{\Gamma_V} [A(u_\varepsilon^1 - u_\varepsilon^2)]^2 d\sigma + \int_{\Gamma_V} (B_1 - B_2 + g_V^2 - g_V^1)\, A(u_\varepsilon^1 - u_\varepsilon^2) d\sigma$$

$$+ \varepsilon \int_{\Gamma_o'} (u_\varepsilon^1 - u_\varepsilon^2)^2 \, d\sigma = 0 \, .$$

Hence

$$\text{(V.17)} \qquad \varepsilon |u_\varepsilon^1 - u_\varepsilon^2|^2_{L^2(\Gamma_o')} \leqslant [\, |B_1 - B_2|_{L^2(\Gamma_V)} + |g_V^2 - g_V^1|_{L^2(\Gamma_V)} \,] |A(u_\varepsilon^1 - u_\varepsilon^2)|_{L^2(\Gamma_V)}$$

and

$$\text{(V.18)} \qquad |A(u_\varepsilon^1 - u_\varepsilon^2)|_{L^2(\Gamma_V)} \leqslant |B_1 - B_2|_{L^2(\Gamma_V)} + |g_V^2 - g_V^1|_{L^2(\Gamma_V)} \, .$$

We deduce

$$\text{(V.19)} \qquad |u_\varepsilon^1 - u_\varepsilon^2)|_{L^2(\Gamma_o')} \leqslant \frac{1}{\sqrt{\varepsilon}} [\, |B_1 - B_2|_{L^2(\Gamma_V)} + |g_V^2 - g_V^1|_{L^2(\Gamma_V)} \,] \, .$$

Since the map $f_V \to B$ defined by (V.10) — (V.11) is continuous linear from $H^1(\Gamma_V)$ into $L^2(\Gamma_V)$, the inequality (V.9) follows from (V.19).

Now suppose there exists $u \in L^2(\Gamma_o')$ such that

$$\text{(V.20)} \qquad Au + B = g_V \, .$$

As u_ε is a solution to $(P^\varepsilon{}_V)$ we have

$$\text{(V.21)} \qquad \int_{\Gamma_V} (A\, u_\varepsilon + B - g_V)^2 \, d\sigma + \varepsilon \int_{\Gamma_o'} |u_\varepsilon|^2 \, d\sigma \leqslant \varepsilon \int_{\Gamma_o'} |u|^2 \, d\sigma \, ,$$

hence :

$$\text{(V.22)} \qquad |u_\varepsilon|_{L^2(\Gamma_o')} \leqslant |u|_{L^2(\Gamma_o')}$$

and

(V.23) $$|A\, u_\varepsilon + B - g_V|_{L^2(\Gamma_V)} \leqslant \sqrt{\varepsilon}\, |u|_{L^2(\Gamma'_o)} .$$

By (V.22) there exists a subsequence of u_ε which converges weakly to $\tilde{u}$ in $L^2(\Gamma'_o)$. Hence $Au_\varepsilon + B - g_V \to A\tilde{u} + B - g_V$ in $L^2(\Gamma_V)$ as $\varepsilon \to 0$, and $A\tilde{u} + B = g_V$ by (V.23).

By Lemma V.1 and (V.20) we have

$$\tilde{u} = u .$$

Using (V.22) we may write

(V.24) $$|u_\varepsilon - u|^2_{L^2(\Gamma'_o)} \leqslant 2\left[\int_{\Gamma'_o} u^2\, d\sigma - \int_{\Gamma'_o} u_\varepsilon\, u\, d\sigma\right]$$

which gives strong convergence of u_ε to u in $L^2(\Gamma'_o)$. □

The optimality conditions for the problem $(P^\varepsilon{}_V)$ are expressed in the following theorem:

<u>Theorem V.2.</u>

The vector $u_\varepsilon \in \mathcal{U}$ is a solution to the problem $(P^\varepsilon{}_V)$ if and only if $(\psi_\varepsilon, p_\varepsilon, u_\varepsilon) \in H^{3/2}(\Omega_o) \times H^{1/2}(\Omega_o) \times L^2(\Gamma'_o)$ is the unique solution to the following system.

(V.25) $$\begin{cases} \psi_\varepsilon = f_V & \text{on} \quad \Gamma_V \\ \dfrac{1}{r}\dfrac{\partial \psi_\varepsilon}{\partial n} = u_\varepsilon & \text{on} \quad \Gamma'_o \\ L\psi_\varepsilon = 0 & \text{in} \quad \Omega_o \end{cases}$$

(V.26) $$\begin{cases} p_\varepsilon = \dfrac{1}{r}\dfrac{\partial \psi_\varepsilon}{\partial n} - g_V & \text{on} \quad \Gamma_V \\ \dfrac{\partial p_\varepsilon}{\partial n} = 0 & \text{on} \quad \Gamma'_o \\ Lp_\varepsilon = 0 & \text{in} \quad \Omega_o \end{cases}$$

(V.27) $$p_\varepsilon = \varepsilon\, u_\varepsilon \quad \text{on} \quad \Gamma'_o\ . \qquad \square$$

Proof of Theorem V.2. Using the operators A and B defined by (V.13) and (V.11) the optimality condition (V.8) can be written

(V.28) $$\int_{\Gamma_V} (Au_\varepsilon + B - g_V)Av\, d\sigma + \varepsilon \int_{\Gamma'_o} u_\varepsilon\, v\, d\sigma = 0\ ,\ \forall v \in \mathcal{U}\ .$$

Since the vector $(1/r\ \partial\psi_\varepsilon/\partial n - g_V)$ belongs to $L^2(\Gamma_V)$ it is clear that (V.26) has a unique solution in $H^{1/2}(\Omega_o)$. In fact if $p_\varepsilon|\Gamma_V \in H^{-1/2}(\Gamma_V)$ then $p_\varepsilon \in L^2(\Omega_o)$; if $p_\varepsilon|\Gamma_V \in H^{1/2}(\Gamma_V)$ then $p_\varepsilon \in H^1(\Omega_o)$, and the fact that $p_\varepsilon \in H^{1/2}(\Omega_o)$ if $p_\varepsilon|\Gamma_V \in L^2(\Gamma_V)$ follows from this by linear interpolation.

Applying Green's formula:

(V.29) $$0 = \langle L\psi_1, p_\varepsilon\rangle_{H^{-\frac12}(\Omega_o)\times H^{\frac12}(\Omega_o)} - \langle Lp_\varepsilon, \psi_1\rangle_{H^{-3/2}(\Omega_o)\times H^{+3/2}(\Omega_o)}$$

$$= -\langle \frac{1}{\mu_o r}\frac{\partial\psi_1}{\partial n}, p_\varepsilon\rangle_{L^2(\Gamma)\times L^2(\Gamma)} + \langle \frac{1}{\mu_o r}\frac{\partial p_\varepsilon}{\partial n}, \psi_1\rangle_{H^{-1}(\Gamma)\times H^1(\Gamma)}$$

where $\Gamma = \Gamma_V \cup \Gamma'_o$ and where ψ_1 is defined by (V.12).

With (V.12), (V.13) and (V.26) we obtain

(V.30) $$\int_{\Gamma'_o} p_\varepsilon\, v\, d\sigma + \int_{\Gamma_V} (A\, u_\varepsilon + B - g_V)A\, v\, d\sigma = 0,\ \forall v \in \mathcal{U}.$$

The relation (V.27) then follows from (V.28) and (V.30). □

Remark V.4. Characterization of u_ε .

By (V.27) we have

(V.31) $$u_\varepsilon = \frac{1}{\varepsilon}\, p_\varepsilon \Big|_{\Gamma'_o}\ .$$

Let $p \in H^{1/2}(\Omega_o)$ be defined from $w \in L^2(\Gamma_V)$ by:

$$(V.32) \qquad \begin{cases} p = w & \text{on} \quad \Gamma_V \\ \dfrac{\partial p}{\partial n} = 0 & \text{on} \quad \Gamma'_o \\ Lp = 0 & \text{in} \quad \Omega_o \,. \end{cases}$$

The map associating $p|\Gamma'_o$ to $p \in H^{1/2}(\Omega_o)$ is continuous linear from $H^{1/2}(\Omega_o)$ to $L^2(\Gamma'_o)$. The map C defined by

$$Cw = p\Big|_{\Gamma'_o}$$

is continuous linear from $L^2(\Gamma_V)$ to $L^2(\Gamma'_o)$. Hence

$$(V.33) \qquad p_\varepsilon\Big|_{\Gamma'_o} = Cp_\varepsilon\Big|_{\Gamma_V}$$

and by (V.26)

$$(V.34) \qquad p_\varepsilon\Big|_{\Gamma_V} = Au_\varepsilon + B - g_V \,.$$

We therefore deduce the following equation for u_ε:

$$(V.35) \qquad (\varepsilon\ I - C\ A)\ u_\varepsilon = C(B - g_V)$$

where I denotes the identity mapping in $L^2(\Gamma'_o)$. The operator CA, which is continuous linear from $L^2(\Gamma'_o)$ to itself, has an unbounded inverse, while by Theorem V.1 the operator $(\varepsilon I - CA)$ does have a bounded inverse. The aim of regularization is thus achieved.

□

V.1.4 Numerical Solution of the Problem (P^ε_V).

V.1.4.1 The Finite Element Method.

We intend to solve equations (V.25) to (V.27) by a finite element method. To do this, we define a triangulation $\mathcal{T}_\ell$ of $\bar{\Omega}_o$ as in I.4.1 using triangles with diameter less than or equal to ℓ. Let Γ'^{ℓ}_o and Γ^ℓ_V denote the polygonal lines that approximate Γ'_o and Γ_V, and let Ω^ℓ_o be the approximation of Ω_o.

We associate to the triangulation $\mathcal{T}_\ell$ the spaces V_ℓ, V_ℓ^o and U_ℓ defined by

$$V_\ell = \{\varphi \in C^0(\overline{\Omega_o^\ell}) \mid \forall T \in \mathcal{T}_\ell , \varphi_{|T} \in P_1 \}$$

$$V_\ell^o = \{\varphi \in C^0(\overline{\Omega_o^\ell}) \mid \forall T \in \mathcal{T}_\ell , \varphi_{|T} \in P_1 \text{ and } \varphi = 0 \text{ on } \Gamma_V^\ell\}$$

$$U_\ell = \{\varphi \in C^o(\Gamma_o'^\ell) \mid \forall C \in \Gamma_o'^\ell , \varphi_{|C} \in P_1\}$$

where P_1 is the vector space of polynomials of degree 1 in r and z, and where C denotes an edge of the polygonal line $\Gamma_o'^\ell$.

Since the function f_V belongs to $H^1(\Gamma_V)$ it belongs to $C^0(\Gamma_V)$ and $f_V(M^j_V)$ can be computed, where the M^j_V denote the k_V vertices of the polygonal line Γ^ℓ_V. The approximation f^ℓ_V to f_V is defined by the values $f_V(M^j_V)$ at the vertices of Γ^ℓ_V and is linear on each edge of Γ^ℓ_V.

The function g_V belongs to $L^2(\Gamma_V)$. We shall consider an approximation g^ℓ_V to g_V in $L^2(\Gamma_V)$ such that g^ℓ_V is continuous on Γ^ℓ_V, defined by its values $g^\ell_V(M^j_V)$ at the vertices of Γ^ℓ_V and linear on each edge of Γ^ℓ_V.

The approximation problem of (V.25), (V.26), (V.27) consists of finding $(\psi^\ell_\varepsilon, p^\ell_\varepsilon, u^\ell_\varepsilon) \in V_\ell \times V_\ell \times U_\ell$ such that

(V.36)
$$\begin{cases} \psi^\ell_\varepsilon = f^\ell_V \quad \text{on} \quad \Gamma^\ell_V \\ a_{\mu_o}(\psi^\ell_\varepsilon, \varphi) = \dfrac{1}{\mu_o} \displaystyle\int_{\Gamma_o'^\ell} u^\ell_\varepsilon \, \varphi \, d\sigma , \quad \forall \varphi \in V^o_\ell \end{cases}$$

(V.37)
$$\begin{cases} p^\ell_\varepsilon = e^\ell_V - g^\ell_V \quad \text{on} \quad \Gamma^\ell_V \\ a_{\mu_o}(p^\ell_\varepsilon, \varphi) = 0 , \quad \forall \varphi \in V^o_\ell \end{cases}$$

(V.38)
$$p^\ell_\varepsilon = \varepsilon \, u^\ell_\varepsilon \quad \text{on} \quad \Gamma_o'^\ell$$

with

$$a_{\mu_o}(\psi, \varphi) = \int_{\Omega_o^\ell} \frac{1}{\mu_o r} \nabla\psi \cdot \nabla\varphi \, dS , \quad \forall (\psi, \varphi) \in V_\ell \times V_\ell$$

where e^{ℓ}_V denotes an approximation to $1/r\ \partial\psi^{\ell}_{\varepsilon}/\partial n$ which is continuous on Γ^{ℓ}_V and linear on each edge of Γ^{ℓ}_V.

For each of the N_{ℓ} vertices M_j of $\mathcal{T}_{\ell}$ (interior to Ω^{ℓ}_0 or belonging to the boundary $\Gamma'^{\ell}_0 \cup \Gamma^{\ell}_V$) we define the function $\varphi_i \in V_{\ell}$ by

$$\varphi_i(M_j) = \delta_{ij} \quad , \qquad 1 \leqslant i,j \leqslant N_{\ell} \quad .$$

The set of these N_{ℓ} functions φ_i constitutes a basis for V_{ℓ}.

Let I be the set of indices $\{1,\ldots,N_{\ell}\}$, with I_V the subset $\{1,\ldots,k^{\ell}_V\}$ corresponding to the k^{ℓ}_V vertices of Γ^{ℓ}_V and I_0 the subset $\{N_{\ell} - k^{\ell}_0 + 1,\ldots,N_{\ell}\}$ corresponding to the k^{ℓ}_0 vertices of Γ'^{ℓ}_0. The set of $(N_{\ell} - k^{\ell}_V)$ functions φ_i, $i \in I - I_V$ is a basis for V^0_{ℓ}.

The basis for U_{ℓ} consists of the k^{ℓ}_0 functions $\varphi'_i \in U_{\ell}$ such that

$$\varphi'_i(M_j) = \delta_{ij} \quad , \qquad 1 \leqslant i,j \leqslant k^{\ell}_0$$

where the M_j are the vertices of Γ'^{ℓ}_0. These functions φ'_i are the restrictions to Γ^{ℓ}_0 of the functions φ_i, for $i \in I_0$.

The system (V.36), (V.37), (V.38) can thus be written:

$$\text{(V.39)} \quad \begin{cases} \psi^{\ell}_{\varepsilon}(M_i) = f^{\ell}_V(M_i) \quad , \quad i \in I_V \\[2ex] \displaystyle\sum_{j=1}^{N_{\ell}} a_{\mu_0}(\varphi_i,\varphi_j)\ \psi^{\ell}_{\varepsilon}(M_j) = \frac{1}{\mu_0} \sum_{j=1}^{k^{\ell}_0} u^{\ell}_{\varepsilon}(M_j) \int_{\Gamma'^{\ell}_0} \varphi_i\varphi_j\, d\sigma \ , \\ \qquad\qquad i \in I - I_V \end{cases}$$

$$\text{(V.40)} \quad \begin{cases} p^{\ell}_{\varepsilon}(M_i) = e^{\ell}_V(M_i) - g^{\ell}_V(M_i) \quad , \quad i \in I_V \\[2ex] \displaystyle\sum_{j=1}^{N_{\ell}} a_{\mu_0}(\varphi_i,\varphi_j)\ p^{\ell}_{\varepsilon}(M_j) = 0 \ , \quad i \in I - I_V \end{cases}$$

$$\text{(V.41)} \quad p^{\ell}_{\varepsilon}(M_i) = \varepsilon\ u^{\ell}_{\varepsilon}(M_i) \quad , \quad i \in I_0 \ .$$

V.1.4.2 Matrix solution.

If $\Psi^{\ell}_{\varepsilon}$ denotes the vector of the $\psi^{\ell}_{\varepsilon}(M_i)$, $i \in I$, with $\mathcal{U}^{\ell}_{\varepsilon}$ the vector of the $U^{\ell}_{\varepsilon}(M_i)$, $i \in I_0$ and $\mathcal{F}^{\ell}_{V}$ the vector of the $f^{\ell}_{V}(M_i)$, $i \in I_V$, then the system (V.39) can be written

$$E\,\Psi^{\ell}_{\varepsilon} = F\,\mathcal{F}^{\ell}_{V} + G\,\mathcal{U}^{\ell}_{\varepsilon} \tag{V.42}$$

where the matrices E,F and G are defined as follow :

$$\begin{cases} E(i,j) = \delta_{ij} & , \quad \forall\, i,j \in I_V \\ E(i,j) = 0 & , \quad \forall\, i \in I_V \quad , \quad j \in I - I_V \\ E(i,j) = a_{\mu_0}(\varphi_i,\varphi_j) & , \quad \forall\, i \in I - I_V \quad , \quad j \in I \end{cases}$$

$$\begin{cases} F(i,j) = \delta_{ij} & , \quad \forall\, i,j \in I_V \\ F(i,j) = 0 & , \quad \forall\, i \in I - I_V \quad , \quad j \in I_V \end{cases}$$

$$G(i,j) = \frac{1}{\mu_0}\int_{\Gamma'^{\ell}_{0}} \varphi_i\,\varphi_j\,d\sigma \quad , \quad \forall\, i \in I \quad , \quad j \in I_0 .$$

Similarly, if $\mathcal{P}^{\ell}_{\varepsilon}$ denotes the vector of the $p^{\ell}_{\varepsilon}(M_i)$, $i \in I$, and $\mathcal{G}^{\ell}_{V}$ the vector of the $g^{\ell}_{V}(M_i)$, $i \in I_V$, then (V.40) can be written :

$$E\,\mathcal{P}^{\ell}_{\varepsilon} = H\,\Psi^{\ell}_{\varepsilon} - F\,\mathcal{G}^{\ell}_{V} \tag{V.43}$$

where H is the matrix associated to the operator $1/r\ \partial/\partial n$ on Γ^{ℓ}_{V}.

Finally, (V.41) becomes

$$I\,\mathcal{P}^{\ell}_{\varepsilon} = \varepsilon\,\mathcal{U}^{\ell}_{\varepsilon} \tag{V.44}$$

with

$$\begin{cases} I(i,j) = 0 & , \quad \forall\, i \in I_0 \quad , \quad j \in I - I_0 \\ I(i,j) = \delta_{ij} & , \quad \forall\, i,j \in I_0 \quad . \end{cases}$$

From (V.42), (V.43), (V.44) we can express $\Psi^{\ell}_{\varepsilon}$ in terms of $\mathcal{F}^{\ell}_{V}$ and $\mathcal{G}^{\ell}_{V}$ by

$$\Psi^{\ell}_{\varepsilon} = J_{\varepsilon}\,\mathcal{F}^{\ell}_{V} + K_{\varepsilon}\,\mathcal{G}^{\ell}_{V} \tag{V.45}$$

where

$$J_\varepsilon = (\varepsilon\, E - G\, I\, E^{-1}\, H)^{-1}\, \varepsilon\, F\,,$$

$$K_\varepsilon = -\, (\varepsilon\, E - G\, I\, E^{-1}\, H)^{-1}\, G\, I\, E^{-1}\, F\,.$$

The plasma boundary Γ_p^ε is then the polygonal line defined as

(V.46) $$\Gamma_p^\varepsilon = \{M \in \Omega_o^\ell \mid \psi_\varepsilon^\ell(M) = \sup_D \psi_\varepsilon^\ell\}\ .$$

Observe in (V.45) that ψ_ε^ℓ is computed by simple multiplication of the vectors $\mathcal{F}^\ell_V$ and $\mathcal{G}^\ell_V$, that represent measurements of ψ and of the tangent field on Γ_V^ℓ, by matrices J_ε and K_ε that can be calculated once and for all. Moreover, as we saw in Section V.1.3, thanks to regularization ψ_ε^ℓ is stable with respect to $\mathcal{F}^\ell_V$ and $\mathcal{G}^\ell_V$. In Section V.3 we shall return to the question of optimal choice of ε and Γ_o'. However, it is henceforth clear that the use of (V.45) allows ultra–rapid identification of the plasma boundary from magnetic measurements, and can help in on–line control of the plasma shape.

Remark V.5.

The matrix H was not made explicit in (V.43); in fact in practice (cf. Section V.3) the tangent field is measured by magnetic probes with rectangular meridian section. Each probe is modelled by the union of two triangles, and the tangent field is then equal to $1/r\ \nabla\psi.\mathbf{n}$ where $\mathbf{n}$ is the unit vector normal to Γ_V and $\nabla\psi$ is constant on each triangle, in conformity with the chosen finite element method P_1. The case of discrete probes will be treated in Section V.3. □

Remark V.6.

Another method for solving the problem (V.1) is the quasi–reversibility method (of R. LATTES – J.L. LIONS). It also allows a stable "elliptic" extension out to a fixed boundary Γ_o'; it has been tested in P. ROUGEVIN–BAVILLE and gives results very close to those of the optimal control method with regularization used above. □

V.2 PARAMETRIC IDENTIFICATION OF THE PLASMA CURRENT DENSITY FROM MAGNETIC MEASUREMENTS.

In this section we shall solve the equation for ψ not only in the vacuum but also in the plasma. We shall exploit the fact that ψ and $\partial\psi/\partial n$ are measured on the vacuum vessel in order to identify a certain number of parameters that characterize the plasma current density.

V.2.1 Formulation of the Problem (P'_V).

Consider the expression (I.21) for the toroidal component of the plasma current density, and put

$$
\text{(V.47)} \quad \begin{cases} g(\psi_N) = (1 - \psi_N)^{\gamma} \\[2ex] h_\beta(r) = \dfrac{r\beta}{R_o} + \dfrac{R_o(1-\beta)}{r} \\[2ex] j_T(r,\psi) = \lambda\, h_\beta(r)\, g(\psi_N) \, . \end{cases}
$$

The plasma current density j_T thus depends on the two parameters β and γ alone. As we saw in Remark I.2, the parameter β is a good approximation to the coefficient β_P ("poloidal beta"), while γ characterizes the "peaked" quality of j_T that is related to the self-inductance ℓ_i of the plasma.

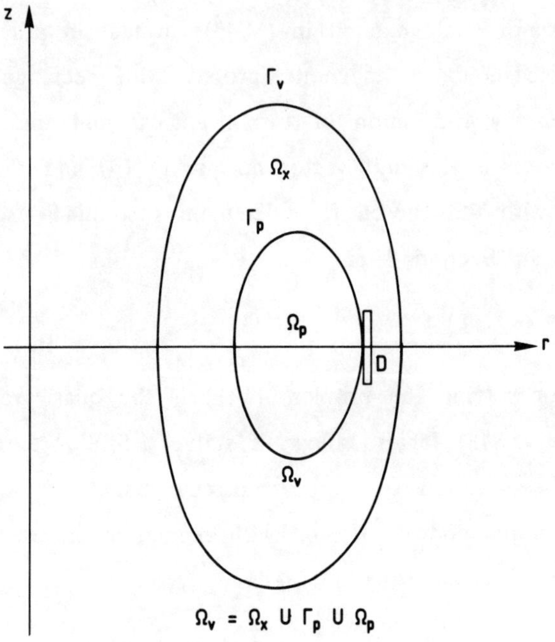

Figure V.2 : Solution domain for the parametric identification of the plasma current density.

Recall that Ω_V denotes the vacuum region, whose boundary is Γ_V (see Fig. V.2). If (β,γ) is given, the equations of state for the pair (ψ,λ) in Ω_V take the form :

(V.48)
$$
\begin{cases}
\psi = f_V & \text{on} \quad \Gamma_V \\
L\psi = 0 & \text{on} \quad \Omega_X \\
L\psi = \lambda \, h_\beta(r)(1-\psi_N)^\gamma & \text{in} \quad \Omega_p \\
I_p = \lambda \displaystyle\int_{\Omega_p} h_\beta(r)(1-\psi_N)^\gamma \, dS & \\
\begin{cases} \psi_e = \psi_i \\ \left[\dfrac{\partial\psi}{\partial n}\right]_e = \left[\dfrac{\partial\psi}{\partial n}\right]_i \end{cases} & \text{on} \quad \Gamma_p
\end{cases}
$$

$$
\text{with} \qquad \Gamma_p = \{M \in \Omega_V \mid \psi(M) = \sup_D \psi\}
$$

$$
\Omega_p = \{M \in \Omega_V \mid \psi(M) > \sup_D \psi\}
$$

where the indices e and i denote quantities calculated on Γ_p outside and inside Ω_p, respectively.

The plasma boundary Γ_p appearing in (V.48) is a free boundary, defined by its contact with the limiter D.

Using the expression for ψ_N given in (I.21) and putting

$$
\lambda' = \frac{\lambda}{(\psi_a - \psi_p)^\gamma}
$$

we can write the system (V.48) for (ψ,λ') as

(V.49)
$$
\begin{cases}
\psi = f_V & \text{on} \quad \Gamma_V \\
L\psi = \lambda' \, h_\beta(r)[(\psi - \psi_p)^+]^\gamma & \text{in} \quad \Omega_V \\
I_p = \lambda' \displaystyle\int_{\Omega_V} h_\beta(r)[(\psi - \psi_p)^+]^\gamma \, dS \, .
\end{cases}
$$

In (V.49) the trace f_V of the flux ψ on Γ_V is used as boundary condition. To determine β and γ "as well as possible" we shall minimize the distance between the tangent field to Γ_V calculated from (V.49) and the measured field g_V. The problem (P'_V) then consists of finding the pair (β,γ) and the vector ψ satisfying (V.49) such that

(V.50)
$$J(\psi) = \inf_{(\beta',\gamma') \in \mathcal{U}_{ad}} J(\psi')$$

with

$$J(\psi') = \frac{1}{2} \int_{\Gamma_V} \left[\frac{1}{r} \frac{\partial \psi'}{\partial n} - g_V\right]^2 d\sigma$$

where ψ' is related to (β',γ') by (V.49) and where $\mathcal{U}_{ad}$ denotes the admissible set of controls (β,γ).

V.2.2 First Order Necessary Conditions for Optimality.

Suppose the hypothesis H1 of Section I.3 is satisfied, namely that $\sup_D \psi$ is attained at a unique point M_0. We then have the following proposition:

Proposition V.1.

If there exists χ' such that

(V.51)
$$\begin{cases} \chi' = \mu_0[\frac{1}{r} \frac{\partial \psi}{\partial n} - g_V] + \text{constant (unknown) on } \Gamma_V \\ L\chi' - \lambda'\gamma\{h_\beta(r)\, 1_{\Omega_p} (\psi - \psi_p)^{\gamma-1}\chi' \\ \quad - \delta_{M_0} \int_{\Omega_p} h_\beta(r)(\psi - \psi_p)^{\gamma-1}\chi' \, ds\} = 0 \\ \int_{\Omega_p} h_\beta(r)\, (\psi - \psi_p)^\gamma\, \chi' \, dS = 0 \end{cases}$$

then the first order necessary condition for optimality of the problem (V.50) is

(V.52)
$$\begin{cases} (\beta' - \beta) \int_{\Omega_p} \left[\frac{r}{R_0} - \frac{R_0}{r}\right] (\psi - \psi_p)^\gamma\, \chi' \, dS \leqslant 0 \\ (\gamma' - \gamma) \int_{\Omega_p} h_\beta(r)(\psi - \psi_p)^\gamma\, \chi' \, \mathrm{Log}(\psi - \psi_p) dS \leqslant 0 \\ \qquad \forall\, (\beta',\gamma') \in \mathcal{U}_{ad} \, . \quad \square \end{cases}$$

Proof (formal).

We proceed as in Section II.2. by introducing a Lagrangian N' defined by

$$N'((\beta,\gamma),\ (\psi,\lambda'),\ (\chi,\nu)) = J(\psi)$$

$$(V.53) \qquad + \int_{\Omega_V} \{L\psi - \lambda'\ h_\beta(r)[(\psi-\psi_p)^+]^\gamma\}\ \chi\ dS$$

$$+ \nu(\lambda' \int_{\Omega_V} h_\beta(r)[(\psi - \psi_p)^+]^\gamma\ dS - I_p)\ .$$

The pair (χ,ν), where χ is defined on Ω_V and where ν is a real number, appears as Lagrange multiplier for the state equations (V.49).

The Lagrangian N' defined by (V.53) can easily be written in the form

$$N' = J(\psi) - \int_{\Gamma_V} \frac{1}{\mu_0\ r} \frac{\partial\psi}{\partial n}\ \chi\ d\sigma + a_{\mu_0}(\psi,\chi)$$

$$(V.54) \qquad - \lambda' \int_{\Omega_p} h_\beta(r)\ (\psi - \psi_p)^\gamma\ \chi\ dS$$

$$+ \nu(\lambda' \int_{\Omega_p} h_\beta(r)(\psi - \psi_p)^\gamma\ dS - I_p)\ .$$

A necessary and sufficient condition for $((\beta,\gamma),\ (\psi,\lambda'))$ to be a solution to (V.50) and (χ,ν) the associated Kuhn–Tucker vector is that $((\beta,\gamma),\ (\psi,\lambda'),\ (\chi,\nu))$ should be a saddle point for N'.

The necessary conditions for optimality can then be written:

$$\left[\frac{\partial N'}{\partial\psi},\ \varphi\right] = 0 = \int_{\Gamma_V} \left[\frac{1}{r}\frac{\partial\psi}{\partial n} - g_V\right] \frac{1}{r}\frac{\partial\varphi}{\partial n}\ d\sigma$$

$$- \int_{\Gamma_V} \frac{1}{\mu_0\ r}\frac{\partial\varphi}{\partial n}\ \chi\ d\sigma + a_{\mu_0}(\varphi,\chi)$$

$$(V.55) \qquad - \lambda'\ \gamma \int_{\Omega_p} h_\beta(r)\ (\psi - \psi_p)^{\gamma-1}\ \chi\ [\varphi-\varphi(M_0)]dS$$

$$+ \nu \lambda' \gamma \int_{\Omega_p} h_\beta(r)(\psi - \psi_p)^{\gamma-1}[\varphi - \varphi(M_o)]dS$$

$$\forall \varphi \text{ such that } \varphi = 0 \text{ on } \Gamma_V .$$

(V.56)

$$\frac{\partial N'}{\partial \lambda'} = 0 = - \int_{\Omega_p} h_\beta(r) \ (\psi - \psi_p)^\gamma \ \chi \ dS$$

$$+ \nu \int_{\Omega_p} h_\beta(r) \ (\psi - \psi_p)^\gamma \ dS$$

Putting $\chi' = \chi - \nu$ equations (V.55) and (V.56) can be written as the partial differential equations (V.51).

Furthermore

(V.57)

$$\left[\frac{\partial N'}{\partial \beta} , \beta' - \beta\right] \geqslant 0 \qquad \forall (\beta',\gamma') \in \mathscr{U}_{ad}$$

$$\left[\frac{\partial N'}{\partial \gamma} , \gamma' - \gamma\right] \geqslant 0 .$$

These inequalities can easily be written in the form (V.52). □

Remark V.7 (open problems).

1. The existence and regularity of ψ and χ' depend on the parameter γ. In practice $\mathscr{U}_{ad} = \{(\beta,\gamma) \in R \times R^+\}$ is often chosen so that only peaked profiles ($\gamma \geqslant 0$) are considered in this model. If $\gamma < 1$ then it is necessary to know the behaviour of $(\psi - \psi_p)$ in the neighbourhood of Γ_p in order to study the existence and regularity of χ'.

2. If we suppose as in Section V.1 that $f_V \in H^1(\Gamma_V)$ and $g_V \in L^2(\Gamma_V)$ then as in J.L. LIONS – E. MAGENES it is legitimate to regard the function ψ defined by (V.49), if it exists, as belonging to $H^{3/2}(\Omega_V)$. The function χ' defined by (V.51), if it exists, probably belongs to $H^{1/2}(\Omega_V)$ (at least for $\gamma \geqslant 1$).

3. If now we suppose $f_V \in W^{2,p}(\Gamma_V)$ and $g_V \in W^{1,p}(\Gamma_V)$ with $p > 2$ then, under suitable hypotheses on γ, the function ψ, if it exists, belongs to $W^{2,p}(\Omega_V)$ and χ belongs to $W^{1,p'}(\Omega_V)$ with $1/p + 1/p' = 1$ (for $\gamma \geqslant 1$). The techniques of proof in Chapter III can be applied to the study of this problem. □

Remark V.8

From the properties of the Lagrangian we can calculate the gradient of J with respect to β and γ:

(V.58)
$$\frac{\partial J}{\partial \beta} = \frac{\partial N'}{\partial \beta} = -\lambda' \int_{\Omega_p} \left[\frac{r}{R_o} - \frac{R_o}{r}\right] (\psi - \psi_p)^{\gamma} \chi' \, dS$$

$$\frac{\partial J}{\partial \gamma} = \frac{\partial N'}{\partial \gamma} = -\lambda' \int_{\Omega_p} h_{\beta}(r) (\psi - \psi_p)^{\gamma} \operatorname{Log}(\psi - \psi_p) \chi' \, dS .$$ □

V.2.3 *Numerical Method of Solution.*

To solve the optimization problem (V.50) we use the same sequential quadratic method as in Section II.3. We linearize the state equations (V.49) with respect to (ψ, λ') and also with respect to the control vector (β, γ), and the algorithm then consists of a sequence of solutions of linear quadratic control problems.

Let $\mathcal{T}_\ell$ be a triangulation of Ω_V, with Γ^ℓ_V denoting the polygonal line approximating the boundary Γ_V of Ω_V and with Ω^ℓ_V the approximation to Ω_V so defined. Let V_ℓ be the space of functions φ that are continuous on Ω^ℓ_V and linear on each triangle T of $\mathcal{T}_\ell$; let V^o_ℓ be the subset of V_ℓ consisting of functions that vanish on Γ^ℓ_V. Let f^ℓ_V, g^ℓ_V be the approximations to f_V, g_V that are linear on each edge of Γ^ℓ_V.

V.2.3.1 *External iterations.*

The algorithm is defined as follows:

Let $((\beta^o, \gamma^o), (\psi^o, \lambda'^o)) \in \mathcal{U}_{ad} \times V_\ell \times R$.

At the nth external iteration we consider the linearized problem at $((\beta^n, \gamma^n), (\psi^n, \lambda'^n))$, that we assume to admit a solution $(\psi, \lambda') \in V_\ell \times R$ when the control vector $(\beta, \gamma) \in \mathscr{U}_{ad}$ is given (thus we are supposing that the matrix for the linearized system is invertible). This system is

$$
(V.59)\quad \left\{
\begin{aligned}
& \psi = f_V^\ell \quad \text{on} \quad \Gamma_V^\ell \\
& a_{\mu_o}(\psi,\varphi) - \lambda'^n \gamma^n \int_{\Omega_p^n} h_{\beta^n}(r)(\psi^n-\psi_p^n)^{\gamma^n-1}[\psi-\psi(M_o^n)]\varphi \, dS \\
& - \lambda' \int_{\Omega_p^n} h_{\beta^n}(r)(\psi^n-\psi_p^n)^{\gamma^n} \varphi \, dS \\
& = \lambda'^n\{(\beta-\beta^n) \int_{\Omega_p^n} \left[\frac{r}{R_o} - \frac{R_o}{r}\right](\psi^n-\psi_p^n)^{\gamma^n} \varphi \, dS \\
& + (\gamma - \gamma^n) \int_{\Omega_p^n} h_{\beta^n}(r)(\psi^n-\psi_p^n)^{\gamma^n} \operatorname{Log}(\psi^n - \psi_p^n) \varphi \, dS \\
& - \gamma^n \int_{\Omega_p^n} h_{\beta^n}(r)(\psi^n-\psi_p^n)^{\gamma^n} \varphi \, dS\} \quad , \forall \varphi \in V_\ell^o \\
& - \lambda'^n \gamma^n \int_{\Omega_p^n} h_{\beta^n}(r)(\psi^n-\psi_p^n)^{\gamma^n-1}[\psi - \psi(M_o^n)] \, dS \\
& - \lambda' \int_{\Omega_p^n} h_{\beta^n}(r)(\psi^n-\psi_p^n)^{\gamma^n} \, dS \\
& = - I_p + \lambda'^n\{-\gamma^n \int_{\Omega_p^n} h_{\beta^n}(r)(\psi^n-\psi_p^n)^{\gamma^n} \, dS \\
& + (\beta-\beta^n) \int_{\Omega_p^n} \left[\frac{r}{R_o} - \frac{R_o}{r}\right](\psi^n-\psi_p^n)^{\gamma^n} \, dS \\
& + (\gamma - \gamma^n) \int_{\Omega_p^n} h_{\beta^n}(r)(\psi^n-\psi_p^n)^{\gamma^n} \operatorname{Log}(\psi^n - \psi_p^n) \, dS\}
\end{aligned}
\right.
$$

where

$$\Gamma^n_p = \{M \in \Omega^\ell_V \mid \psi^n(M) = \psi^n_p = \sup_D \psi^n = \psi^n(M^n_o)\}$$

$$\Omega^n_p = \{M \in \Omega^\ell_V \mid \psi^n(M) > \psi^n_p\}$$

and the point M^n_o is the point of contact (assumed to be unique) between Γ^n_p and D.

The problem $(P^n{}_V)$ consists of finding $((\beta^{n+1}, \gamma^{n+1}), (\psi^{n+1}, \lambda'^{n+1})) \in \mathcal{U}_{ad} \times V_\ell \times R$ to satisfy (V.59) and such that

(V.60)
$$J_\ell(\psi^{n+1}) = \inf_{\substack{(\beta,\gamma) \in \mathcal{U}_{ad} \\ (\psi,\lambda') \in V_\ell \times R \\ \text{related by (V.59)}}} J_\ell(\psi)$$

with

$$J_\ell(\psi) = \frac{1}{2} \int_{\Gamma^\ell_V} [e^\ell_V(\psi) - g^\ell_V]^2 \, d\sigma$$

where $e^\ell{}_V$ denotes an approximation to $1/r \; \partial\psi/\partial n$ continuous on $\Gamma^\ell{}_V$ and linear on each of its edges. Analogously to Proposition (V.1) we can prove the following :

<u>Proposition V.2.</u>

The first order necessary conditions for optimality for the problem $(P^n{}_V)$ are:

. equations (V.59) where

$$\psi = \psi^{n+1}, \quad \lambda' = \lambda'^{n+1}, \quad \beta = \beta^{n+1}, \quad \gamma = \gamma^{n+1};$$

. the following equations for the adjoint state $\chi'^{n+1} \in V_\ell$:

$$
(V.61)\quad \begin{cases}
\chi'^{n+1} = \mu_0\,[e^{\ell}_V(\psi^{n+1}) - g^{\ell}_V] \;+\; \text{constant on } \Gamma^{\ell}_V \\[2ex]
a_{\mu_0}(\chi'^{n+1},\varphi) - \lambda'^n\,\gamma^n \displaystyle\int_{\Omega^n_p} h_{\beta^n}(r)\,(\psi^n-\psi^n_p)^{\gamma^n-1}\chi'^{n+1}[\varphi-\varphi(M^n_o)]\,dS \\[2ex]
= 0\;,\;\forall\,\varphi \in V^{\ell}_o \\[2ex]
\displaystyle\int_{\Omega^n_p} h_{\beta^n}(r)\,(\psi^n-\psi^n_p)^{\gamma^n}\,\chi'^{n+1}\,dS = 0\;;
\end{cases}
$$

. the following inequalities

$$
(V.62)\quad \begin{cases}
(\beta'-\beta^{n+1}) \displaystyle\int_{\Omega^n_p} \left[\frac{r}{R_o} - \frac{R_o}{r}\right](\psi^n-\psi^n_p)^{\gamma^n}\,\chi'^{n+1}\,dS \leqslant 0 \\[2ex]
(\gamma' - \gamma^{n+1}) \displaystyle\int_{\Omega^n_p} h_{\beta^n}(r)\,(\psi^n-\psi^n_p)^{\gamma^n}\chi'^{n+1}\,\mathrm{Log}(\psi^n - \psi^n_p)\,dS \leqslant 0 \\[1ex]
\qquad\qquad\qquad\qquad\forall\,(\beta',\gamma') \in \mathcal{U}_{ad}\;. \quad \square
\end{cases}
$$

V.2.3.2 *Internal Iterations.*

First of all consider the problem without constraints ($\mathcal{U}_{ad} = R^2$). To solve $(P^n{}_V)$, we use a conjugate gradient algorithm as in Section II.3.2.

We initialize by

$$(\beta^n_o,\ \gamma^n_o) = (\beta^n,\ \gamma^n)\ .$$

The jth iteration thus goes :

. calculate $(\psi^n{}_j,\ \lambda'^n{}_j) \in V_{\ell} \times R$ from $(\beta^n{}_j,\ \gamma^n{}_j)$ by equations (V.59) where

$$\psi = \psi^n_j\ ,\ \lambda' = \lambda'^n_j\ ,\ \beta = \beta^n_j,\ \gamma = \gamma^n_j\ ;$$

. calculate the adjoint state $\chi'^n_j \in V_{\ell}$ by :

(V.63)
$$\begin{cases} \chi'^n_j = \mu_o \, [e^\ell_V(\psi^n_j) - g^\ell_V] + \text{constant on } \Gamma^\ell_V \\[2ex] a_{\mu_o}(\chi'^n_j,\varphi) - \lambda'^n \gamma^n \displaystyle\int_{\Omega^n_p} h_{\beta^n}(r)(\psi^n-\psi^n_p)^{\gamma^n-1} \chi'^n_j [\varphi-\varphi(M^n_o)] \, dS \\[2ex] = 0 \;, \; \forall \varphi \in V^\ell_o \;; \\[2ex] \displaystyle\int_{\Omega^n_p} h_{\beta^n}(r)(\psi^n-\psi^n_p)^{\gamma^n} \chi'^n_j \, dS = 0 \end{cases}$$

. the gradient $G^n{}_j$ of the functional J_ℓ at the jth iteration has as components the following variables:

(V.64)
$$\begin{cases} (G^n_j)_1 = - \lambda'^n \displaystyle\int_{\Omega^n_p} \left[\frac{r}{R_o} - \frac{R_o}{r}\right] (\psi^n-\psi^n_p)^{\gamma^n} \chi'^n_j \, dS \\[2ex] (G^n_j)_2 = - \lambda'^n \displaystyle\int_{\Omega^n_p} h_{\beta^n}(r)(\psi^n-\psi^n_p)^{\gamma^n} \chi'^n_j \, \text{Log}(\psi^n - \psi^n_p) \, dS \;; \end{cases}$$

. we put

(V.65)
$$\begin{cases} w^n_o = G^n_o \\ w^n_j = G^n_j + \kappa^n_j \, w^n_{j-1} \qquad \text{for } j > 0 \end{cases}$$

and define the sequence $((\beta^n{}_j, \gamma^n{}_j), (\psi^n{}_j, \lambda'^n{}_j)) \in R^2 \times (V_\ell \times R)$ by

(V.66)
$$\begin{cases} \beta^n_{j+1} = \beta^n_j - \rho^n_j (w^n_j)_1 \\[1ex] \gamma^n_{j+1} = \gamma^n_j - \rho^n_j (w^n_j)_2 \\[1ex] \psi^n_{j+1} = \psi^n_j - \rho^n_j \, \hat{\psi}^n_j \\[1ex] \lambda'^n_{j+1} = \lambda'^n_j - \rho^n_j \, \hat{\lambda}^n_j \end{cases}$$

where $(\hat{\psi}^n{}_j, \hat{\lambda}^n{}_j) \in V^o{}_\ell \times R$ are defined by

$$a_{\mu_0}(\hat{\psi}^n_j,\varphi)-\lambda'^n\gamma^n\int_{\Omega^n_p} h_{\beta^n}(r)(\psi^n-\psi^n_p)^{\gamma^n-1}[\hat{\psi}^n_j-\hat{\psi}^n_j(M^n_o)]\,\varphi\,dS$$

$$-\hat{\lambda}^n_j\int_{\Omega^n_p} h_{\beta^n}(r)\ (\psi^n-\psi^n_p)^{\gamma^n}\ \varphi\,dS$$

$$\text{(V.67)}\qquad =\lambda'^n[(w^n_j)_1\int_{\Omega^n_p}\left[\frac{r}{R_o}-\frac{R_o}{r}\right](\psi^n-\psi^n_p)^{\gamma^n}\ \varphi\,dS$$

$$+(w^n_j)_2\int_{\Omega^n_p} h_{\beta^n}(r)\ (\psi^n-\psi^n_p)^{\gamma^n}\ \mathrm{Log}(\psi^n-\psi^n_p)\ \varphi\,dS]$$

$$\forall\ \varphi\in V^o_\ell \qquad \text{and} \qquad \text{for}\quad \varphi\equiv 1\quad \text{on}\quad \Omega^\ell_V$$

and where $\rho^n{}_j$ and $\kappa^n{}_j$ are such that

$$\text{(V.68)}\qquad J_\ell(\psi^n_{j+1})=\inf_{\rho>0} J_\ell(\psi^n_j-\rho\,\hat{\psi}^n_j)$$

and

$$\text{(V.69)}\qquad \int_{\Gamma^\ell_V} e^\ell_V(\hat{\psi}^n_j)\cdot e^\ell_V(\hat{\psi}^n_{j-1})\,d\sigma=0\ .$$

Since the number of control parameters equals 2, the conjugate gradient algorithm converges in 2 iterations and

$$\text{(V.70)}\qquad \begin{cases}\beta^n_2=\beta^{n+1}\ , & \gamma^n_2=\gamma^{n+1}\ ,\\ \psi^n_2=\psi^{n+1}\ , & \lambda'^n_2=\lambda'^{n+1}\ .\end{cases}$$

Remark V.9.

In the case when $\mathcal{U}_{ad}=R\times R^+$ we used the above conjugate gradient algorithm and took

$$\text{(V.71)}\qquad (\beta^{n+1},\gamma^{n+1})=\mathrm{Proj}_{\mathcal{U}_{ad}}\ (\beta^n_2,\gamma^n_2)\ .$$

If in the limit of the external iterations the parameter γ is zero, this means that the current density profile is concave and cannot be represented by (V.47). Polynomial functions in ψ_N have been used in J. BLUM – J.C. GILBERT – B. THOORIS to express the plasma current density, and these enable concave profiles to be simulated. □

V.2.3.3 *Structure and Convergence of the Algorithm*

The flow chart for the program is as follows:

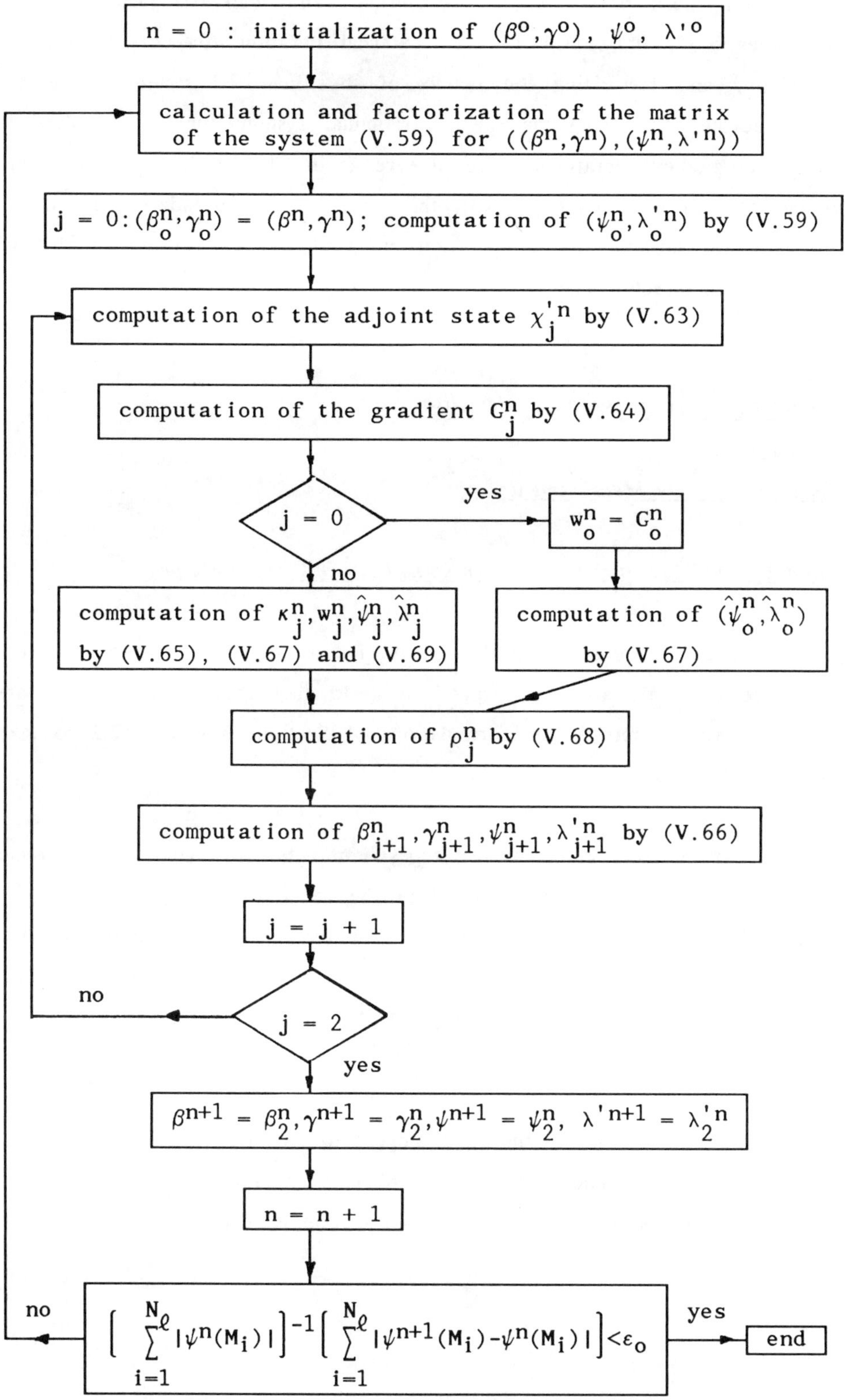

According to the particular case, this algorithm converges in from 5 to 10 external iterations for $\varepsilon_0 = 10^{-5}$. As each external iteration is of the order of 0.3s on CRAY 1 for the JET mesh consisting of 870 vertices and 1680 triangles, one case requires of the order of 2s of CPU. In each external iteration 70% of the time is devoted to calculating and factorizing the matrix of the linearized system, while the remaining 30% is devoted to solving the various triangular systems that are involved in the internal conjugate gradient iterations. We observe as in Chapter II that since there is a small number of control parameters (two), this algorithm is well adapted to solving the control problem, the cost of the computation being of the same order as that of Newton's method for the direct problem.

V.3 APPLICATIONS TO TORE SUPRA AND JET

V.3.1 The Case of Discrete Measurements

V.3.1.1 Position of the Magnetic Measurements in JET and TORE Supra

In Sections V.1 and V.2 we supposed that ψ and $1/r\ \partial\psi/\partial n$ were measured at every point of Γ_V and equal to f_V and g_V respectively. In fact these measurements are discrete, and so we have to modify the formulations of Sections V.1 and V.2 to take account of their discrete nature.

The flux loops that measure ψ are of two types: complete loops going right around the torus, and saddle–type loops on each octant of the vacuum vessel (cf. L.de KOCK – G.TONETTI). These loops are shown in Fig. V.3 for one octant of the vacuum vessel of JET. As we see in Fig. V.4, which is a meridian section of the torus, they enable ψ to be measured at every instant at 14 points M_i situated on the outer boundary of the vacuum vessel of JET.

The magnetic probes measuring the field tangent to Γ_V (i.e. $1/r\ \partial\psi/\partial n$) are also shown on Figs. V.3. and V.4. They are 18 in number, situated on the inner boundary of the vacuum vessel at points N_j, and their meridian section is rectangular.

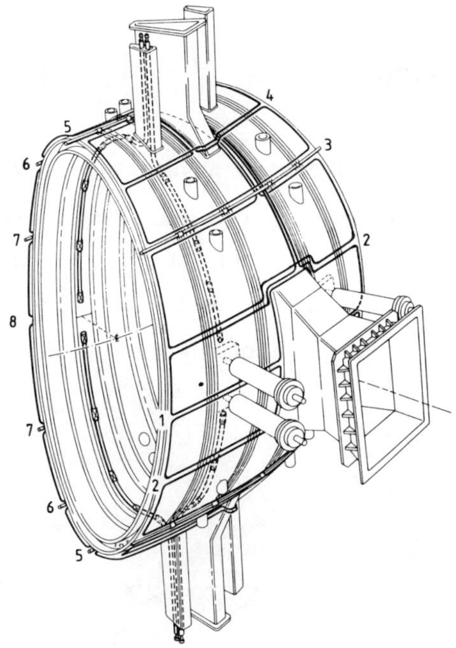

Figure V.3 : Flux loops and magnetic probes on one octant of the vacuum vessel of JET.

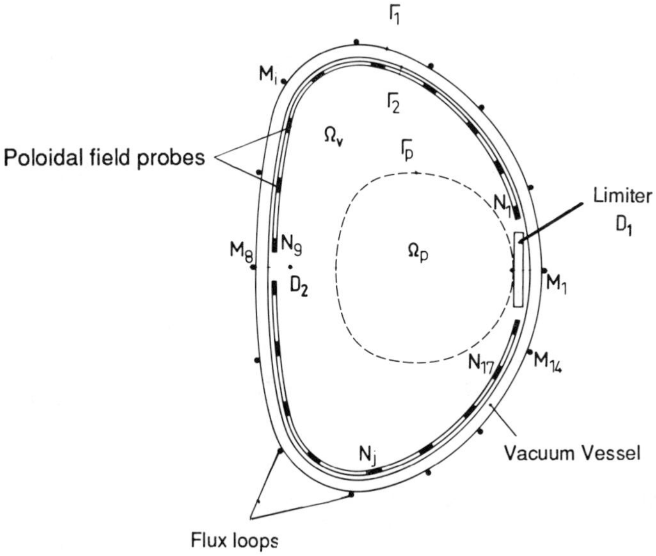

Figure V.4 : Position of the flux loops and magnetic probes in a meridian section of JET.

Figure V.5 shows the 14 flux loops and the 16 magnetic probes in TORE Supra. They are all situated on the inner boundary of the vacuum vessel.

V.3.1.2 *Interpolation of* ψ *on* Γ_V

The boundary Γ_V of the domain being studied will be the outer boundary of the vacuum vessel in JET (cf. Fig. V.4) and the inner boundary in TORE Supra (cf. Fig. V.5). To reconstruct f_V on the whole of Γ_V we shall interpolate between the flux measurements at the points M_i. Let s be the arc length coordinate of a point M on Γ_V, with the outermost point M_1 corresponding to $s = 0$. We distinguish two types of interpolation.

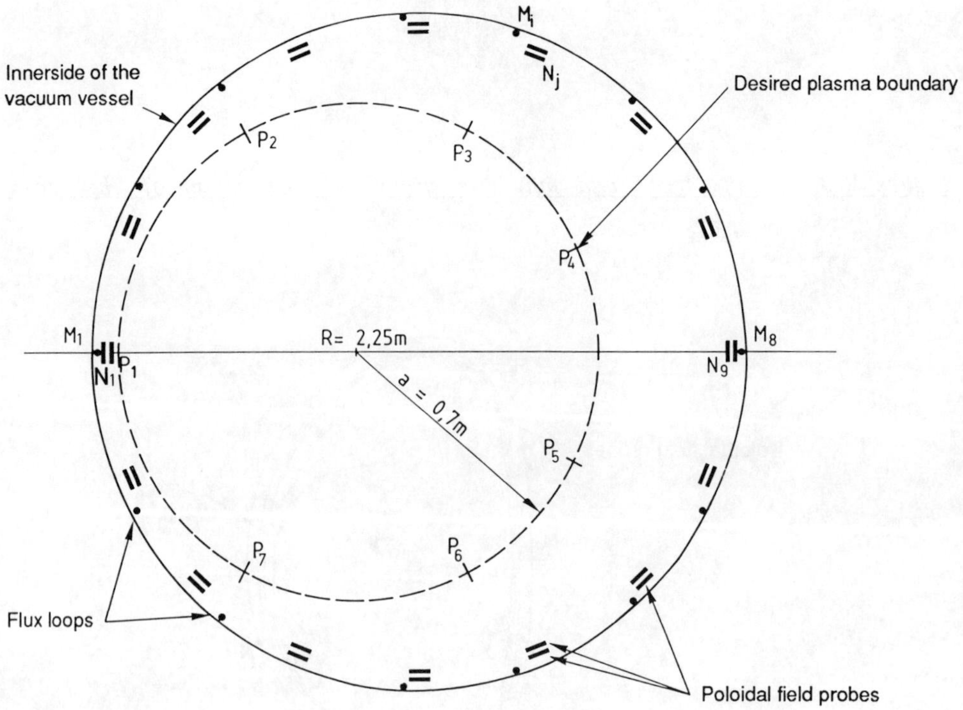

Figure V.5 : Position of flux loops and magnetic probes in a meridian section of TORE Supra

i) <u>linear interpolation</u>

Let M have coordinate s on Γ_V; then we put

$$f_V(s) = \frac{s-s_i}{s_{i+1}-s_i} f_V^{i+1} + \frac{s_{i+1} - s}{s_{i+1} - s_i} f_V^i , \qquad \text{(V.72)}$$

$$\forall s \in [s_i , s_{i+1}] , \ i \in \{1,\ldots,13\}$$

where s_i denotes the coordinate of the vertex M_i corresponding to the ith flux loop and f^i_V the experimental measurement of ψ by this loop.

ii) <u>Fourier series decomposition</u>

We define θ from s by

$$\theta = 2\pi \frac{s}{s_o}$$

where s_o denotes the perimeter of Γ_V, and put

$$f_V(\theta) = \sum_{m=0}^{7} f_m \cos m\theta + \sum_{m=1}^{6} f'_m \sin m\theta \qquad \text{(V.73)}$$

where the f_m and f'_m are calculated by solving the following linear system:

$$f_V(\theta_i) = f_V^i , \ i \in \{1,\ldots,14\} , \qquad \text{(V.74)}$$

the θ_i corresponding to the positions of the 14 flux loops.

By one or other of these interpolations we reconstruct the Dirichlet condition

$$\psi = f_V \ \text{on} \ \Gamma_V . \qquad \text{(V.75)}$$

V.3.1.3 *<u>Discrete Cost Function</u>*

We saw that the field tangent to Γ_V was measured on k rectangles forming the magnetic probes (k = 18 for JET, k = 16 for TORE Supra). The cost function J defined by (V.3) or (V.50) ought therefore to be replaced by the following discrete sum:

(V.76) $$J' = \frac{1}{2} \sum_{j=1}^{k} [(\frac{1}{r} \nabla\psi . n)(N_j) - g^j_V]^2$$

where g^j_V denotes the experimental measurement of the tangent field on the probe centered at N_j, and where **n** denotes the unit normal vector to Γ_V at N_j. In the mesh each rectangular probe is made up of the union of two triangles and the vector $\nabla\psi$ will be taken equal to the average of this quantity calculated on each of the 2 triangles.

Analogously to (V.4) we define for the rapid identification method a regularizing functional

(V.77) $$J'_\varepsilon(v) = J'(v) + \frac{\varepsilon}{2} \int_{\Gamma'_o} v^2 \, d\sigma$$

where ψ is related to v by (V.2), with f_V calculated by (V.72) or (V.73). The problem of minimizing J'_ε with respect to v is called the problem $(\tilde{P}^\varepsilon_V)$. Likewise, for the second method the problem of minimizing J', given by (V.76), with respect to β and γ when ψ is related to (β, γ) by equations (V.49) is called $(\tilde{P}'_V)$.

V.3.1.4 *Numerical Solution of the Problem* $(\tilde{P}^\varepsilon_V)$

We use the methods developed in Section V.1.4. The only modification concerns the definition of the adjoint state in conditions (V.36) – (V.38) necessary for optimality. Explicitly, instead of equation (V.37) the adjoint state p^ℓ_ε is now defined by

(V.78) $$\begin{cases} p^\ell_\varepsilon \in V^\ell_o \\ a_{\mu_o}(p^\ell_\varepsilon , \varphi) = - \sum_{j=1}^{k} \{ [(\frac{1}{r} \nabla\psi^\ell_\varepsilon . n)(N_j) - g^j_V] \\ \qquad (\frac{1}{r} \nabla\varphi . n)(N_j) \} , \quad \forall \varphi \in V^\ell_o \end{cases}$$

where we recall that $\nabla\psi$ and $\nabla\varphi$ are calculated by taking the average of these quantities on the two triangles T^1_j and T^2_j that comprise the jth magnetic probe centered at N_j.

The system (V.40) then becomes

(V.79)
$$\begin{cases} p_\varepsilon^\ell(M_i) = 0 \quad , \quad i \in I_V \\ \sum_{j=1}^{N_\ell} a_{\mu_0}(\varphi_i,\varphi_j)\, p_\varepsilon^\ell(M_j) = \\ \quad - \sum_{j'=1}^{k} \{ [\sum_{j=1}^{N_\ell} \psi_\varepsilon^\ell(M_j)(\frac{1}{r}\nabla\varphi_j.n)(N_j) - g_V^{j'}] \\ \qquad (\frac{1}{r}\nabla\varphi_i.n)(N_{j'}) \} \quad , \quad i \in I - I_V . \end{cases}$$

As far as the matrix solution is concerned, the system (V.42) remains valid and the system (V.43) now becomes

(V.80)
$$E\mathscr{P}_\varepsilon^\ell = H'\,\Psi_\varepsilon^\ell - F'\,\mathscr{G}_V$$

where $\mathscr{G}_V$ denotes the vector of the $g^j{}_V$, $j \in \{1,\ldots,k\}$ and where H' and F' are defined as follows:

$$H'(i,j) = - \sum_{j'=1}^{k} (\frac{1}{r}\nabla\varphi_i.n)(N_{j'})(\frac{1}{r}\nabla\varphi_j.n)(N_{j'}),\ \forall i \in I-I_V, j \in I$$

$$H'(i,j) = 0 \quad , \quad \forall\, i \in I_V \quad , \quad j \in I$$

$$F'(i,j) = - (\frac{1}{r}\nabla\varphi_i.n)(N_j) \quad , \quad \forall\, i \in I - I_V \ , \ j \in \{1,\ldots,k\}$$

$$F'(i,j) = 0 \quad , \quad \forall\, i \in I_V \quad , \quad j \in \{1,\ldots,k\} .$$

Equation (V.44) remains valid. Finally, whether it be by linear interpolation (V.72) or Fourier series decomposition (V.73) – (V.74), the vector $\mathscr{F}_V^\ell$ is deduced linearly from the measurements f^i_V of ψ by the flux loops and can be written

(V.81)
$$\mathscr{F}_V^\ell = M\,\mathscr{F}_V$$

where $\mathscr{F}_V$ denotes the vector of the $f^i{}_V$, $i \in \{1,\ldots,14\}$ and M the interpolation matrix.

From (V.42), (V.44), (V.80), (V.81) we can express $\Psi^\ell{}_\varepsilon$ in terms of $\mathscr{F}_V$ and $\mathscr{G}_V$ by

(V.82) $$\Psi^{\ell}_{\varepsilon} = J'_{\varepsilon}\, \mathscr{F}_V + K'_{\varepsilon}\, \mathscr{G}_V$$

with

$$J'_{\varepsilon} = (\varepsilon\, E - G\, I\, E^{-1}\, H')^{-1}\, \varepsilon\, F'\, M$$

$$K'_{\varepsilon} = -\,(\varepsilon\, E - G\, I\, E^{-1}\, H')^{-1}\, G\, I\, E^{-1}\, F'\ .$$

The matrices J'_{ε} and K'_{ε} are calculated once and for all; the function $\psi^{\ell}{}_{\varepsilon}(M_i)$, where M_i is any vertex of the grid, is a linear combination of the magnetic measurements $f^j{}_V$ and $g^j{}_V$. The calculation time for $\psi^{\ell}{}_{\varepsilon}$ at all the vertices of the mesh, in multiplying the vectors $\mathscr{F}_V$ and $\mathscr{G}_V$ by the matrices J'_{ε} and K'_{ε}, is 0.25 ms on CRAY 1. The boundary Γ_p is then determined by (V.46).

V.3.1.5 *Numerical Solution of the Problem* ($\tilde{P}'_V$)

The method of solution is identical to that of Section V.2.3, except that $J_{\ell}(\psi)$, defined by (V.60), is replaced by the functional J' given by (V.76).

At each internal iteration the adjoint state, previously given by (V.63), is now defined by:

(V.83)
$$\begin{cases} \chi'^{n}_{j} = \text{constant (unknown) on } \Gamma^{\ell}_{V} \\[2ex] a_{\mu_o}(\chi'^{n}_{j},\varphi) - \lambda'^{n}\gamma^{n} \displaystyle\int_{\Omega^n_p} h_{\beta^n}(r)(\psi^n-\psi^n_p)^{\gamma^n-1}\chi'^{n}_{j}[\varphi-\varphi(M^n_o)]\,dS \\[2ex] = -\displaystyle\sum_{j'=1}^{k} [(\frac{1}{r}\nabla\psi^n_j.n)(N'_j)-g^{j'}_V](\frac{1}{r}\nabla\varphi.n)(N_{j'})\ ,\ \forall\varphi\in V^{\ell}_o \\[2ex] \displaystyle\int_{\Omega^n_p} h_{\beta^n}(r)(\psi^n-\psi^n_p)^{\gamma^n}\chi'^{n}_{j}\,dS = 0\ . \end{cases}$$

The determination of β and γ by the algorithm of Section V.2.3 thus modified takes 2 to 3s of CPU time on CRAY 1.

V.3.2 Identification and Control of the Plasma Shape in TORE Supra

V.3.2.1 Identification of the Plasma Boundary

Consider a case of equilibrium in TORE Supra calculated as in Section II.4.3.2 and corresponding to a "quasi-circular" plasma of major radius R = 2.25m and minor radius a = 0.70m (with $\beta = 1$). We then take the values of the flux ψ on the flux loops and of the tangent field $1/r\ \partial\psi/\partial n$ on the magnetic probes. By the method of Section V.1.4 modified as in Section V.3.1.4 we shall attempt to reconstruct the plasma boundary Γ_p from these magnetic measurements (cf.G. ROBILLARD).

The inner boundary Γ_o' of Ω_o is chosen to be circular of major radius R_o = 2.25m and minor radius a_o = 0.35 m. We use the formula (V.73) for interpolation by Fourier series in order to reconstruct f_V on Γ_V.

Figure V.6 shows the comparison between the real boundary of the plasma (obtained in Chapter II) and the boundary obtained by minimizing J_ε' for various values of ε : $\varepsilon = 10^{-4}, 10^{-6}, 10^{-8}, 10^{-10}$. For this unperturbed case note that only the value $\varepsilon = 10^{-4}$ gives an inaccurate result.

We then perturb the data f^i_V and g^i_V randomly within a limit of 3% relative error. Figure V.7 shows the comparison between the real plasma boundary and that obtained by minimizing J_ε' for the same ε-values as before. Note that for $\varepsilon = 10^{-10}$ the regularization is insufficient. For $\varepsilon = 10^{-4}$ we see as before that it is too strong. The value $\varepsilon = 10^{-6}$ seems optimal.

Likewise, we can modify the interior contour Γ_o' and we observe that Γ_p^ε does not depend on the choice of Γ_o', provided that Γ_o' is inside the plasma and is not reduced to a point.

From these few cases we may draw the following conclusions:

- The solution to the problem is independent of the choice of Γ_o';
- there is an optimal value of ε which renders the problem stable relative to the magnetic measurements without "deforming" the plasma boundary (this value of course depends on the size of the errors in the magnetic measurements);
- this method is ultra-rapid : it enables the plasma boundary to be identified in less than 1 ms.

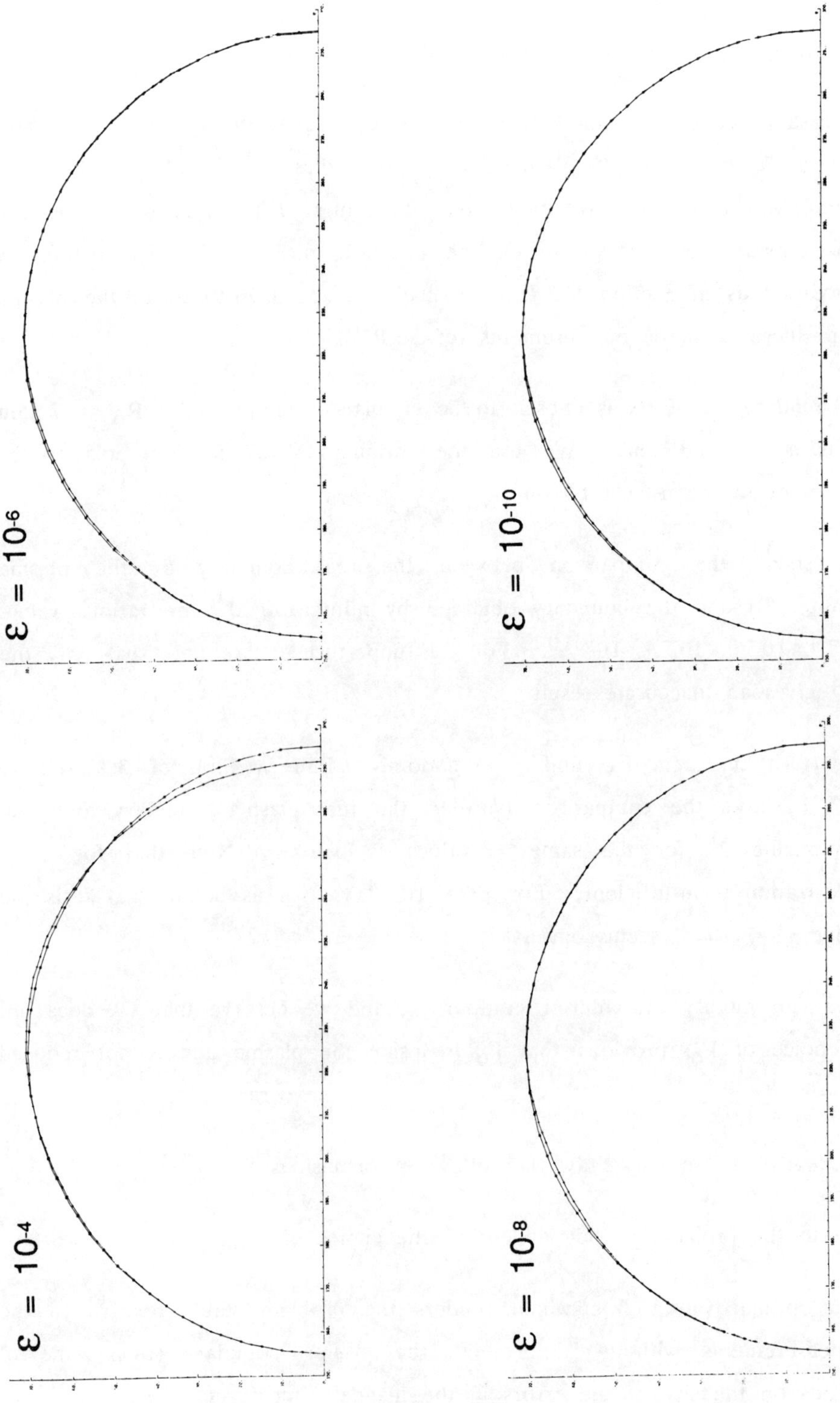

Figure V.6 : Comparison between the real plasma boundary and that obtained by rapid identification from the magnetic measurements, for various values of ε.

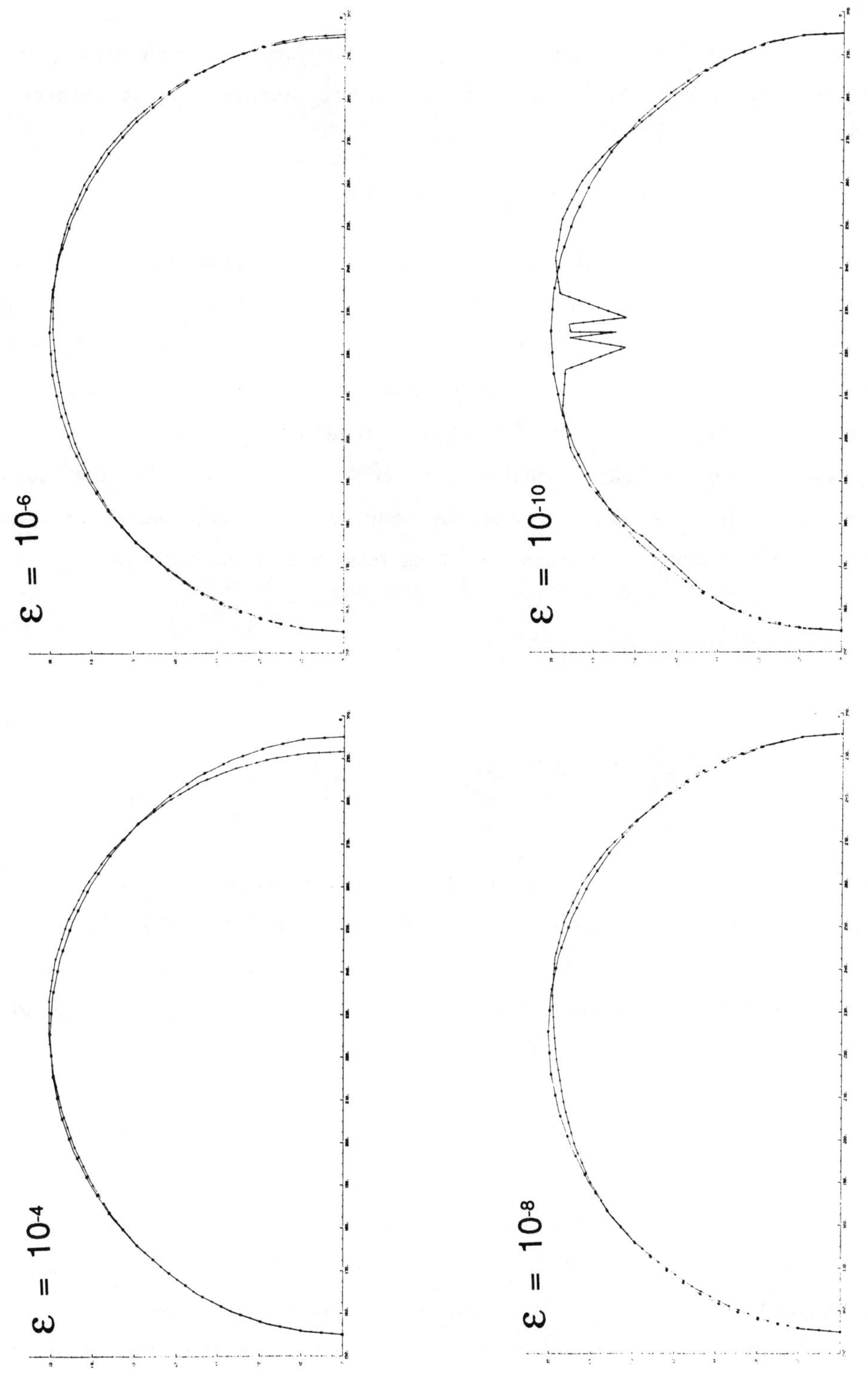

Figure V.7 : As in Fig. V.6. but with data perturbed within a 3% limit.

V.3.2.2 Application to Control of the Plasma Shape

As can be seen in Fig. V.5 the desired plasma boundary is a circle passing through the seven points P_1 to P_7. We try to make the plasma boundary Γ_p pass through these seven points, so

$$\psi(P_i) = \psi(P_1) \quad , \quad \forall\, i \in \{2,\ldots,7\} . \tag{V.84}$$

Once the mesh and the parameter ε are fixed we can suppose that the function $\psi^{\ell}_{\varepsilon}$ defined by (V.82) is the sought for poloidal flux function ψ. The computation of $\psi^{\ell}_{\varepsilon}$ at the seven points P_i, $i \in \{1,\ldots,7\}$, is by simple linear combination of the $f^i{}_V$ and $g^j{}_V$ with coefficients being those of the seven rows of the matrices J'_{ε} and K'_{ε} that correspond to the vertices P_i. This computation, requiring 210 multiplications, can be done on line. The variables initiating the feedback will then be the quantities $[\psi^{\ell}{}_{\varepsilon}(P_i), - \psi^{\ell}{}_{\varepsilon}(P_1)]$, $i \in \{2,\ldots,7\}$, and the voltages V_j of the generators will be calculated by a proportional – derivative feedback relative to these quantities:

$$V_j = \sum_{i=2}^{7} G_{i,j}[\psi^{\ell}_{\varepsilon}(P_i) - \psi^{\ell}_{\varepsilon}(P_1)] + \tag{V.85}$$

$$+ \sum_{i=2}^{7} G'_{i,j} \frac{d}{dt} [\psi^{\ell}_{\varepsilon}(P_i) - \psi^{\ell}_{\varepsilon}(P_1)]$$

where the matrices of the gains $G_{i,j}$ and $G'_{i,j}$ are determined in such a way as to control the plasma shape "as well as possible" (cf. Chapter VI). The voltages of the generators, of which there are nine in TORE Supra, also serve to control the total plasma current and the distribution of the currents in the coils (see Chapter VI and cf. J. BLUM – J. LE FOLL – C. LELOUP).

Remark V.10.

It can be seen in Fig V.5. that the points P_1 to P_7 are chosen at the intersection of the desired circle with the normals to the vacuum vessel passing through the points M_{2i-1} corresponding to the flux loops with odd numbers. One very simple method consists of calculating ψ at the points P_i by a Taylor expansion from the points M_{2i-1} along the normal to Γ_V (cf. F. SCHNEIDER);

$$\psi(P_i) = \psi(M_{2i-1}) + d_i(\frac{\partial\psi}{\partial n})(M_{2i-1}) + \frac{1}{2}\, d_i^2\, (\frac{\partial^2\psi}{\partial n^2})(M_{2i-1}), \quad i \in \{1,\ldots,7\} \tag{V.86}$$

where d_i denotes the distance from M_{2i-1} to P_i. The values of $(\partial\psi/\partial n)(M_{2i-1})$ can be deduced from the $g^j{}_V$ by interpolation. The values of $(\partial^2\psi/\partial n^2)(M_{2i-1})$ can be deduced from double probes measuring the gradient of $(1/r\ \partial\psi/\partial n)$ at the points N_j, if these measurements are established to be of sufficient quality. The formula (V.86) is easier to employ then (V.82) since it requires only 28 multiplications. In cases when the measurements of the gradient of the tangent field might be too inaccurate and we would be forced to restrict ourselves to the first order in (V.86), then the use of (V.82) to calculate the $\psi(P_i)$ could be indispensable. □

V.3.3 Determination of the Plasma Boundary and the Parameters β_P and ℓ_i in JET

We consider the two test-cases of Section I.5.4, one elliptic and the other circular. The values of $f^i{}_V$ and $g^j{}_V$ are taken from these simulations and regarded as data for the identification of Γ_p by the methods of Sections V.1 and V.2 modified as in V.3.1 to take account of the discrete character of the measurements. The plasma boundary obtained by these two methods, as well as the parameters β and γ (obtained by the method of Section V.2), will be compared with the "real" plasma of the simulation of Section I.5.4. We shall likewise study the influence of perturbations of the $f^i{}_V$ and $g^j{}_V$ on the solution obtained.

As the two configurations studied are symmetric relative to the equatorial plane, we solve the problem in the upper half-plane, using the values of the $f^i{}_V$ at the eight flux loops and the $g^j{}_V$ at the nine magnetic probes in this half-plane.

V.3.3.1 The Elliptic Case

Recall that this case corresponds to $\beta = 1.5$, $\gamma = 1$. Fig. V.8 represents the function ψ on the outer boundary Γ_V of the vacuum vessel as a function of the arc-length coordinate s on Γ_V in three cases:

1. as it is obtained in the simulation of Section I.5.4;

2. after linear interpolation according to the formula (V.72) between the $f^i{}_V$ obtained in Section I.5.4;

3. after linear interpolation between values of $f^i{}_V$ perturbed by -5%, $+3\%$, -3%, $+8\%$, $+4\%$, -2%, 0%, $+4\%$ for i going from 1 to 8.

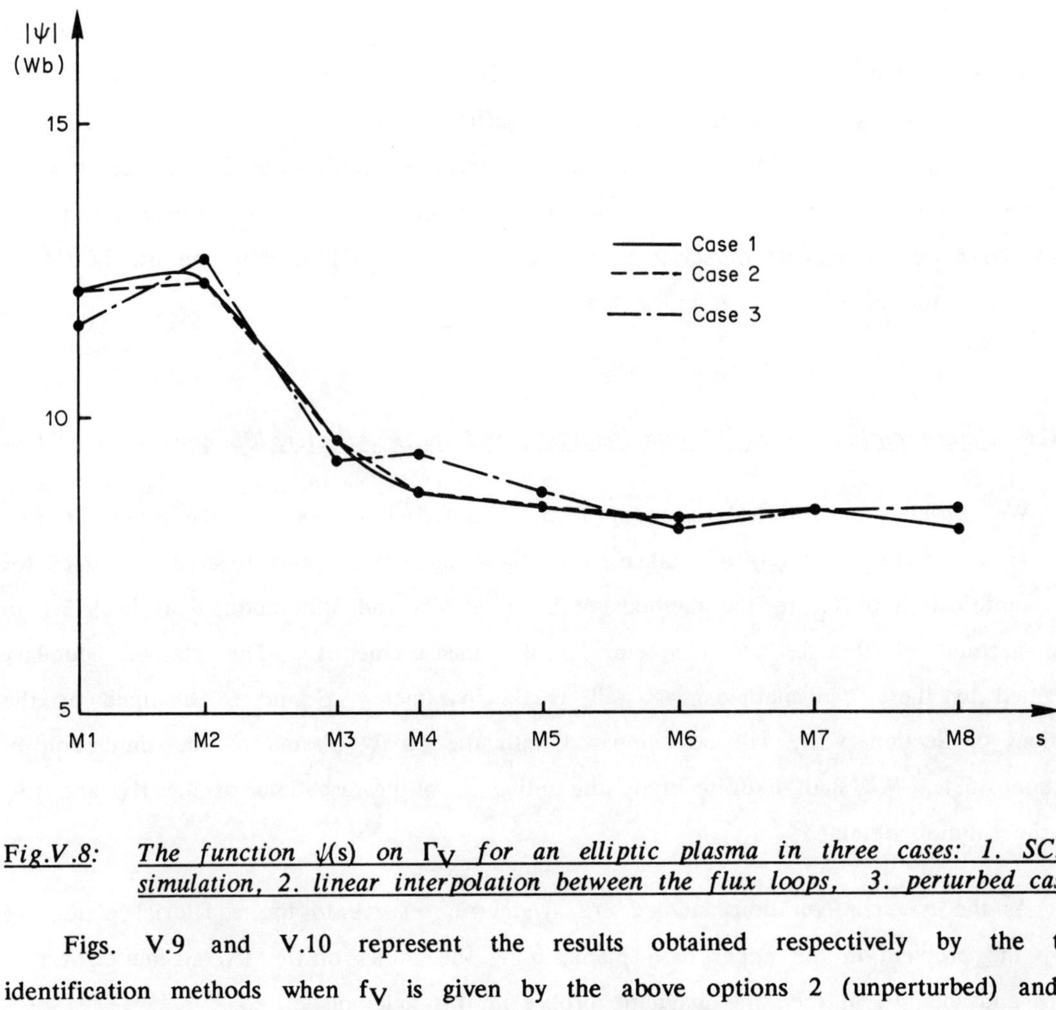

Fig.V.8: The function $\psi(s)$ on Γ_V for an elliptic plasma in three cases: 1. SCED simulation, 2. linear interpolation between the flux loops, 3. perturbed case.

Figs. V.9 and V.10 represent the results obtained respectively by the two identification methods when f_V is given by the above options 2 (unperturbed) and 3 (perturbed) and for the g^j_V taken from the simulation of Section I.5.4. The index a) for each figure corresponds to the method of Section V.1. modified in Section V.3.1.4 and the value of the regularization parameter ε is indicated. The index b) for each figure corresponds to the method of Section V.2 modified in Section V.3.1.5 and the values of β and γ obtained at convergence are indicated. The boundary Γ_2 denotes the inner boundary of the vacuum vessel (the edge of the solution domain is the outer boundary Γ_V of the vessel). The continuous line represents the plasma boundary obtained by the identification method, while the broken line represents the "real" boundary obtained in Section I.5.4.

Fig. V.11 represents the results obtained for a function f_V on Γ_V given by option 2 (linear interpolation between unperturbed measurements), while the g^j_V are perturbed by -7%, $+9\%$, -3%, -5%, -4%, $+8\%$, $+5\%$, $+7\%$, -6%, respectively, for j going from 1 to 9.

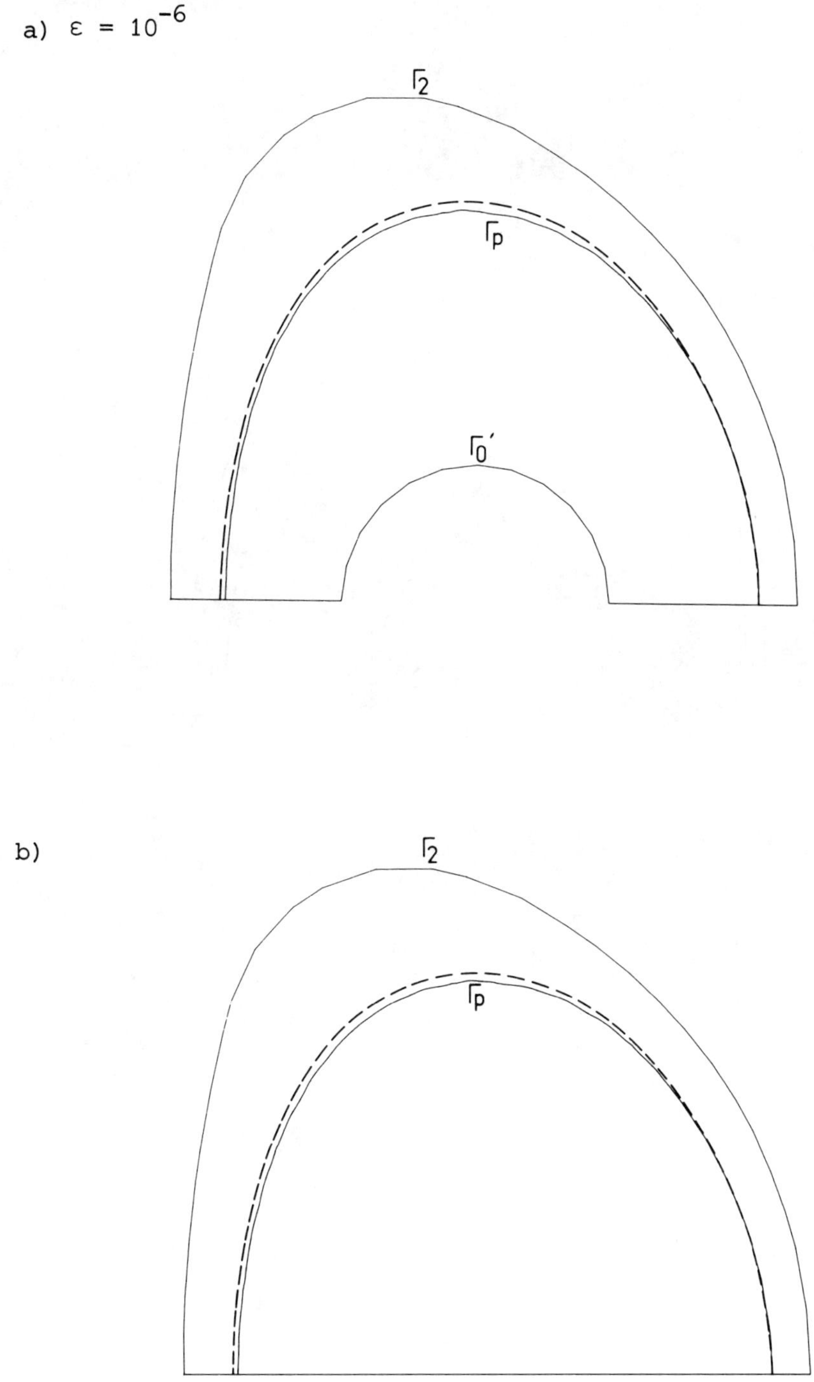

$\beta = 1.52,\ \gamma = 0.98$

Figure V.9 : *Identification of* Γ_p *by the two methods with* f_V *corresponding to case 2 of Fig V.8. (unperturbed)*

a) $\varepsilon = 10^{-4}$:

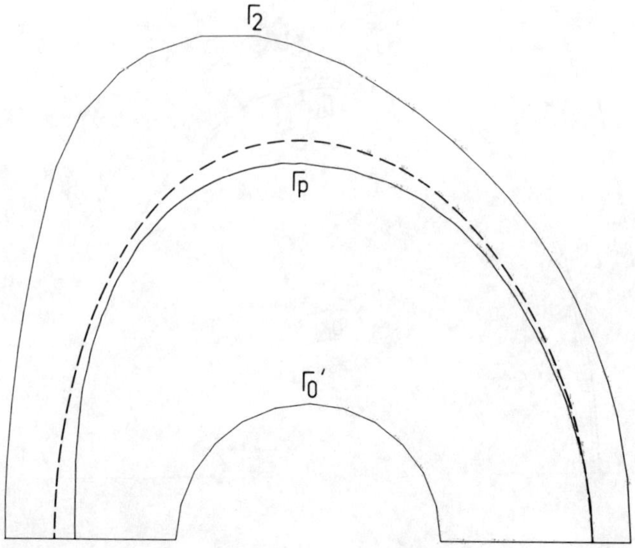

b)

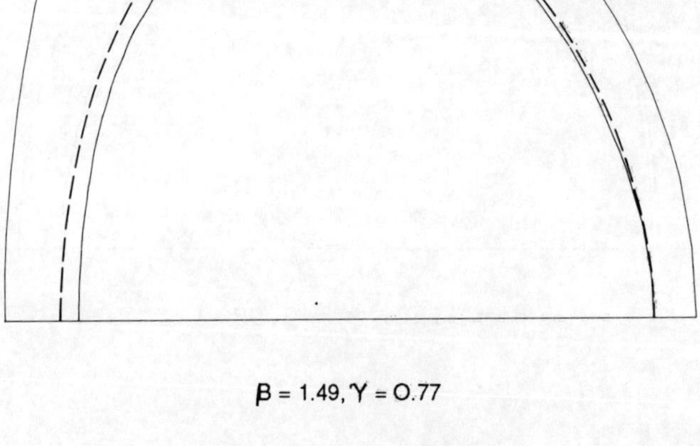

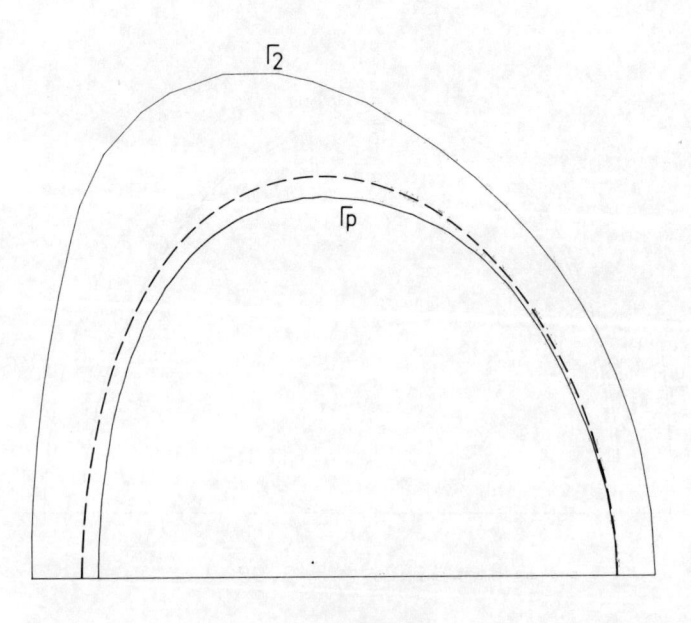

$\beta = 1.49, \gamma = 0.77$

Figure V.10 : Identification of Γ_p by the two methods with f_V corresponding to case 3 of Fig V.8 (perturbed)

a) $\varepsilon = 10^{-4}$

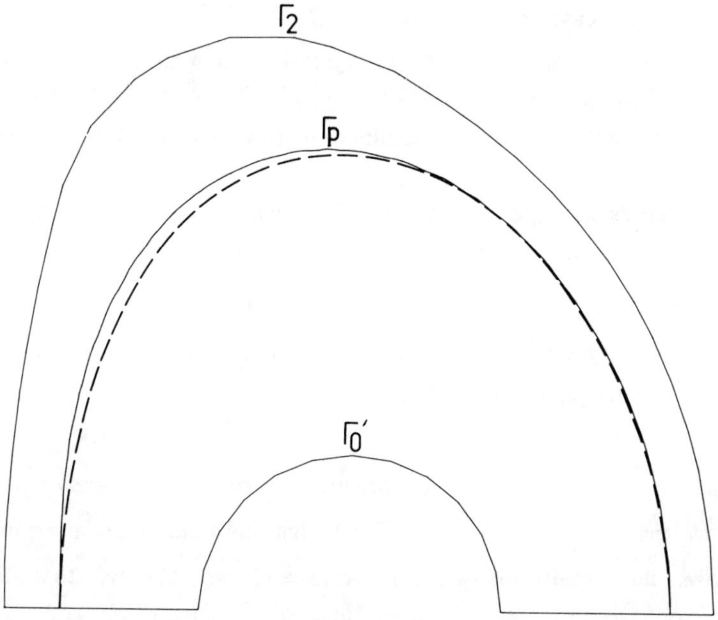

b)

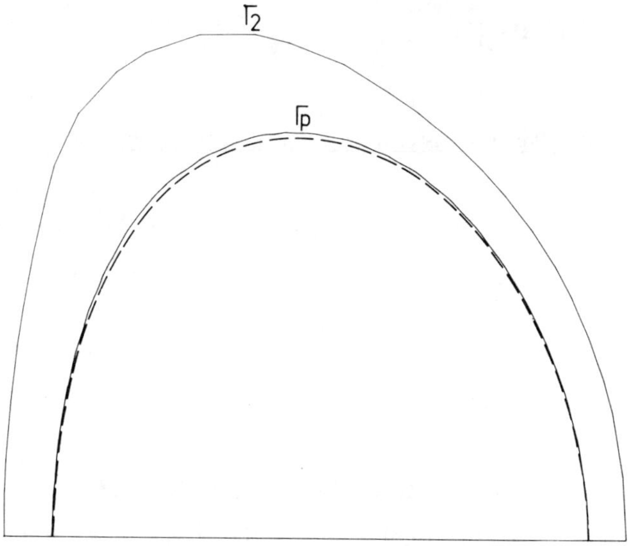

$\beta = 1.50,\ \gamma = 0.83$

Figure V.11 : *Identification of* Γ_p *with* f_V *corresponding to case 2 of Fig.V.8 and the data* g_V^{i} *perturbed.*

V.3.3.2 The Circular Case

Recall that this case corresponds to $\beta = 2$, $\gamma = 1$. The function f_V is determined by linear interpolation between unperturbed measurements f^j_V.

Fig. V.12 represents the results obtained with unperturbed measurements g^j_V.

Fig. V.13 represents the results of identification obtained with measurements g^j_V perturbed by -7%, $+9\%$, -3%, -5%, -4%, $+8\%$, $+5\%$, $+7\%$, -6%, respectively, for j going from 1 to 9. Observe that in this case the second method does not converge with two parameters β and γ. We therefore fixed γ at 1, and the value of β obtained at convergence is given in Fig V.13b.

The value of γ is, however, in practice unknown; to study the sensitivity of Γ_p with respect to the choice of γ we fixed this parameter at a wrong value: $\gamma = 3$. Fig V.14 shows the results of identification, and we observe that the value of β is wrong; it is the quantity $\beta_p + \ell_i/2$ that is identical to the reference case (see Section I.5.2 for the definitions of β_p and ℓ_i).

Other simulations for the elliptic and circular case, with more substantial perturbations or with breakdowns in certain loops or probes, are given in J. BLUM – J. LE FOLL – B. THOORIS [3].

V.3.3.3 Comparisons of the Two Methods of Identification

As we have seen, for the elliptic case as for the circular case the two methods of identification give the same plasma boundary, even when perturbations in the measurements induce errors in the plasma shape.

The regularisation parameter ε in the first identification method has to be optimized as a function of the size of the perturbations. For an unperturbed case the value $\varepsilon = 10^{-6}$ seems optimal. By contrast, for perturbations of the order of several per cent, $\varepsilon = 10^{-4}$ seems to be the right value. Fig. V.15 shows a perturbed case where the too small value of ε does not suffice to regularize the problem.

The second method of identification enables the values of β and γ to be determined for an elliptical plasma in a way which is stable with respect to perturbations in the measurements. The parameter β_p (poloidal beta) defined by (I.95) is very close to β. On the other hand there is a correspondence between the self-inductance ℓ_i of the plasma, defined by (I.96), and the exponent γ which characterizes the "peaked" nature of the plasma current density (for $\gamma = 0$, $\ell_i \simeq 0.5$: "flat" current; for $\gamma = 1$, $\ell_i \simeq 1$; for $\gamma = 3$, $\ell_i \simeq 1.5$: "peaked" currents). In the case of an elliptical plasma β_p and ℓ_i can therefore be identified separately.

This is no longer the case for a circular plasma; indeed, as we saw in the previous section, as soon as there is the least perturbation in f_V^i and g_V^i it is no longer possible to identify β and γ and thus β_p and ℓ_i; only the parameter $(\beta_p + \ell_i/2)$ can be identified correctly.

The final point of comparison between the two methods concerns the computation time: the first needs less than 1ms on CRAY 1, while the second takes 2 to 3s. This predestines each of these methods to its own type of use: the first for the on-line control of plasma shape, the second for the analysis of a discharge at various intervals of time between two shocks. The second method has been used on JET since it came into service, as the first link in the chain of diagnostic codes (cf. M. BRUSATI et al).

<u>*Remark V.11*</u>

In this section we have assumed the position of the flux loops and the magnetic probes to be fixed. In fact this program has been used for optimizing the position and number of these probes and loops. Criteria which have emerged from the study of numerous configurations are:

. the necessity of putting flux loops in the equatorial plane

. spacing the loops and probes as regularly as possible

. if possible having a flux loop and a magnetic probe opposite each other (i.e. on the same normal to Γ_V).

The solution settled upon is a compromise between these criteria and the technological constraints (avoiding access ports,...). □

a) $\varepsilon = 10^{-6}$

Γ_2

Γ_p

Γ_0'

b)

Γ_2

Γ_p

$\beta = 1.98, \gamma = 0.96$

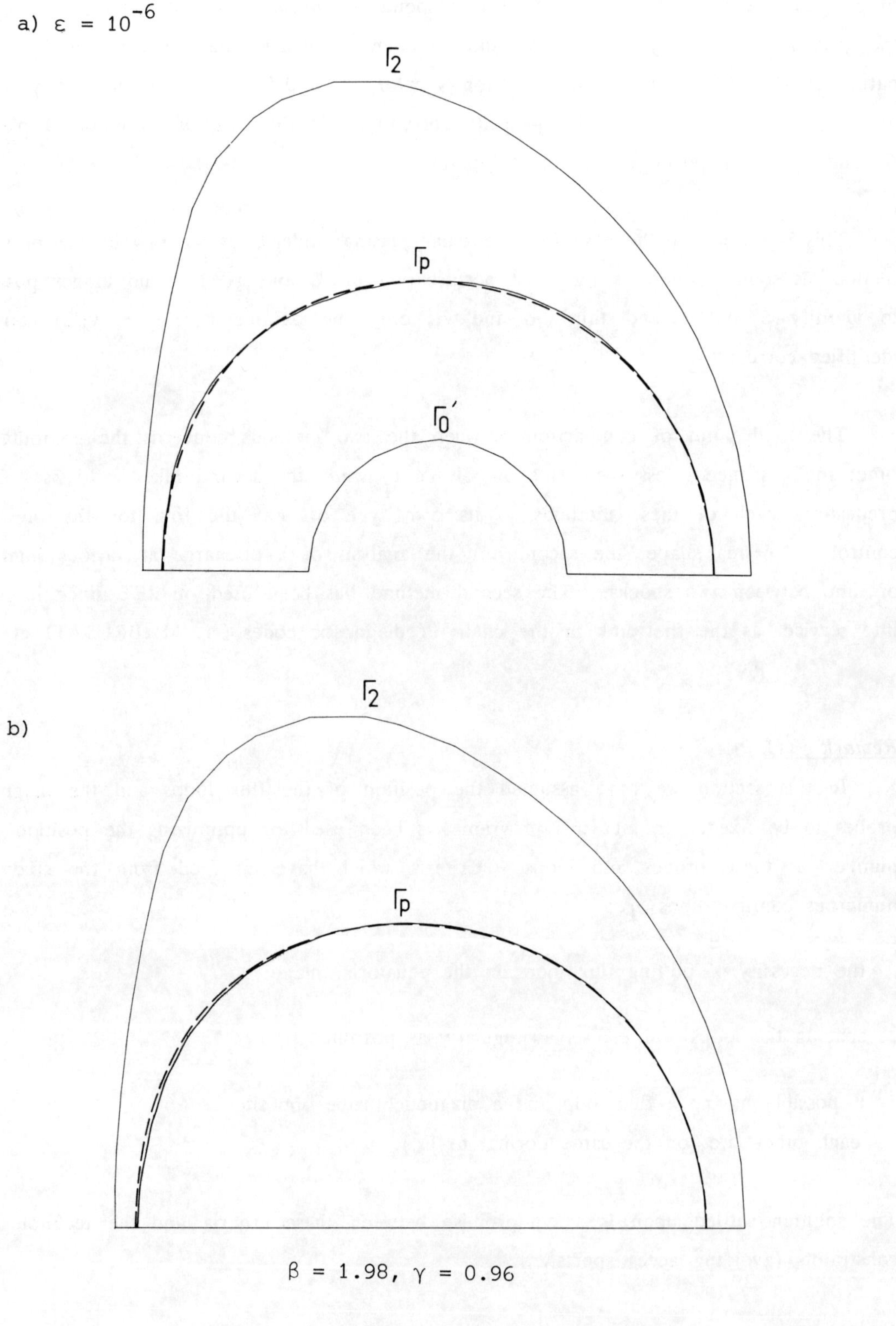

Figure V.12 : *Identification of Γ_p by the two methods for a circular plasma with unperturbed data f_V^i and g_V^i*

a) $\varepsilon = 10^{-4}$

b)

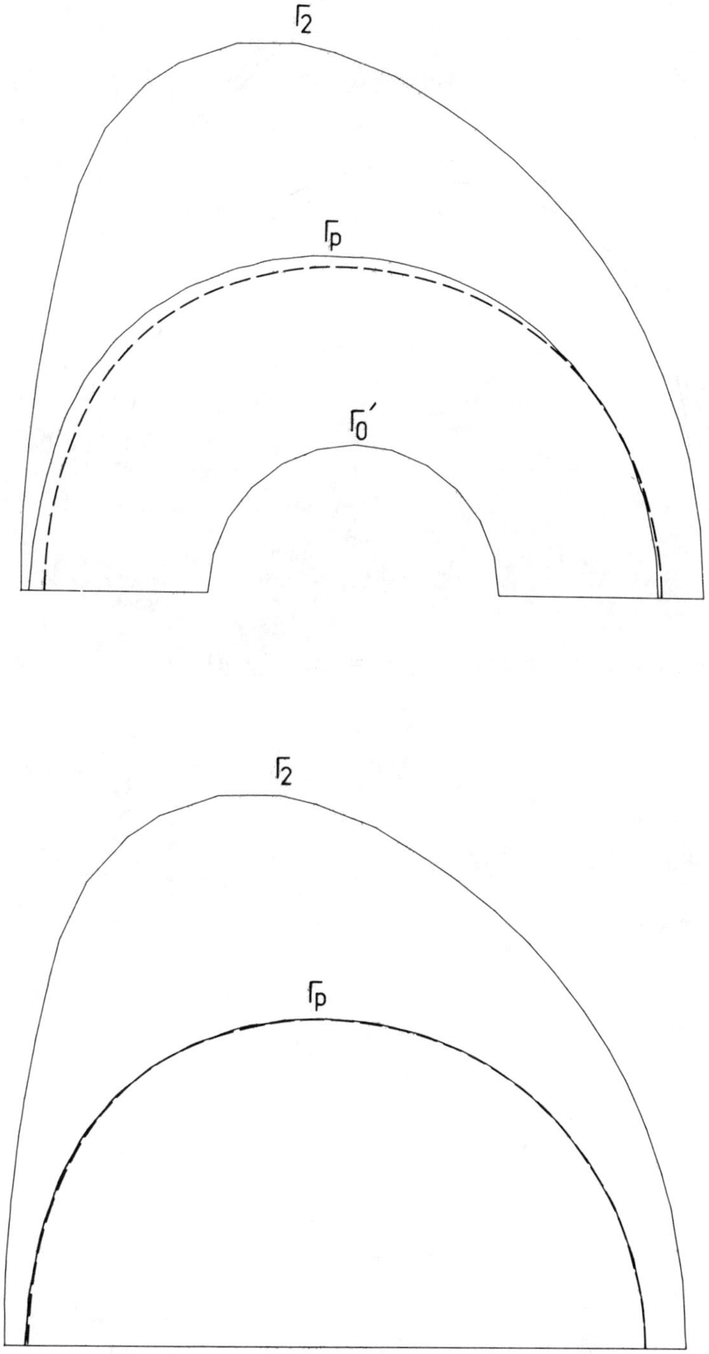

$\beta = 1.91$, $\gamma = 1$ (fixed)

Figure V.13 : *Identification of* Γ_p *in the circular case with perturbed data* g_V^i.

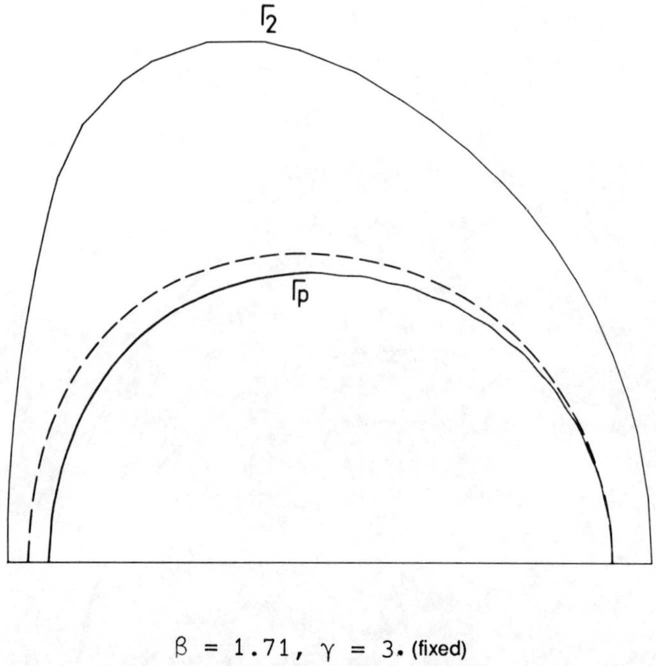

Figure V.14 : *Same as Fig V.13b but with* $\gamma = 3.$ *(fixed)*

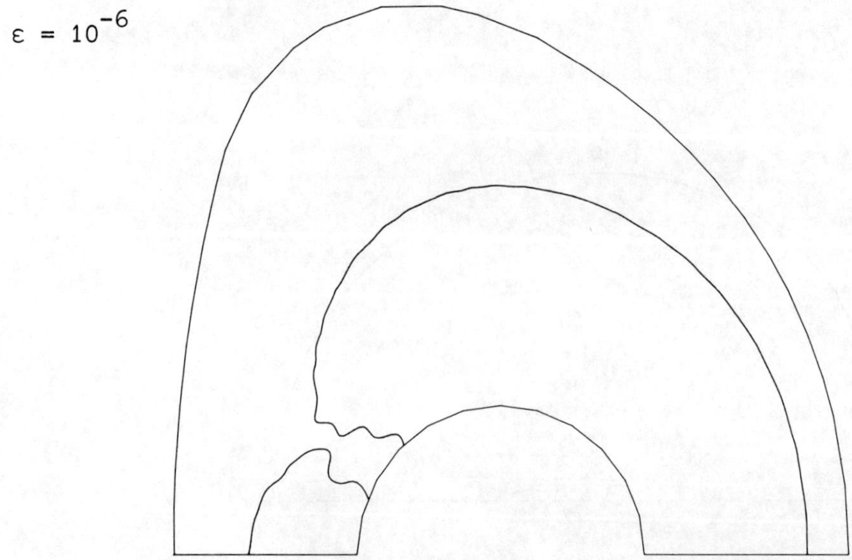

Figure V.15 : *Identification of* Γ_p *for an elliptical plasma with perturbed data by the first method, with regularization parameter* ε *too small*.

V.3.4. Some Comments on Identification of the Plasma Current Density

V.3.4.1 Sensitivity to the Choice of the Function $g(\psi_N)$.

In this chapter we chose to write the current density in the form (V.47).

By (I.17) the toroidal component of the plasma current density is

(V.87) $$j_T(r,\psi) = r\,A(\psi) + \frac{1}{r}\,B(\psi)$$

with

$$A(\psi) = \frac{\partial p}{\partial \psi}\ ,$$

$$B(\psi) = \frac{1}{2\mu_o}\frac{\partial f^2}{\partial \psi}\ .$$

In J. BLUM – J.C. GILBERT – B. THOORIS the functions $A(\psi)$ and $B(\psi)$ are chosen as polynomials in the variable ψ_N. It has been shown in a large number of test–cases that it is not possible to identify $A(\psi)$ and $B(\psi)$ separately from magnetic measurements. However, if we write j_T in the form

(V.88) $$j_T(r,\psi) = \lambda\left[\frac{r\beta}{R_o} + \frac{R_o(1-\beta)}{r}\right]\left[(1-\psi_N)+a_1(1-\psi_N)^2\right]$$

then for elliptical plasmas β and a_1 can be identified separately and the values of β_P and ℓ_i so obtained are identical to those obtained by using the expression (V.47). For circular plasmas with small β_P it has not been possible to identify β and a_1 separately, but only $\beta_P + \ell_i/2$.

The identification of the parameters β_P and ℓ_i thus does not depend on the choice of the function $g(\psi_N)$. The same type of conclusion has been put forward recently in L.L. LAO et al [1].

Note that if there is experimental information on the pressure profile in the equatorial plane then it is possible to identify $A(\psi)$ and $B(\psi)$ in (V.87) separately, as shown in J. BLUM – J.C. GILBERT – B. THOORIS. It is likewise possible to identify the function $g(\psi_N)$, not in parametric form but as a sufficiently smooth function. This is done in J. BLUM – J.C. GILBERT – J. LE FOLL – B. THOORIS by adding a smoothing term of type $\int_0^1 (\partial^2 g/\partial \psi^2_N)^2\, d\psi_N$ to the cost function.

V.3.4.2 *The Circular Case*

In all cases we have observed that the plasma boundary Γ_p is identified correctly from magnetic measurements. The poloidal field B_p on Γ_p is then also known and is denoted g_p.

The problem of identifying the current density in the plasma is then stated as follows: let Γ_p be the plasma boundary, assumed known. We have

$$\text{(V.89)} \qquad \begin{cases} \psi = 0 & \text{on } \Gamma_p \\ \dfrac{1}{r}\dfrac{\partial\psi}{\partial n} = g_p & \text{on } \Gamma_p \\ L\psi = j_T(r,\psi) & \text{in } \Omega_p . \end{cases}$$

Given g_p, what information can be obtained about $j_T(r, \psi)$? First we make a cylindrical approximation in which L is replaced by $(-\Delta)$ where (V.89) becomes

$$\text{(V.90)} \qquad \begin{cases} \psi = 0 & \text{on } \Gamma_p \\ \dfrac{\partial\psi}{\partial n} = g_p & \text{on } \Gamma_p \\ -\Delta\psi = j_T(\psi) & \text{in } \Omega_p . \end{cases}$$

If Γ_p is a circle then the only solutions are radial and g_p has to be constant on Γ_p. The only information we obtain is the total plasma current I_p, since

$$\text{(V.91)} \qquad I_p = -\int_{\Gamma_p} g_p \, d\sigma = \int_{\Omega_p} j_T \, dS .$$

Now let us return to the toroidal case, and carry out an expansion to first order with respect the inverse ε of the aspect ratio, i.e. $\varepsilon = a/R$ where a and R are the minor and major radii of the torus, respectively. This is the object of the equilibrium theory of V.D. SHAFRANOV [3] and [4] which will be developed in Chapter VII. The magnetic field g_p on the boundary Γ_p (all the time assumed circular) is

$$\text{(V.92)} \qquad g_p = g_0(1 + \varepsilon\, \xi \cos\theta)$$

with

$$\xi = \beta_p + \frac{\ell_i}{2} - 1 \ , \quad g_0 = -\frac{\mu_o I_p}{2\pi a}$$

where θ is the polar angle defined relative to the centre of Γ_p. The expression (V.92) shows that in the case of a Tokamak with high aspect ratio and a plasma with circular section the only information that can be deduced is the value of $(\beta_P + \ell_i/2)$.

V.3.4.3 The Limit Between "Circular" and "Elliptical" Plasma

The minimum elongation from which it is possible to separate β_P and ℓ_i depends on the accuracy of the magnetic measurements. It seems that in JET this minimum elongation is about 1.2.

The possibility of identifying β_P and ℓ_i separately for elliptical plasmas has also been shown by J.L. LUXON – B.B. BROWN by statistical methods.

An expression for $(\beta_P + 1/2\ \ell_i)$ involving integrals over Γ_V of quantities depending only on the tangential and normal components of the poloidal field B_P to Γ_V is given by V.D. SHAFRANOV [2] for a plasma of arbitrary section. L.L. LAO et al. [2] give an additional integral relation that allows β_P and ℓ_i to be calculated separately for non–circular plasmas: in particular, for SOLOVEV equilibria, for which an analytic solution is known, they show that for an elongation greater than 1.1 (in the absence of noise in the magnetic measurements) β_P and ℓ_i can be meaningfully separated.

In conclusion, β_P and ℓ_i can be separated more easily, the more the plasma is elliptical with small aspect ratio (i.e. compact) and $(\beta_P + \ell_i/2)$ is large.

As regards the determination of $j_T(r, \psi)$ in (V.89), J.P. CHRISTIANSEN – J.B. TAYLOR have shown that if the shape of the magnetic surfaces is known then $j_T(r, \psi)$ can generally be deduced from it (an exception is given by BISHOP – TAYLOR). However, the determination of $j_T(r, \psi)$ from g_P (or from f_V and g_V) is an open problem.

Finally, we mention the excellent review article by B.J. BRAAMS on methods of interpreting magnetic diagnostics in a Tokamak.

6. Evolution of the equilibrium at the diffusion time scale

In the preceding chapters we have been concerned with the stationary equilibrium problem. Now we shall study how this equilibrium evolves on a "slow" time–scale, i.e. on the time–scale of diffusion of the heat and of the current inside the plasma, which is long compared with the Alfven time. On this slow time–scale the plasma can be regarded as being in equilibrium at every instant, the MHD instabilities having been stabilized. The plasma thus traverses a succession of equilibrium states related to each other by the induction of the currents in the external coils, by penetration of the magnetic field into the vacuum vessel, and by diffusion phenomena within the plasma.

H. GRAD – J. HOGAN have shown, for the classical theory of diffusion, the need to develop a self–consistent description of the evolution of the axisymmetric equilibrium of the plasma on the resistivity time–scale. The method of magnetic surface averaging was introduced by H. GRAD – P.N. HU – D.C. STEVENS for adiabatic evolutions of equilibrium, and by H. GRAD – P.N. HU – D.C. STEVENS – E. TURKEL for resistive evolutions. The numerical method proposed by H. GRAD for solving the equilibrium equations consists of iterating between the solution of the 2–dimensional elliptic equation for equilibrium, and the averaged equation which is a second order differential equation. These techniques have also been used by F.L. HINTON – R.D. HAZELTINE for the neoclassical theory of transport. In view of the fact that the speed of diffusion parallel to the magnetic surfaces is much greater than the perpendicular diffusion speed, these methods allow us to write the set of equations of conservation of particles, of energy and of flux as a system of 1–dimensional diffusion – convection equations in space with respect to a variable indexing the magnetic surfaces.

In the first part of this chapter we set up the equations for the model. These are on the one hand the equilibrium evolution equations, and on the other hand the equations for diffusion within the plasma. We shall show the elements of coupling between the two systems.

In the second part we present numerical methods for solving these two systems, and describe how they are coupled.

In the third part we apply this model to the simulation of the various phases of a discharge in TORE Supra.

VI.1 THE EQUATIONS OF EQUILIBRIUM AND TRANSPORT

The resistive magnetohydrodynamic (MHD) equations are written (cf. S.I. BRAGINSKII):

(VI.1)

$$\dot{n} + \nabla\cdot(n\,u) = s$$
(conservation of particles)

$$mn(\dot{u} + u.\nabla\, u) + \nabla\, p = j \times B$$
(conservation of momentum)

$$3/2\,(\dot{p} + u.\nabla\, p) + 5/2\, p\, \nabla.u + \nabla.Q = s'$$
(conservation of particle energy)

$$\nabla \times E = -\dot{B}$$ (Faraday's Law)

$$\nabla.\, B = 0$$ (conservation of B)

$$E + u \times B = \eta j$$ (Ohm's Law)

$$\nabla \times H = j$$ (Ampère's Law)

$$B = \mu H$$ (magnetic permeability)

$$p = n\,k\,T$$ (law of perfect gases).

where n denotes the density of the particles, m their mass, **u** their mean velocity, p their pressure, T their temperature, Q the heat flux, η the resistivity tensor, s and s' the source terms, k the Boltzmann constant, and $\dot{A}$ denotes the time derivative of A for an arbitrary quantity A. We shall give here the equations for a model with two fluids (electrons and ions) as in S.I. BRAGINSKII.

To simplify the system (VI.1) we need to define a few characteristic time constants of the plasma. The Alfven time constant τ_A is

$$\tau_A = \frac{a(\mu_0 mn)^{\frac{1}{2}}}{B_0}$$

where a is the minor radius of the plasma and B_0 is the toroidal magnetic field. It is of the order of a microsecond for present Tokamaks.

The diffusion time constant of the particle density n is

$$\tau_n = \frac{a^2}{D}$$

where D is the particle diffusion coefficient. Likewise, the time constants for diffusion of heat of the electrons and of the ions are

$$\tau_e = \frac{n_e a^2}{K_e}$$

$$\tau_i = \frac{n_i a^2}{K_i}$$

where n_e, n_i are the density of electrons and ions, respectively, and K_e, K_i are their thermal conductivities. These constants τ_n, τ_e, τ_i are of the order of a millisecond on Tokamaks currently operating.

Finally, the resistive time constant for the diffusion of current density and magnetic field in the plasma is given by

$$\tau_r = \frac{\mu_0 a^2}{\eta}$$

and is of the order of a second.

If a global time constant for plasma diffusion is defined by

$$\tau_p = \inf(\tau_n, \tau_e, \tau_i, \tau_r)$$

we note that

$$\tau_A \ll \tau_p$$

On the diffusion time-scale τ_p the term $(\dot{u} + u.\nabla u)$ is small compared with ∇p (cf. E.K. MASCHKE – J. PANTUSO SUDANO, D.B. NELSON – H. GRAD) and the equilibrium equation (I.4) is thus satisfied at every instant.

For an exhaustive study of time-scales in a Tokamak refer to S.C. JARDIN [1].

Our aim here is to rewrite the system (VI.1) on the time-scale τ_p and for an axisymmetric configuration.

In Section VI.1.1. we shall see what the equilibrium equations for an axisymmetric configuration are like when account is taken of the external circuits and of diffusion of the currents induced in the vacuum vessel.

In Section VI.1.2. the equations of transport for electrons and ions will be rewritten in the light of fact that the speed of diffusion parallel to the magnetic surfaces is very large compared with that of perpendicular diffusion.

In Section VI.1.3 we shall study the coupling elements between these two systems.

VI.1.1 *The System of Equilibrium Equations*

We suppose that the configuration is axisymmetric and aim to write down the equations that govern the behaviour of the function $\psi(r,z,t)$.

The equations (I.1) to (I.3) are satisfied at each instant; the same is true for equation (I.4) if we work on the diffusion time-scale τ_p, as discussed above.

As a result of the axial symmetry assumption, this also holds for equations (I.11) and (I.12). By (I.12) we have

(VI.2) $$L\psi = j_T$$

This stationary equation has to be satisfied at each moment of the discharge, but the current density j_T diffuses in the vacuum vessel and the plasma and varies in the poloidal field coils.

We shall now rewrite equation (VI.2) in each region of the domain Ω shown in Fig. I.4, namely the iron Ω_f, the air Ω_a (including the coils B_i and the vacuum vessel Ω_{cv}) and the plasma Ω_p. Let Ω_a' denote the region

$$\Omega_a' = \Omega_a - \bigcup_{i=1}^{k} \bar{B}_i - \bar{\Omega}_{cv} ,$$

i.e. the air but not the coils nor the vacuum vessel.

. In the magnetic circuit Ω_f and in the region Ω_a', we have

(VI.3) $$L\psi = 0 \qquad \text{in } \Omega_f \cup \Omega_a' .$$

The operator L, given by (I.13), is nonlinear in Ω_f and linear in Ω'_a.

. In each coil B_i we have

(VI.4) $$L\psi = j_i(t)$$

where the current density j_i is assumed homogeneous in each coil. Suppose further that the poloidal field system consists of independent circuits, each made up of a certain number of windings of the coil B_i. If V_i is the voltage applied to this circuit, R_i its resistance, n_i the number of turns and S_i the area of B_i then the equation for this circuit can be written:

(VI.5) $$V_i(t) = R_i S_i \; j_i(t) + \frac{n_i}{S_i} \int_{B_i} \dot{\psi} \; dS$$

where $\dot{\psi}$ is the time derivative of ψ.

We can deduce from (VI.4) and (VI.5) the integro-differential evolution equation for ψ in each coil B_i:

(VI.6) $$\frac{n_i}{R_i\ S_i^2}\int_{B_i} \dot{\psi}\ dS + L\psi = \frac{V_i}{R_i\ S_i} \quad \text{in } B_i\ .$$

. In the vacuum vessel Ω_{cv} Ohm's law is

(VI.7) $$j_T = \sigma_v\ E_T$$

where σ_v is the conductivity of the vacuum vessel and E_T is the toroidal component of the electric field. By Faraday's Law we have in an axisymmetric configuration:

(VI.8) $$E_T = -\frac{1}{r}\dot{\psi}\ .$$

From (VI.2), (VI.7) and (VI.8) we deduce the diffusion equation for ψ in the vacuum vessel:

(VI.9) $$\frac{\sigma_v}{r}\dot{\psi} + L\psi = 0 \quad \text{in } \Omega_{cv}\ .$$

. In the plasma the Grad–Shafranov equation (I.17) is satisfied at each instant and we have

(VI.10) $$L\psi = r\frac{\partial p}{\partial \psi} + \frac{1}{2\mu_0 r}\frac{\partial f^2}{\partial \psi} \quad \text{in } \Omega_p\ .$$

. The boundary condition is the homogeneous Dirichlet condition on the boundary Γ of Ω:

(VI.11) $$\psi = 0 \quad \text{on} \quad \Gamma\ .$$

In the case of configuration symmetric with respect to the equatorial plane we take the boundary conditions (I.25).

. The transmission conditions at the interfaces between the air, the coils, the vacuum and the plasma are those of continuity of ψ and its normal derivative $\partial\psi/\partial n$. At the air/iron interface Γ_{af} we have conditions (I.26), namely

(VI.12) $$\begin{cases} \left(\frac{1}{\mu}\frac{\partial \psi}{\partial n}\right)_f = \left(\frac{1}{\mu_0}\frac{\partial \psi}{\partial n}\right)_a & \text{on } \Gamma_{af} \\ (\psi)_f = (\psi)_a\ . \end{cases}$$

Recall finally that the free boundary Γ_p of the plasma is defined as follows:

(VI.13) $$\Gamma_p = \{M \in \Omega_V \mid \psi(M) = \sup_D \psi\} .$$

To summarize, the equations for the flux ψ in Ω are the stationary elliptic equation (VI.3) in the iron and the air and the parabolic equations (VI.6) and (VI.9)in the coils and the vacuum vessel, while the equation (VI.10) for ψ in the plasma is of elliptic nature if the profiles $p(\psi)$ and $f(\psi)$ are known at every instant. In the next section we shall establish the equations for $p(\psi,t)$ and $f(\psi,t)$.

The initial condition for this system is

(VI.14) $$\psi(0) = \psi_0$$

where ψ_0 is a solution to the equilibrium problem corresponding to currents $j_T(0) = j_0$ in the coils and to the functions $p(\psi,0)$ and $f(\psi,0)$ in the plasma.

As in Chapter I, we can give a weak formulation of the system of equations for ψ. Let $L^{p_0}(0,T;V^{p_0}(\Omega))$ be the space of functions ψ such that

$$\int_0^T |\psi|^{p_o}_{V^{p_o}(\Omega)} dt<\infty$$

where T is the duration of the discharge. The weak formulation of the problem consists of finding a function $\psi \in L^{p_0}(0,T;V^{p_0}(\Omega))$ which satisfies:

$$\frac{d}{dt}\int_{\Omega_{cv}} \frac{\sigma_V}{r} \psi\, \phi\, dS + \sum_{i=1}^{k} \frac{n_i}{R_i\, S_i^2} \frac{d}{dt}\int_{B_i} \psi\, dS \int_{B_i} \phi\, dS$$

(VI.15) $$+ a_\mu(\psi,\phi) = \int_{\Omega_p} j_T(r,\psi)\phi\, dS + \sum_{i=1}^{k} \frac{V_i}{R_i\, S_i}\int_{B_i} \phi\, dS ,$$

$$\forall\phi \in V^{p_0'}(\Omega) , \forall t \in [0,T]$$

with

$$\psi(r,z;0) = \psi_0(r,z) ,$$

$$\Omega_p = \{M \in \Omega_V \mid \psi(M) > \sup_D \psi\} ,$$

$$a_\mu(\psi,\phi) = \int_\Omega \frac{1}{\mu r} \nabla\psi.\nabla\phi \; dS \;,$$

$$j_T(r,\psi) = r\,\frac{\partial p}{\partial\psi} + \frac{1}{2\mu_0 r}\,\frac{\partial f^2}{\partial\psi}$$

$$\frac{1}{p_0} + \frac{1}{p'_0} = 1$$

VI.1.2 The System of Transport Equations.

We shall show how by applying the method of averaging over the magnetic surfaces we are able to write the transport equations in the form of a system of 1−dimensional equations in space with respect to a variable indexing the magnetic surfaces. The proofs of the propositions that follow are formal.

VI.1.2.1 The method of Averaging Over the Magnetic Surfaces.

Let ρ be an arbitrary coordinate indexing the magnetic surfaces; any quantity constant on each magnetic surface could *a priori* be chosen as the variable ρ. We shall return later to the question of this choice.

Define the average $<A>$ of an arbitrary quantity A on a magnetic surface S by

$$<A> = \frac{\partial}{\partial V} \int_V A \; dV \qquad \text{(VI.16)}$$

where V is the volume inside S. This notion of average has the following properties.

Proposition VI.1

For an arbitrary function A and any vector field **W** we have

$$<A> = \frac{1}{V'} \int_S \frac{A dS}{|\nabla\rho|} = \frac{\displaystyle\int_c \frac{A d\ell}{B_p}}{\displaystyle\int_c \frac{d\ell}{B_p}} \qquad \text{(VI.17)}$$

$$<\nabla.W> = \frac{\partial}{\partial V} <W.\nabla V> = \frac{1}{V'} \frac{\partial}{\partial\rho} [V' <W.\nabla\rho>] \qquad \text{(VI.18)}$$

(VI.19) $$\frac{\partial}{\partial t}(V' \langle A\rangle) = V' \langle \dot{A}\rangle + \frac{\partial}{\partial \rho}[V'\langle A u_\rho . \nabla\rho\rangle]$$

with

$$V' = \frac{\partial V}{\partial \rho} = \int_S \frac{dS}{|\nabla\rho|} = 2\pi \int_C \frac{r d\ell}{|\nabla\rho|}$$

and

$$\dot{\rho} + u_\rho . \nabla\rho = 0 \ .$$

The contour C is the line of flux that is the intersection of the magnetic surface S and the Orz plane of the meridian section. The notation $\partial/\partial t$ represents the partial derivative with respect to t keeping ρ fixed, while $\dot{A}$ represents the time derivative of A at a fixed point (r,z). The vector u_ρ is the velocity of the surfaces of constant ρ. □

Proof

The equalities (VI.17) follow from the definition of $\langle A\rangle$ with

$$dV = \frac{dS\ d\rho}{|\nabla\rho|} = \frac{2\pi\ r\ d\ell\ d\rho}{|\nabla\rho|} = \frac{2\pi\ r\ d\ell\ d\psi}{|\nabla\psi|}$$

and the expression for the amplitude of the poloidal field B_P given by (I.11) :

$$B_P = \frac{1}{r} |\nabla\psi|$$

The expression for V' comes from (VI.17) by putting $A = 1$ in those relations.

We prove (VI.18) as follows :

$$\langle \nabla . W\rangle = \frac{\partial}{\partial V}\int_V \nabla . W\ dV = \frac{\partial}{\partial V}\int_S W . dS$$

$$= \frac{\partial}{\partial V}\int_S \frac{W . \nabla\rho}{|\nabla\rho|} dS = \frac{\partial}{\partial V}[V' \langle W . \nabla\rho\rangle]$$

To establish (VI.19) we may write

$$\frac{\partial}{\partial t}\int_V A\ dV = \int_V \dot{A}\ dV + \int_S A u_\rho . dS$$

from which on differentiating with respect to ρ we obtain

$$\frac{\partial}{\partial t}(V'\langle A\rangle) = V'\langle\dot{A}\rangle + \frac{\partial}{\partial\rho}\int_S \frac{A\, u_\rho.\nabla\rho}{|\nabla\rho|}\, dS .$$

This proves (VI.19), using (VI.17). □

Remark VI.1

Letting $A = 1$ in (VI.19) gives

$$\frac{\partial V'}{\partial t} = \frac{\partial}{\partial\rho}\langle u_\rho.\nabla V\rangle \tag{VI.20}$$ □

We shall now average the transport equations over each magnetic surface (cf.F.L. HINTON – R.D. HAZELTINE) in order to obtain a system of 1−dimensional equations in ρ.

VI.1.2.2 The Equation of Conservation of Electrons

This equation is :

$$\dot{n}_e + \nabla.(n_e u_e) = s_1 \tag{VI.21}$$

where n_e is the electron density and $\mathbf{u_e}$ the electron velocity, while s_1 is a source term representing the flux of particles entering or leaving.

As the speed of diffusion parallel to the magnetic surfaces is much greater than the speed of perpendicular diffusion, we can assume that the electron density is constant on each magnetic surface :

$$n_e = n_e(\rho, t) .$$

Proposition VI.2

The equation of conservation of electrons (VI.21) can be written

$$\frac{\partial}{\partial t}(n_e V') + \frac{\partial}{\partial\rho}(\Gamma_e V') = \langle s_1\rangle V' \tag{VI.22}$$

where the electron flux Γ_e across a surface of constant ρ is defined by

$$\Gamma_e = \langle n_e (u_e - u_\rho) . \nabla\rho \rangle \qquad \square$$

Proof

Averaging (VI.21) over each magnetic surface and multiplying by V' we obtain

$$\text{(VI.23)} \qquad V' \langle \dot{n_e} \rangle + V' \langle \nabla . (n_e \, u_e) \rangle = V' \langle s_1 \rangle \; .$$

Now using (VI.18) and (VI.19) the equation (VI.23) can be written

$$\frac{\partial}{\partial t} (\langle n_e \rangle \, V') + \frac{\partial}{\partial \rho} [V' \, \langle n_e (u_e - u_\rho) . \nabla\rho \rangle] = V' \langle s_1 \rangle \; .$$

Since we assumed that n_e is constant on each magnetic surface, the equation obtained is none other than (VI.22). □

VI.1.2.3 The Equations of Conservation of Energy.

If viscosity terms are neglected, the equation of conservation of energy for electrons may be written (cf. F.L. HINTON – R.D. HAZELTINE) :

$$\text{(VI.24)} \qquad 3/2 \, \dot{p_e} + \nabla . (Q_e + 5/2 \, p_e \, u_e) = j.E - Q_\Delta - u_i . \nabla p_i + s_2 .$$

where p_e, p_i denote the pressure of the electrons and the ions, respectively, Q_e the heat flux of the electrons, j.E the Joule effect term, Q_Δ the equipartition term between electrons and ions, u_i the ion velocity and s_2 a source term.

Likewise the equation of conservation of energy for ions is

$$\text{(VI.25)} \qquad 3/2 \, \dot{p_i} + \nabla . (Q_i + 5/2 \, p_i \, u_i) = Q_\Delta + u_i . \nabla p_i + s_3$$

where Q_i is the heat flux of the ions and s_3 is a source term.

We have moreover the following relations:

$$\text{(VI.26)} \qquad \begin{cases} p_e = n_e\, k\, T_e \\ p_i = n_i\, k\, T_i \\ p = p_e + p_i \\ n_e = Z n_i \\ Q_\Delta = \dfrac{3m_e}{m_i} \dfrac{n_e}{t_e} (T_e - T_i) \end{cases}$$

where T_e, T_i denote the electronic and ionic temperatures, respectively, k is Boltzmann's constant, Z the charge number, m_e and m_i the mass of an electron and an ion, respectively, and t_e the collision time for electrons (cf. S.I. BRAGINSKII).

We saw in Chapter I that the plasma pressure p is a function constant on the magnetic surfaces. We saw in Section VI.1.2.2 that n_e could be regarded as constant on lines of flux. From (VI.26) we can deduce from this that T_e and T_i are approximately uniform on the magnetic surfaces. We then have the following:

Proposition VI.3

The equations of conservation of energy for electrons (VI.24) and ions (VI.25) can be expressed as

$$\frac{3}{2V'^{2/3}} \frac{\partial}{\partial t} (p_e\, V'^{5/3}) + \frac{\partial}{\partial \rho} [(q_e + 5/2\, k\, T_e\, \Gamma_e)V']$$

$$\text{(VI.27)} \qquad = [\langle j.E\rangle - \langle u_\rho.\nabla p_e\rangle - \langle u_i.\nabla p_i\rangle - Q_\Delta + \langle s_2\rangle]V'$$

$$\text{(VI.28)} \qquad \frac{3}{2V'^{2/3}} \frac{\partial}{\partial t} (p_i\, V'^{5/3}) + \frac{\partial}{\partial \rho} [(q_i + 5/2\, \frac{k\, T_i\, \Gamma_e}{Z})V']$$

$$- \frac{\Gamma_e V'}{n_e} \frac{\partial p_i}{\partial \rho} = Q_\Delta\, V' + \langle s_3\rangle\, V'$$

with

$$q_e = \langle Q_e.\nabla\rho\rangle, \quad q_i = \langle Q_i.\nabla\rho\rangle$$

and where p_e, p_i, n_i and Q_Δ are given by (VI.26). □

Proof

Averaging (VI.24), multiplying by V' and by using (VI.18) and (VI.19) we obtain

$$3/2 \frac{\partial}{\partial t} (p_e V') - 3/2 \frac{\partial}{\partial \rho} [V' \langle p_e\, u_\rho.\nabla\rho\rangle]$$

$$+ \frac{\partial}{\partial \rho} [V' \langle (Q_e + 5/2\, p_e\, u_e).\nabla\rho\rangle]$$

$$= [\langle j.E\rangle - Q_\Delta - \langle u_i.\nabla p_i\rangle + \langle s_2\rangle] V' \; .$$

Using (VI.20) and the definitions of Γ_e and q_e this can be re-expressed as

$$3/2 \frac{\partial}{\partial t} (p_e V') + p_e \frac{\partial V'}{\partial t} + V' \langle u_\rho.\nabla\rho\rangle \frac{\partial p_e}{\partial \rho}$$

$$+ \frac{\partial}{\partial \rho} [q_e + 5/2\, kT_e\Gamma_e)V'] = [\langle j.E\rangle - Q_\Delta - \langle u_i.\nabla p_i\rangle + \langle s_2\rangle]\, V'$$

which is none other than equation (VI.27).

Likewise by averaging (VI.25), multiplying by V' and using the relations (VI.18) and (VI.19) we obtain

$$\frac{3}{2} \frac{\partial}{\partial t} (p_i V') - \frac{3}{2} \frac{\partial}{\partial \rho} [V' \langle p_i\, u_\rho.\nabla\rho\rangle] + \frac{\partial}{\partial \rho} [V' \langle (Q_i + \frac{5}{2} p_i u_i).\nabla\rho\rangle]$$

$$= [Q_\Delta + \langle u_i.\nabla p_i\rangle + \langle s_3\rangle]\, V' \; .$$

Using (VI.20) as well as the property

$$\text{(VI.29)} \qquad \Gamma_i = \langle n_i (u_i - u_\rho).\nabla\rho\rangle = \frac{\Gamma_e}{Z}$$

we obtain

$$3/2 \frac{\partial}{\partial t} (p_i V') + p_i \frac{\partial V'}{\partial t} - V' \frac{\partial p_i}{\partial \rho} <(u_i - u_\rho).\nabla\rho>$$

$$+ \frac{\partial}{\partial \rho} [(q_i + 5/2 \frac{k\, T_i\, \Gamma_e}{Z})V'] = (Q_\Delta + <s_3>)V'$$

which gives (VI.28). □

VI.1.2.4 The Equations of Flux

In Chapter I we saw the physical significance of the poloidal flux ψ (see equation (I.8)). We now define the toroidal flux ϕ by

(VI.30) $$\phi = \int_{D'} B.dS = \frac{1}{2\pi} \int_V \frac{B_T}{r} dV = \frac{1}{2\pi} \int_V \frac{f}{r^2} dV$$

where D' is the domain bounded by the line of flux C' which is the intersection of the meridian section of the torus with the magnetic surface S, enclosing the volume V.

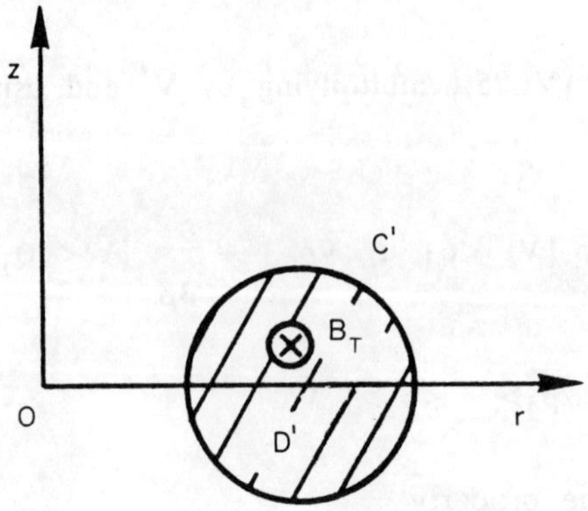

Further, we define the safety factor q by

(VI.31) $$q = - \frac{1}{2\pi} \frac{\partial \phi}{\partial \psi} .$$

The quantity q is the number of turns in the toroidal sense that a field line makes in going once around the magnetic axis in the poloidal sense. This number q is called the safety factor because for reasons of MHD stability it is necessary to have $q > 1$ (c.f. C. MERCIER).

From (VI.30) and (VI.31) we have

(VI.32) $$q = -\frac{1}{4\pi^2}\frac{\partial V}{\partial \psi} f \left\langle \frac{1}{r^2} \right\rangle .$$

We can then establish the following proposition:

<u>Proposition VI.4</u>

The evolution equations for ψ and ϕ are

(VI.33) $$\frac{\partial \psi}{\partial t} = (u_\rho - u_\psi).\nabla\psi$$

(VI.34) $$\frac{\partial \phi}{\partial t} = \frac{1}{2\pi}\frac{\partial V}{\partial \psi} \langle E.B \rangle - 2\pi q (u_\rho - u_\psi).\nabla\psi$$

where the velocity $\mathbf{u}_\psi$ of the surfaces of constant ψ is defined by $\dot{\psi} + \mathbf{u}_\psi.\nabla\psi = 0$. □

<u>Proof</u>

The time derivatives $\dot{\psi}$ of ψ at a fixed point and $\partial\psi/\partial t$ with ρ held fixed are related by:

$$\dot{\psi} = \frac{\partial \psi}{\partial t} + \frac{\partial \psi}{\partial \rho}\dot{\rho} .$$

Hence by using the definitions of $\mathbf{u}_\rho$ and $\mathbf{u}_\psi$ given in Propositions (VI.1) and (VI.4) we obtain

$$\frac{\partial \psi}{\partial t} = - u_\psi.\nabla\psi + \frac{\partial \psi}{\partial \rho} u_\rho.\nabla\rho$$

which can be rewritten in the form of equation (VI.33).

Let us differentiate the expression (VI.30) for ϕ with respect to time, keeping ρ fixed. We obtain

$$2\pi \frac{\partial\phi}{\partial t} = \int_V \frac{\dot{B}_T}{r}\, dV + \int_S \frac{B_T}{r} \frac{u_\rho . \nabla\rho}{|\nabla\rho|}\, dS \; . \tag{VI.35}$$

Faraday's Law states

$$\nabla \times E = - \dot{B} \tag{VI.36}$$

Projecting this equation onto the unit vector e_T and dividing by r gives

$$\nabla \times E . \frac{e_T}{r} = \nabla \cdot (E \times \frac{e_T}{r}) = - \frac{\dot{B}_T}{r} \; . \tag{VI.37}$$

The first term on the right hand side of (VI.35) is then

$$\begin{aligned} \int_V \frac{\dot{B}_T}{r}\, dV &= - \int_V \nabla . (E \times \frac{e_T}{r})\, dV \\ &= - \int_S (\frac{e_T}{r} \times E) \cdot \frac{\nabla\psi}{|\nabla\psi|}\, dS = \int_S \frac{E . B_P}{|\nabla\psi|}\, dS \; . \end{aligned} \tag{VI.38}$$

The second term on the right hand side of (VI.35) is

$$\begin{aligned} \int_S \frac{B_T}{r} \frac{u_\rho . \nabla\rho}{|\nabla\rho|}\, dS &= \int_S \frac{B_T}{r} \frac{u_\psi . \nabla\psi}{|\nabla\psi|}\, dS + \int_S \frac{B_T}{r} \frac{(u_\rho - u_\psi) \cdot \nabla\rho}{|\nabla\rho|}\, dS \\ &= \int_S \frac{B_T . E_T}{|\nabla\psi|}\, dS + \int_S \frac{f}{r^2} \frac{(u_\rho - u_\psi) . \nabla\rho}{|\nabla\rho|}\, dS \end{aligned} \tag{VI.39}$$

since $u_\psi . \nabla \psi = -\dot{\psi} = r\, E_T$.

From (VI.35), (VI.38) and (VI.39) we deduce

$$\begin{aligned} 2\pi \frac{\partial\phi}{\partial t} &= \int_S \frac{E . B}{|\nabla\psi|}\, dS + f(u_\rho - u_\psi) . \nabla\rho \int_S \frac{dS}{r^2 |\nabla\rho|} \\ &= \frac{\partial V}{\partial\psi} \langle E . B \rangle + f \frac{\partial V}{\partial\psi} (u_\rho - u_\psi) . \nabla\psi \langle \frac{1}{r^2} \rangle \; . \end{aligned}$$

Using the expression (VI.32) for q we obtain (VI.34). □

VI.1.2.5 The Averaged Equilibrium Equation.

The axisymmetric equilibrium equation (VI.10) is a 2–dimensional equation in the meridian plane section of the torus. By averaging this equation over each magnetic surface we obtain an ordinary differential equation in the variable ρ which will later enable us to calculate $f(\rho)$ and $V'(\rho)$ at each instant.

Proposition VI.5

The averaged Grad–Shafranov equation is

$$-\frac{\partial\psi}{\partial\rho}\frac{\partial}{\partial\rho}\left(C_2\frac{\partial\psi}{\partial\rho}\right) = \mu_0 V'\frac{\partial p}{\partial\rho} + \frac{C_3}{2}\frac{\partial f^2}{\partial\rho} \tag{VI.40}$$

with

$$C_2 = V'\left\langle\frac{|\nabla\rho|^2}{r^2}\right\rangle, \quad C_3 = V'\left\langle\frac{1}{r^2}\right\rangle. \qquad \square$$

Proof

The Grad–Shafranov equation (VI.10) can be rewritten

$$-\nabla\cdot\left(\frac{1}{\mu_0 r^2}\nabla\psi\right) = \frac{\partial p}{\partial\psi} + \frac{1}{2\mu_0 r^2}\frac{\partial f^2}{\partial\psi} \tag{VI.41}$$

where the differential operators $\nabla.$ and ∇ are here taken in (x,y,z)–space. Averaging (VI.41) and using (VI.18) we obtain

$$-\frac{1}{V'}\frac{\partial}{\partial\rho}\left[V'\left\langle\frac{1}{\mu_0 r^2}\nabla\psi\cdot\nabla\rho\right\rangle\right] = \frac{\partial p}{\partial\psi} + \frac{1}{2\mu_0}\left\langle\frac{1}{r^2}\right\rangle\frac{\partial f^2}{\partial\psi}$$

and hence

$$-\frac{\partial}{\partial\rho}\left(\frac{\partial\psi}{\partial\rho}V'\left\langle\frac{|\nabla\rho|^2}{r^2}\right\rangle\right) = \mu_0 V'\frac{\partial p}{\partial\psi} + \frac{V'}{2}\left\langle\frac{1}{r^2}\right\rangle\frac{\partial f^2}{\partial\psi}$$

which implies (VI.40). □

Remark VI.2

We deduce from the expression (VI.10) for the toroidal current density j_T in the plasma and from the averaged equilibrium equation (VI.40) that

$$<\frac{j_T}{r}> = -\frac{1}{\mu_0 V'}\frac{\partial}{\partial \rho}(C_2 \frac{\partial \psi}{\partial \rho}) \quad . \qquad \text{(VI.42)}$$ □

VI.1.2.6 The Choice of the Variable ρ

Define the quantities N_e, σ_e and σ_i by

$$\begin{cases} N_e = n_e \ V' \\ \sigma_e = p_e \ V'^{5/3} \\ \sigma_i = p_i \ V'^{5/3} \quad . \end{cases} \qquad \text{(VI.43)}$$

The quantity N_e is the total number of particles between two adjacent magnetic surfaces, while σ_e and σ_i are related to the entropy of the electrons and ions, respectively. It is clear from (VI.22), (VI.27) and (VI.28) that N_e, σ_e and σ_i remain invariant during adiabatic movements, i.e. movements that are sufficiently rapid for dissipation and transport phenomena to be negligible. In fact in these cases the equations are

$$\frac{\partial N_e}{\partial t} = \frac{\partial \sigma_e}{\partial t} = \frac{\partial \sigma_i}{\partial t} = 0 \ .$$

The quantities N_e, σ_e and σ_i are called adiabatic variables.

Equations (VI.22), (VI.27), (VI.28), (VI.33) and (VI.34) constitute a set of 1-dimensional evolution equations in the spatial variable ρ for the five unknown functions N_e, σ_e, σ_i, ψ and ϕ. We shall show that if we choose the variable ρ suitably this system reduces to a set of four diffusion equations.

The simplest idea is to take ψ to be the independent variable. This was the choice made by R.N. BYRNE – H.H. KLEIN in the code G2M, and by J.A. HOLMES – Y.K. PENG – S.J. LYNCH for the study of F.C.T. (Flux Conserving Tokamaks). The problem is that in conventional Tokamaks the poloidal flux ψ varies with the plasma current and with the inducing voltages in the primary. It therefore seems preferable to choose for ρ the normalized flux ψ_N defined by (I.21), which remains in the interval

[0,1]. This choice was made by D.E. SHUMAKER et al. for the study of reversed field mirrors and by A.D. TURNBULL – D.G. STORER in the cylindrical case.

However, in present day Tokamaks the toroidal flux ϕ varies much less than the poloidal flux ψ. In fact the toroidal field that is applied outside the plasma is large compared with the poloidal field and is constant throughout the discharge, so that the toroidal flux is almost invariant during the discharge. This is why a mean radius of the section of each magnetic surface is defined from the toroidal flux by

(VI.44) $$\rho = \sqrt{\frac{\phi}{\pi B_o}}$$

where B_o is a constant, namely the applied toroidal magnetic field at the centre of the vacuum vessel. This is the choice made by F.L.HINTON – R.D. HAZELTINE, and it has been used by J. BLUM – J. LE FOLL – B. THOORIS [1] in the code SCED and by R.L. MILLER for the study of Doublet configurations. We shall likewise adopt it in what follows, and rewrite the diffusion equations with ρ defined by (VI.44).

<u>Proposition VI.6</u>

If the variable ρ is defined by the relation (VI.44) then equation (VI.33) for the poloidal flux can be written

(VI.45) $$\frac{\partial \psi}{\partial t} - \frac{\eta_{\shortparallel}\rho}{\mu_o C_3^2}\frac{\partial}{\partial \rho}\left(\frac{C_2 C_3}{\rho}\frac{\partial \psi}{\partial \rho}\right) = 0$$

where $\eta_{\shortparallel}$ is the component of the resistivity tensor that is parallel to the magnetic surfaces.

Equation (VI.27) of conservation of energy for the electrons is

(VI.46) $$\frac{3}{2V'^{2/3}}\frac{\partial \sigma_e}{\partial t} + \frac{\partial}{\partial \rho}\left[\left(q_e + \frac{5}{2}k\,T_e\Gamma_e\right)V'\right] = -\frac{\Gamma_e V'}{n_e}\frac{\partial p_i}{\partial \rho}$$

$$+ \frac{\eta_{\shortparallel}\rho}{\mu_o^2 C_3^2}\frac{\partial}{\partial \rho}\left(C_2\frac{\partial \psi}{\partial \rho}\right)\frac{\partial}{\partial \rho}\left(\frac{C_2 C_3}{\rho}\frac{\partial \psi}{\partial \rho}\right) - Q_\Delta V' + \langle s_2\rangle V' \,.$$ □

Proof

From the definition (VI.44) of ρ we have $\partial\phi/\partial t = 0$ and so (VI.34) and (VI.32) give

(VI.47) $$(u_\rho - u_\psi)\cdot\nabla\psi = -\frac{<E\cdot B>}{f<\frac{1}{r^2}>} .$$

and (VI.33) gives

(VI.48) $$\frac{\partial\psi}{\partial t} = -\frac{<E\cdot B>}{f<\frac{1}{r^2}>} .$$

From Ohm's Law in a plasma (c.f. S.I.BRAGINSKII):

(VI.49) $$E + u \times B = \eta j$$

we can deduce

(VI.50) $$<E\cdot B> = \eta_{\parallel} <j\cdot B>$$

where $\eta_{\parallel}$ is the component of resistivity parallel to the magnetic surfaces.

Furthermore

$$<j\cdot B> = <j_P\cdot B_P> + <j_T\cdot B_T>$$

and from the expressions (I.11) and (I.12) for **B** and j, with the help of (VI.42) we have

$$<j_P\cdot B_P> = \frac{1}{\mu_0}\frac{\partial f}{\partial\rho}\frac{\partial\psi}{\partial\rho}<\frac{|\nabla\rho|^2}{r^2}>$$

$$<j_T\cdot B_T> = -\frac{f}{\mu_0 V'}\frac{\partial}{\partial\rho}(C_2\frac{\partial\psi}{\partial\rho})$$

Hence

(VI.51) $$<j\cdot B> = -\frac{f^2}{\mu_0 V'}\frac{\partial}{\partial\rho}(\frac{C_2}{f}\frac{\partial\psi}{\partial\rho}) .$$

By (VI.50) and (VI.51) we may write (VI.48) as

(VI.52) $$\frac{\partial\psi}{\partial t} - \frac{\eta_{\shortparallel} f}{\mu_0 C_3} \frac{\partial}{\partial\rho}\left(\frac{C_2}{f}\frac{\partial\psi}{\partial\rho}\right) = 0 \; .$$

From the definitions (VI.30) and (VI.44) of ϕ and ρ we have

$$\frac{\partial\phi}{\partial\rho} = \frac{fV'}{2\pi} \left\langle\frac{1}{r^2}\right\rangle = 2\pi\,\rho\,B_0 \; .$$

and hence

(VI.53) $$f = \frac{4\pi^2\,\rho\,B_0}{C_3} \; .$$

By replacing f by the expression (VI.53) we can now write equation (VI.52) in the form of the parabolic equation (VI.45) which characterizes the resistive diffusion of the poloidal flux ψ in terms of ρ.

We now calculate the Joule effect term $\langle j.E \rangle$. From (I.11) and (I.12) and the equilibrium equation (I.17) we deduce

(VI.54) $$j \cdot E = \frac{1}{\mu_0}\frac{\partial f}{\partial\psi} E \cdot B + \frac{\partial p}{\partial\psi} rE_T$$

Using the averaged equilibrium equation (VI.40) we have

(VI.55) $$\frac{\partial f}{\partial\psi} = -\frac{1}{C_3 f}\frac{\partial}{\partial\rho}\left(C_2\frac{\partial\psi}{\partial\rho}\right) - \frac{\mu_0}{f\left\langle\frac{1}{r^2}\right\rangle}\frac{\partial p}{\partial\psi}$$

and hence by (VI.54)

(VI.56) $$\langle j.E\rangle = -\frac{\langle E.B\rangle}{\mu_0 f C_3}\frac{\partial}{\partial\rho}\left(C_2\frac{\partial\psi}{\partial\rho}\right) + \frac{\partial p}{\partial\psi}\left[\langle rE_T\rangle - \frac{\langle E.B\rangle}{f\left\langle\frac{1}{r^2}\right\rangle}\right]$$

Now by (VI.8), (VI.47) and the definition of $\mathbf{u}_\psi$ we have

(VI.57) $$\langle rE_T\rangle - \frac{\langle E.B\rangle}{f\left\langle\frac{1}{r^2}\right\rangle} = \langle u_\rho . \nabla\psi\rangle$$

With (VI.50), (VI.51) and (VI.57) the expression (VI.56) then becomes

$$\langle j.E\rangle = \frac{\eta_{\shortparallel} f}{\mu_0^2 V' C_3}\frac{\partial}{\partial\rho}\left(C_2\frac{\partial\psi}{\partial\rho}\right)\frac{\partial}{\partial\rho}\left(\frac{C_2}{f}\frac{\partial\psi}{\partial\rho}\right) + \langle u_\rho \cdot \nabla p\rangle$$

or, using (VI.53)

$$(VI.58)\quad <j.E> = \frac{\eta'' \rho}{\mu_o^2 V' C_3^2} \frac{\partial}{\partial \rho} (C_2 \frac{\partial \psi}{\partial \rho}) \frac{\partial}{\partial \rho} (\frac{C_2 C_3}{\rho} \frac{\partial \psi}{\partial \rho}) + <u_\rho \cdot \nabla p>$$

From this we deduce

$$<j.E> - <u_\rho \cdot \nabla p_e> - <u_i \cdot \nabla p_i>$$

$$= \frac{\eta'' \rho}{\mu_o^2 V' C_3^2} \frac{\partial}{\partial \rho} (C_2 \frac{\partial \psi}{\partial \rho}) \frac{\partial}{\partial \rho} (\frac{C_2 C_3}{\rho} \frac{\partial \psi}{\partial \rho}) - <(u_i - u_\rho) \cdot \nabla p_i>$$

and from the expression (VI.29) for Γ_i the equation (VI.27) can be written in the form (VI.46). □

Equations (VI.22) and (VI.28) in the unknowns N_e and σ_i remain unchanged, so that the system of equations (VI.22), (VI.46), (VI.28) and (VI.45) constitute a set of four one–dimensional diffusion equations in ρ for the unknowns N_e, σ_e, σ_i and ψ.

To ensure that this is a closed system, the expressions for the electronic particle flux Γ_e, the heat fluxes q_e, q_i and the resistivity η'' have to be given as functions of N_e, σ_e, σ_i and ψ. This is done in F.L. HINTON – R.D. HAZELTINE for the neo–classical theory of transport : the fluxes Γ_e, q_e and q_i are quasilinear combinations of $\partial n_e/\partial \rho$, $\partial p_e/\partial \rho$, $\partial p_i/\partial \rho$ and $<E.B>$. By (VI.50), (VI.51) there is a second derivative of ψ appearing in $<E.B>$. To obtain a set of 2nd order partial differential equations in ρ we therefore have to take $\psi' = \partial \psi/\partial \rho$ as the unknown instead of ψ. The equation for ψ' is obtained by differentiating (VI.45) with respect to ρ:

$$(VI.59)\qquad \frac{\partial \psi'}{\partial t} - \frac{1}{\mu_o} \frac{\partial}{\partial \rho} [\frac{\eta'' \rho}{C_3^2} \frac{\partial}{\partial \rho} (\frac{C_2 C_3}{\rho} \psi')] = 0 .$$

The system of equations (VI.22),(VI.46),(VI.28) and (VI.59) is the system of equations of transport for the unknowns N_e, σ_e, σ_i and ψ' with respect to the variable ρ. It is convenient to adjoin also the differential equation (VI.40) and the relation (VI.53) in order to calculate $f(\rho)$ and $V'(\rho)$.

Remark VI.3

The rotational transform χ is defined as being the inverse of the safety factor q. The derivative ψ' of ψ with respect to ρ is then related to χ by the expression

$$\psi' = - B_0 \rho \chi .$$

From (VI.59) the diffusion equation for χ is :

$$\text{(VI.60)} \qquad \frac{\partial \chi}{\partial t} - \frac{1}{\mu_0 \rho} \frac{\partial}{\partial \rho} \left[\frac{\eta'' \rho}{C_3^2} \frac{\partial}{\partial \rho} (C_2 C_3 \chi) \right] = 0 .$$

□

Remark VI.4: The Adiabatic Case.

The relations (VI.59) and (VI.60) show that ψ' and hence χ and q are preserved during adiabatic movements. The variables σ_e and σ_i are likewise adiabatic variables, and we deduce that

$$\text{(VI.61)} \qquad \pi(\psi) = p(\psi) \left(\frac{\partial V}{\partial \psi}\right)^{5/3}$$

is an adiabatic quantity. From (VI.61) we have

$$\text{(VI.62)} \qquad \frac{\partial p}{\partial \psi} = \frac{\partial \pi}{\partial \psi} \left(\frac{\partial \psi}{\partial V}\right)^{5/3} + 5/3\ \pi(\psi) \left(\frac{\partial \psi}{\partial V}\right)^{-1/3} \frac{\partial^2 \psi}{\partial V^2}$$

and from (VI.32) we deduce

$$\text{(VI.63)} \qquad \frac{1}{2} \frac{\partial f^2}{\partial \psi} = \frac{16\pi^4 q(\psi)}{<r^{-2}>} \left[\left(\frac{\partial \psi}{\partial V}\right)^2 \frac{\partial}{\partial \psi} \left(\frac{q}{<r^{-2}>}\right) + \frac{q}{<r^{-2}>} \frac{\partial^2 \psi}{\partial V^2} \right]$$

Since $\pi(\psi)$ and $q(\psi)$ are given adiabatic functions, the Grad–Shafranov equation (VI.10) then takes the form:

$$\text{(VI.64)} \qquad L\psi = F\left(r, \psi, \frac{\partial \psi}{\partial V}, \frac{\partial^2 \psi}{\partial V^2}\right)$$

This type of equation, valid for adiabatic movements of the plasma, was introduced by H. GRAD – P.N. HU – D.C. STEVENS under the name "generalized differential equation" and studied mathematically in C. VIGFUSSON, R. TEMAM [3] and

J. MOSSINO – R. TEMAM where the notion of monotone rearrangement of the function $\psi(V)$ is used. □

Remark VI.5

Certain authors (F.J. HELTON – R.L. MILLER – J.M. RAWLS, J.T. HOGAN) have chosen as ρ the actual mean radius of the section of a magnetic surface, defined from the volume by

$$\rho = \sqrt{\frac{V}{2\pi^2 R_o}} \tag{VI.65}$$

The volume V is not an adiabatic variable, so that in J.T. HOGAN each phase of the evolution is separated into two stages: a stage in which the geometry of the magnetic surfaces is fixed and where ρ defined by (VI.65) is taken as a spatial variable in the system of diffusion equations, and an adiabatic stage in which a new equilibrium is computed while the adiabatic variables N_e, σ_e, σ_i and q are kept constant. □

VI.1.2.7. An Empirical Model

Since the neoclassical theory of transport is rendered invalid by experiment as far as the diffusion of electrons is concerned, we use an empirical model which limits itself to transport coefficients that are orthogonal to the magnetic surfaces. This "diagonal" model consists of assuming that the particle flux and heat flux are proportional to the gradients of n_e, T_e and T_i. These latter quantities are taken to be constant on the magnetic surfaces, so the particle flux and heat flux are orthogonal to the lines of flux. More precisely we assume

$$\begin{cases} n_e(u_e - u_\rho) = -D\,\nabla n_e \\ Q_e = -K_e\,\nabla T_e \\ Q_i = -K_i\,\nabla T_i \end{cases} \tag{VI.66}$$

where D is the coefficient of electron diffusion, and K_e, K_i are the electronic and ionic thermal conductivities, respectively. We then have:

(VI.67)
$$\begin{cases} \Gamma_e = -D \langle \nabla^2 \rho \rangle \dfrac{\partial n_e}{\partial \rho} \\ q_e = -K_e \langle \nabla^2 \rho \rangle \dfrac{\partial T_e}{\partial \rho} \\ q_i = -K_i \langle \nabla^2 \rho \rangle \dfrac{\partial T_i}{\partial \rho} \end{cases}$$

<u>Proposition VI.7</u>

Under the hypothesis (VI.66) the equations (VI.22), (VI.46) and (VI.28) for conservation of particles and energy can be written as follows.

(VI.68)
$$\frac{\partial N_e}{\partial t} - \frac{\partial}{\partial \rho}(DC_1 \frac{\partial n_e}{\partial \rho}) = \langle s_1 \rangle V'$$

$$\frac{3}{2V'^{2/3}} \frac{\partial \sigma_e}{\partial t} - \frac{\partial}{\partial \rho}(K_e C_1 \frac{\partial T_e}{\partial \rho}) - \frac{5k}{2} \frac{\partial}{\partial \rho}(DC_1 T_e \frac{\partial n_e}{\partial \rho})$$

(VI.69)
$$= \frac{\eta'' \rho}{\mu_o^2 C_3^2} \frac{\partial}{\partial \rho}(C_2 \psi') \frac{\partial}{\partial \rho}(\frac{C_2 C_3}{\rho} \psi') + \frac{DC_1}{n_e} \frac{\partial n_e}{\partial \rho} \frac{\partial p_i}{\partial \rho}$$

$$- Q_\Delta V' + \langle s_2 \rangle V'$$

(VI.70)
$$\frac{3}{2V'^{2/3}} \frac{\partial \sigma_i}{\partial t} - \frac{\partial}{\partial \rho}(K_i C_1 \frac{\partial T_i}{\partial \rho}) - \frac{5k}{2} \frac{\partial}{\partial \rho}(\frac{DC_1 T_i}{Z} \frac{\partial n_e}{\partial \rho})$$

$$+ \frac{DC_1}{n_e} \frac{\partial n_e}{\partial \rho} \frac{\partial p_i}{\partial \rho} = Q_\Delta V' + \langle s_3 \rangle V' \quad (\text{with } C_1 = V' \langle \nabla^2 \rho \rangle).$$ □

The proof of this proposition follows trivially from (VI.67).

The set of equations (VI.68), (VI.69), (VI.70) and (VI.59) constitutes the system of diffusion equations for N_e, σ_e, σ_i and ψ' as functions of ρ. It is necessary to add the averaged equilibrium equation (VI.40) and the relation (VI.53) for the functions $f(\rho)$ and $V'(\rho)$.

Remark VI.6

The transport coefficients corresponding to collision theory have been computed in E.K. MASCHKE, assuming the magnetic field to be stationary; their validity in the context of the study of evolution of equilibrium has been analyzed in E.K. MASCHKE – J. PANTUSO SUDANO or in D.B. NELSON – H. GRAD. These coefficients involve the averages over each magnetic surface of the quantities j.B, B^2 and $1/B^2$. The non–uniformity of n_e, T_e, T_i on the magnetic surfaces contributes to the computation of the heat fluxes for the collision theory.

The transport coefficients for the neoclassical theory are given in the survey article by F.L. HINTON – R.D. HAZELTINE in the different regimes (collision, banana, plateau). □

VI.1.2.8. Boundary Conditions

The variable ρ belongs to the interval $[0,\rho_{max}]$ where ρ_{max} denotes the value of ρ on the plasma boundary. Equations (VI.68), (VI.69), (VI.70) and (VI.59) degenerate for $\rho = 0$ because $V'(0) = 0$ and so $C_1(0) = C_2(0) = C_3(0) = 0$.

For $\rho = \rho_{max}$ we impose the conditions

$$\text{(VI.71)} \qquad \begin{cases} n_e(\rho_{max}) = n_e^o \\ T_e(\rho_{max}) = T_e^o \\ T_i(\rho_{max}) = T_i^o \end{cases}$$

where n_e^o, T_e^o, T_i^o are given.

By integrating (VI.42) over the whole plasma we obtain :

$$I_p = \int_{\Omega_p} j_T \, dS = \frac{1}{2\pi} \int_{V_p} \langle \frac{j_T}{r} \rangle \, dV = - \frac{1}{2\pi\mu_o} \left(C_2 \frac{\partial\psi}{\partial\rho}\right)(\rho_{max})$$

where V_p denotes the plasma volume. Therefore

(VI.72) $$\psi'(\rho_{max}) = - \frac{2\pi\mu_0 I_p}{C_2(\rho_{max})}$$

In the case when the total plasma current I_p is given, the relation (VI.72) is the boundary condition for the equation (VI.59). We shall see in the next Section how to use this condition for the equation for ψ' in the case when $I_p(t)$ is an unknown in the problem and it is the voltages $V_i(t)$ applied to the external circuits that are given.

Finally, (VI.40) is a first order equation for f and its boundary condition is

(VI.73) $$f(\rho_{max}) = R_0 B_0$$

VI.1.3. Coupling Between the Systems of Equilibrium Equations and Transport Equations

Denote by **Eq** the system of equilibrium equations written in variational form (VI.15), and by **Dif** the system of diffusion equations (VI.68), (VI.69), (VI.70), (VI.59), to which the average equilibrium equation (VI.40) and the relation (VI.53) are added.

The Grad–Shafranov equation included in **Eq** requires knowledge of the functions $\partial p/\partial\psi$ and $\partial f^2/\partial\psi$ which are deduced from **Dif**. Conversely, **Dif** requires knowledge of the geometric coefficients $<|\nabla\rho|^2>$, $<|\nabla\rho|^2/r^2>$ and $<r^{-2}>$ which are deduced from the solution of the equilibrium equations **Eq**. In the case when the plasma boundary is fixed and identified with a superconducting shell ($\psi = 0$), and the total plasma current I_p is given, the coupling between the systems **Dif** and **Eq** reduces to these elements.

However, if the data in the problem are the voltages $V_i(t)$ applied to the external circuits, then the plasma boundary is a free boundary which evolves with time, and the time–variation of the plasma current $I_p(t)$ is an unknown. In this case we have to consider a compatibility condition between the two systems, namely that the values of I_p and of ρ_{max} should agree:

(VI.74) $$\begin{cases} (I_p)_{Dif} = (I_p)_{Eq} \\ \\ (\rho_{max})_{Dif} = (\rho_{max})_{Eq} \end{cases}$$

The coupling between the two systems can be visualized as follows :

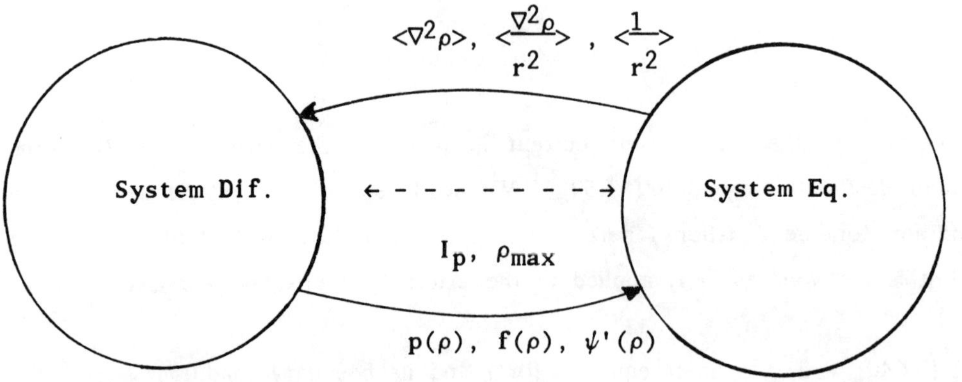

VI.2. NUMERICAL METHODS

VI.2.1. The System Eq.

We aim to solve the system (VI.15) numerically. To do this we use the finite element method defined in Section I.4.1 with an implicit time scheme and a Newton or quasi– Newton iteration at each time step.

If we denote by ψ^n the function ψ at the instant $n\Delta t$, belonging to the finite dimensional space V_ℓ defined in Section I.4.1, an implicit time scheme consists of the computation of $\psi^{n+1} \in V_\ell$ from ψ^n as follows

$$a_{\mu^{n+1}}(\psi^{n+1},\varphi) + \sum_{i=1}^{k} \frac{n_i}{R_i\, S_i^2} \int_{B_i} \frac{\psi^{n+1}-\psi^n}{\Delta t}\, dS \int_{B_i} \varphi dS$$

$$\text{(VI.75)} \qquad + \int_{\Omega_{cv}} \frac{\sigma_V}{r} \frac{\psi^{n+1}-\psi^n}{\Delta t}\, \varphi\, dS = \int_{\Omega_p^{n+1}} j_T(r,\psi^{n+1})\varphi\, dS$$

$$+ \sum_{i=1}^{k} \frac{V_i^{n+1}}{R_i\, S_i} \int_{B_i} \varphi\, dS\ ,\ \forall\varphi \in V_\ell$$

where

$$\Omega_p^{n+1} = \{M \in \Omega_V \mid \psi^{n+1}(M) > \sup_D \psi^{n+1}\}$$

Since this system is nonlinear we shall use the Newton linearization method as in Section I.4.4, and at each time step carry out a single Newton iteration so that instead of solving (VI.75) we have to solve the following linear system:

$$\sum_{i=1}^{k} \frac{n_i}{R_i S_i^2} \int_{B_i} \psi^{n+1}\, dS \int_{B_i} \varphi\, dS + \int_{\Omega_{cv}} \frac{\sigma_v}{r} \psi^{n+1} \varphi\, dS$$

$$+ \Delta t \Big[a_{\mu^n} (\psi^{n+1}, \varphi) + a'_{u^n} (\psi^{n+1}, \varphi)$$

$$- \int_{\Omega_p^n} j'_T(r, \psi^n) \psi^{n+1} \varphi\, dS - \int_{\Gamma_p^n} j_T(r, \psi^n) \frac{\psi^{n+1}(M_o^n) - \psi^{n+1}}{\frac{\partial \psi^n}{\partial n}} \varphi\, d\Gamma \Big]$$

(VI.76)

$$= \sum_{i=1}^{k} \frac{n_i}{R_i S_i^2} \int_{B_i} \psi^n\, dS \int_{B_i} \varphi\, dS + \int_{\Omega_{cv}} \frac{\sigma_v}{r} \psi^n \varphi\, dS$$

$$+ \Delta t \Big[a'_{u^n} (\psi^n, \varphi) + \int_{\Omega_p^n} j_T(r, \psi^n) \varphi\, dS$$

$$+ \sum_{i=1}^{k} \frac{V_i^{n+1}}{R_i S_i} \int_{B_i} \varphi\, dS - \int_{\Omega_p^n} j'_T(r, \psi^n) \psi^n \varphi\, dS \Big] \quad , \forall \varphi \in V_\ell$$

with

$$a'_{u^n}(\psi, \varphi) = - \int_{\Omega_f} \frac{2\mu'^n}{(\mu^n)^2 r^3} (\nabla\psi^n \cdot \nabla\varphi)(\nabla\psi^n \cdot \nabla\psi)\, dS$$

and

$$j'_T(r, \psi) = r \frac{\partial^2 p}{\partial \psi^2} + \frac{1}{2\mu_o r} \frac{\partial^2 f^2}{\partial \psi^2} \ .$$

The method used consists therefore of calculating at each time step the trajectory tangent to the evolution of $\psi(r,z,t)$ at the instant $n\Delta t$, and deducing the function ψ at the instant $(n+1)\Delta t$ by inverting this linearized system.

In order to gain machine time a quasi–Newton method can also be used (cf. Section I.4.4.2), that is to compute the Jacobian matrix of (VI.76) only when the linearized system has changed in some significant way.

Remark VI.7

S.P. HIRSHMAN – S.C. JARDIN differentiate the Grad – Shafranov equation with respect to time to obtain an equation for the speed of the surfaces of constant ϕ. A.D. TURNBULL – R.G. STORER obtain an equation for (rE_T) by differentiating the equilibrium equation. In these two papers an equation obtained by differentiating the equations of state is therefore used. This can be compared with the method used here, which consists of computing the trajectory tangent to the evolution system. The aim is essentially the same : to solve a linear system at each time step, while the equations of state are nonlinear. □

VI.2.2. The System Dif

Here we use a fractional step method (c.f. YANENKO) to solve the diffusion equations. We shall solve the equations for N_e, σ_e, σ_i and ψ' successively by an implicit scheme in time and a discretization by centred finite differences in space. The variable belongs to the interval $[0,\rho_{max}]$: here ρ_{max} varies with time. We thus use the reduced variable $\bar{\rho} = \rho/\rho_{max}$ which belongs to [0,1], and it is therefore appropriate in equations (VI.68), (VI.69), (VI.70) and (VI.59) to replace $\partial/\partial\rho$ by $(1/\rho_{max})\partial/\partial\bar{\rho}$, and $\partial/\partial t$ by $\partial/\partial t - (\bar{\rho}/\rho_{max})(d\rho_{max}/dt)\partial/\partial\bar{\rho}$. A similar normalization has been carried out by G. CENACCHI – A. TARONI for the code GETTO.

We now write down the discretization of each of these equations.

Equation for N_e :

If $(N_e)_j^n$ corresponds to the value of N_e for $\bar{\rho} = (j- 1/2)\Delta\rho$ and $t = n\Delta t$, a discretization of (VI.68) can be written :

$$\frac{(N_e)_j^{n+1}-(N_e)_j^n}{\Delta t} - \frac{(j-\frac{1}{2})}{\rho^n_{max}} \left(\frac{d\rho_{max}}{dt}\right) \frac{(N_e)_{j+1}^{n+1}-(N_e)_{j-1}^{n+1}}{2}$$

(VI.77)
$$- \frac{1}{(\rho^n_{max}\Delta\rho)^2} \left\{ (DC_1)^n_{j+1/2}\left[\frac{(N_e)^{n+1}_{j+1}}{V'^n_{j+1}} - \frac{(N_e)^{n+1}_j}{V'^n_j}\right]\right.$$

$$\left. - (DC_1)^n_{j-1/2}\left[\frac{(N_e)^{n+1}_j}{V'^n_j} - \frac{(N_e)^{n+1}_{j-1}}{V'^n_{j-1}}\right]\right\} = [\langle s_1\rangle\ V']^n_j$$

<u>Equations for</u> σ_e <u>and</u> σ_i :

Define $\tau_e = T_e V'^{2/3}$ and $\tau_i = T_i V'^{2/3}$ so that $\sigma_e = N_e k\tau_e$ and $\sigma_i = N_i k\tau_i$. Equation (VI.69) is discretized as follows for the $(\tau_e)^n{}_j$:

$$\frac{3k}{2(V'^n_j)^{2/3}} \left[\frac{(N_e)^{n+1}_j(\tau_e)^{n+1}_j - (N_e)^n_j(\tau_e)^n_j}{\Delta t}\right.$$

$$\left. - \frac{(j-1/2)}{\rho^n_{max}} \frac{d\rho_{max}}{dt} \frac{(N_e)^{n+1}_{j+1}(\tau_e)^{n+1}_{j+1} - (N_e)^{n+1}_{j-1}(\tau_e)^{n+1}_{j-1}}{2}\right]$$

$$- \frac{1}{(\rho^n_{max}\Delta\rho)^2} \left\{ (K_eC_1)^n_{j+1/2} \left[\frac{(\tau_e)^{n+1}_{j+1}}{(V'^n_{j+1})^{2/3}} - \frac{(\tau_e)^{n+1}_j}{(V'^n_j)^{2/3}}\right]\right.$$

(VI.78)
$$\left. - (K_eC_1)^n_{j-1/2} \left[\frac{(\tau_e)^{n+1}_j}{(V'^n_j)^{2/3}} - \frac{(\tau_e)^{n+1}_{j-1}}{(V'^n_{j-1})^{2/3}}\right]\right\} - \frac{5k}{4(\rho^n_{max}\Delta\rho)^2}$$

$$\left\{ (DC_1)^n_{j+1/2} \left[\frac{(\tau_e)^{n+1}_{j+1}}{(V'^n_{j+1})^{2/3}} + \frac{(\tau_e)^{n+1}_j}{(V'^n_j)^{2/3}}\right]\left[\frac{(N_e)^{n+1}_{j+1}}{V'^n_{j+1}} - \frac{(N_e)^{n+1}_j}{V'^n_j}\right]\right.$$

$$\left. - (DC_1)^n_{j-1/2} \left[\frac{(\tau_e)^{n+1}_j}{(V'^n_j)^{2/3}} + \frac{(\tau_e)^{n+1}_{j-1}}{(V'^n_{j-1})^{2/3}}\right]\left[\frac{(N_e)^{n+1}_j}{V'^n_j} - \frac{(N_e)^{n+1}_{j-1}}{V'^n_{j-1}}\right]\right\}$$

$$= \frac{\eta^n_j (j-1/2)}{4\mu_o^2 [(C_3)^n_j]^2 (\rho^n_{max}\Delta\rho)^2} \left[(C_2)^n_{j+1}\psi'^n_{j+1} - (C_2)^n_{j-1}\psi'^n_{j-1} \right]$$

$$\left[\frac{(C_2C_3)^n_{j+1}\psi'^n_{j+1}}{j+1/2} - \frac{(C_2C_3)^n_{j-1}\psi'^n_{j-1}}{j-3/2} \right]$$

$$+ \frac{(DC_1V')^n_j}{4(N_e)^{n+1}_j (\rho^n_{max}\Delta\rho)^2} \left[\frac{(\sigma_i)^n_{j+1}}{(V'^n_{j+1})^{5/3}} - \frac{(\sigma_i)^n_{j-1}}{(V'^n_{j-1})^{5/3}} \right] \left[\frac{(N_e)^{n+1}_{j+1}}{V'^n_{j+1}} - \frac{(N_e)^{n+1}_{j-1}}{V'^n_{j-1}} \right]$$

$$- \frac{3m_e}{m_i} \frac{(N_e)^{n+1}_j}{(t_e)^n_j} \left[(\tau_e)^{n+1}_j - (\tau_i)^n_j \right] (V'^n_j)^{-2/3} + \langle s_2 \rangle^n_j V'^n_j$$

In the same way we deduce the equation for the $(\tau_i)^n_j$ from the discretization of (VI.70).

$$\frac{3k}{2Z(V'^n_j)^{2/3}} \left[\frac{(N_e)^{n+1}_j(\tau_i)^{n+1}_j - (N_e)^n_j(\tau_i)^n_j}{\Delta t} \right.$$

$$\left. - \frac{(j-1/2)}{\rho^n_{max}} \frac{d\rho_{max}}{dt} \frac{(N_e)^{n+1}_{j+1}(\tau_i)^{n+1}_{j+1} - (N_e)^{n+1}_{j-1}(\tau_i)^{n+1}_{j-1}}{2} \right]$$

$$- \frac{1}{(\rho^n_{max}\Delta\rho)^2} \left\{ (K_iC_1)^n_{j+1/2} \left[\frac{(\tau_i)^{n+1}_{j+1}}{(V'^n_{j+1})^{2/3}} - \frac{(\tau_i)^{n+1}_j}{(V'^n_j)^{2/3}} \right] \right.$$

(VI.79)

$$\left. - (K_iC_1)^n_{j-1/2} \left[\frac{(\tau_i)^{n+1}_j}{(V'^n_j)^{2/3}} - \frac{(\tau_i)^{n+1}_{j-1}}{(V'^n_{j-1})^{2/3}} \right] \right\} - \frac{5k}{4Z(\rho^n_{max}\Delta\rho)^2}$$

$$\left\{ (DC_1)^n_{j+1/2} \left[\frac{(\tau_i)^{n+1}_{j+1}}{(V'^n_{j+1})^{2/3}} + \frac{(\tau_i)^{n+1}_j}{(V'^n_j)^{2/3}} \right] \left[\frac{(N_e)^{n+1}_{j+1}}{V'^n_{j+1}} - \frac{(N_e)^{n+1}_j}{V'^n_j} \right] \right.$$

$$- (DC_1)^n_{j-1/2} \left[\frac{(\tau_i)^{n+1}_j}{(V'^n_j)^{2/3}} + \frac{(\tau_i)^{n+1}_{j-1}}{(V'^n_{j-1})^{2/3}} \right] \left[\frac{(N_e)^{n+1}_j}{V'^n_j} - \frac{(N_e)^{n+1}_{j-1}}{V'^n_{j-1}} \right]$$

$$+ \frac{k(DC_1V')^n_j}{4Z(N_e)^{n+1}_j(\rho^n_{max}\,\Delta\rho)^2}\left[\frac{(N_e)^{n+1}_{j+1}}{V'^n_{j+1}} - \frac{(N_e)^{n+1}_{j-1}}{V'^n_{j-1}}\right]\left[\frac{(N_e\tau_i)^{n+1}_{j+1}}{(V'^n_{j+1})^{5/3}} - \frac{(N_e\tau_i)^{n+1}_{j-1}}{(V'^n_{j-1})^{5/3}}\right]$$

$$= \langle s_3\rangle^n_j\, V'^n_j + \frac{3m_e}{m_i}\,\frac{(N_e)^{n+1}_j}{(t_e)^n_j}\,\frac{(\tau_e)^{n+1}_j - (\tau_i)^{n+1}_j}{(V'^n_j)^{2/3}}$$

Equation for ψ':

Equation (VI.59) can be discretized as follows :

$$\frac{\psi'^{n+1}_j - \psi'^n_j}{\Delta t} - \frac{(j-1/2)}{2\rho^n_{max}}\left(\frac{d\rho_{max}}{dt}\right)(\psi'^{n+1}_{j+1} - \psi'^{n+1}_{j-1})$$

(VI.80)

$$- \frac{1}{\mu_0(\rho^n_{max}\Delta\rho)^2}\left\{\frac{\eta^n_{j+1/2}\,j}{[(C_3)^n_{j+1/2}]^2}\left[\frac{(C_2C_3)^n_{j+1}\,\psi'^{n+1}_{j+1}}{j+1/2} - \frac{(C_2C_3)^n_j\,\psi'^{n+1}_j}{j-1/2}\right]\right.$$

$$\left. - \frac{\eta^n_{j-1/2}(j-1)}{[(C_3)^n_{j-1/2}]^2}\left[\frac{(C_2C_3)^n_j\,\psi'^{n+1}_j}{j-1/2} - \frac{(C_2C_3)^n_{j-1}\psi'^{n+1}_{j-1}}{j-3/2}\right]\right\} = 0$$

The variable j goes from 1 to N, where N is the number of steps in the spatial discretization. The discretization schemes (VI.77), (VI.78), (VI.79) and (VI.80) have to be modified for j = 1 and j = N to take account of the fact that C_1, C_2 and C_3 vanish on the magnetic axis ($\bar{\rho} = 0$) and of the boundary conditions (VI.71) on the plasma boundary ($\bar{\rho} = 1$). For the equation for ψ' we consider the boundary condition (VI.72) if I_p is given, and in the case when it is the voltages in the external circuits that are given we use the conditions (VI.74). We shall return to this last point in the next Section. Solving each of these four discretized equations therefore comes down to solving a tridiagonal N $\times$ N system.

Equations for f and V'

It remains now to calculate $f(\rho)$ and $V'(\rho)$ by means of the equation (VI.40) and the relation (VI.53).

Equation (VI.40), discretized as follows, enables the f_j^{n+1} to be computed:

$$\frac{(C_3)^n_{j+\frac{1}{2}}}{2}\left[\left[f^{n+1}_{j+1}\right]^2-\left[f^{n+1}_{j}\right]^2\right] = -\frac{\psi'^{n+1}_{j+1}+\psi'^{n+1}_{j}}{2}\left[(C_2)^n_{j+1}\,\psi'^{n+1}_{j+1}-(C_2)^n_j\,\psi'^{n+1}_j\right]$$

(VI .81)

$$-\mu_0 k\,\frac{V'^n_{j+1}+V'^n_j}{2}\left\{\frac{(N_e)^{n+1}_{j+1}\left[(\tau_e)^{n+1}_{j+1}+\frac{(\tau_i)^{n+1}_{j+1}}{Z}\right]}{\left[V'^n_{j+1}\right]^{5/3}}-\frac{(N_e)^{n+1}_{j}\left[(\tau_e)^{n+1}_{j}+\frac{(\tau_i)^{n+1}_{j}}{Z}\right]}{\left[V'^n_{j}\right]^{5/3}}\right\}$$

with the boundary condition (VI.73): $f(1) = R_0B_0$.

The $V_j'^{n+1}$ can be computed from (VI.53) by:

(VI.82)
$$V'^{n+1}_j = \frac{4\pi^2 B_0 (j-1/2)\rho^n_{max}\,\Delta\rho}{f^{n+1}_j\left\langle\frac{1}{r^2}\right\rangle^n_j}$$

Since (VI.81) depends on V' (via C_2 and C_3 in particular) we iterate by a fixed point algorithm between (VI.81) and (VI.82) until the f_j^{n+1} and $V_j'^{n+1}$ have converged.

In the discretizations (VI.77) – (VI.82) certain nonlinearities have been taken explicitly, i.e. at the instant $n\Delta t$. A predictor–corrector algorithm could be used to obtain a discretization that is purely implicit in time. The discretizations (VI.77)–(VI.82) constitute the predictor of the method; the correctors consist of solving these equations once again with the explicit terms replaced by their implicit values obtained at the previous iteration.

Remark VI.8

A method that is different from the fractional step method consists of solving simultaneously the four diffusion equations; this method is more implicit and leads to the solution of a tridiagonal block system. It was used by R.N. BYRNE – H.H. KLEIN, S.C. JARDIN [2], D.E. SHUMAKER et al.

The nonlinearities can be treated by a method of predictor corrector type as above. However, it is possible also to resort to a Newton method simultaneously on the diffusion equations and the averaged equilibrium equation as in H.H. KLEIN – R.N. BYRNE. □

VI.2.3 *Coupling of the Systems Eq and Dif*

If G denotes the vector (N_e, σ_e, σ_i, ψ', f) the system **Dif** can be written symbolically as follows :

$$K \frac{\partial G}{\partial t} - \frac{\partial}{\partial \rho} \left(K' \frac{\partial G}{\partial \rho}\right) + K'' \frac{\partial G}{\partial \rho} + K'''G = S' \qquad \text{(VI.83)}$$

where K is a diagonal matrix (with K(5,5) = 0), and K', K", K"' are the 5 × 5 matrices of transport coefficients that depend nonlinearly on G. These coefficients depend also on ψ via the geometrical coefficients $<|\nabla\rho|^2>$, $<|\nabla\rho|^2/r^2>$ and $<1/r^2>$.

The system **Eq** can be described as

$$A\dot{\psi} + L\psi = B(\psi, G) \qquad \text{(VI.84)}$$

where A is a linear operator (identically zero on the whole of Ω except in the coils and the vacuum vessel), L is the nonlinear elliptic operator defined by (I.13) and B is a function of ψ depending on G via p and f .

Using the predictor–corrector method of Section VI.2.2, the system (VI.83) is discretized in time implicitly for G and explicitly for ψ as:

$$K(G^{n+1})\frac{G^{n+1}-G^{n}}{\Delta t} - \frac{\partial}{\partial\rho}\left[K'(G^{n+1},\psi^{n})\frac{\partial G^{n+1}}{\partial\rho}\right]$$

(VI.85)

$$+ K''(G^{n+1},\psi^{n})\frac{\partial G^{n+1}}{\partial\rho} + K'''(G^{n+1},\psi^{n})G^{n+1} = S'^{n+1}$$

By the method of Section VI.2.1 the system (VI.84) is discretized in time as follows:

$$A\frac{\psi^{n+1}-\psi^{n}}{\Delta t} + (L^{n}+L'^{n})\psi^{n+1} - B'(\psi^{n},G^{n+1})\psi^{n+1}$$

(VI.86)

$$= B(\psi^{n},G^{n+1}) - B'(\psi^{n},G^{n+1})\psi^{n} + L'^{n}\psi^{n}$$

where the operator L' is defined by (I.71) and B' is the derivative of B with respect to ψ.

The algorithm (VI.85), (VI.86) consists of advancing the system **Dif** implictly at each time step, and then carrying out the same time step $[n\Delta t, (n+1)\Delta t]$ for the system **Eq** using the profiles of p and f obtained at the instant $(n+1)\Delta t$ by the system **Dif**. The coupling between the two systems is thus explicit.

For solving the problem with fixed boundary and given I_p this algorithm is stable, but if we wish to solve the free boundary problem with voltages in the external circuits as data, the algorithm is unstable and it is necessary to couple the systems **Dif** and **Eq** implicitly in order to satisfy the consistency conditions (VI.74). This comes down to replacing ψ^n by ψ^{n+1} in (VI.85) and we then have to iterate between (VI.85) and (VI.86) at each time step.

More precisely, since we are working on a linearized system we have to solve the system **Dif** and the system **Eq** four times at each time step in order to satisfy the two consistency relations (VI.74).

VI.2.4 ***Flowchart***

We can summarize the structure of such an evolutive code as this by means of the flow diagram:

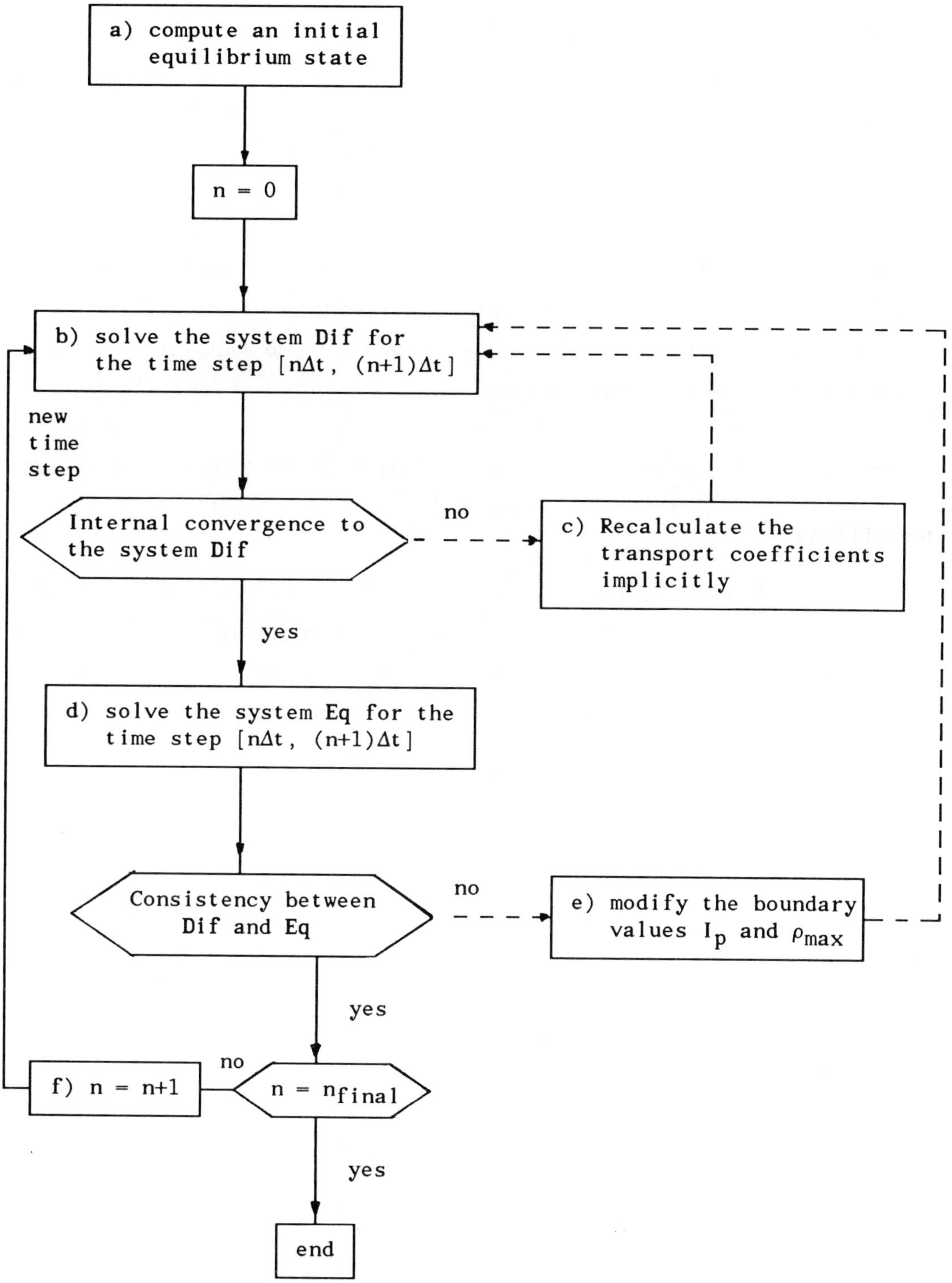
a) compute an initial equilibrium state
n = 0
b) solve the system Dif for the time step [nΔt, (n+1)Δt]
new time step
Internal convergence to the system Dif
no
c) Recalculate the transport coefficients implicitly
yes
d) solve the system Eq for the time step [nΔt, (n+1)Δt]
Consistency between Dif and Eq
no
e) modify the boundary values I_p and ρ_{max}
yes
no
f) n = n+1
n = n_{final}
yes
end

Its stages are as follows:

(a) compute an initial equilibrium state and the functions $n_e(\rho,0)$, $T_e(\rho,0)$, $T_i(\rho,0)$, $\psi'(\rho,0)$, $f(\rho,0)$, $V'(\rho,0)$ associated to this equilibrium, alternating between the solution of the 2−dimensional equilibrium equations and that of the 1−dimensional averaged Grad−Shafranov equation

(b) advance the quantities N_e, τ_e, τ_i, ψ', f and V' from the instant $n\Delta t$ to the instant $(n+1)\Delta t$, solving the diffusion equations and the averaged equilibrium equation (system **Dif**)

(c) (optional, according to precision desired): recalculate the nonlinearities of the diffusion equations and in particular the transport coefficients at the instant $(n+1)\Delta t$ so that the diffusion equations are solved totally implicitly; return to stage (b) until N_e, σ_e, σ_i, ψ', f and V' have converged at the instant $(n+1)\Delta t$

(d) solve the equilibrium equations for the time step $[n\Delta t, (n+1)\Delta t]$ using the profiles of p and f determined at the instant $(n+1)\Delta t$ in stage (b) (system **Eq**)

(e) modify the values of I_p and ρ_{max} in the system **Dif** so that they are consistent with the analogous quantities for the system **Eq** (this stage is irrelevant in the case where the plasma boundary is fixed and the current I_p is given). Return to stage (b) until the consistency relations are satisfied

(f) the time step ends. Return to (b) for a new time step.

This type of code has been called a $1\frac{1}{2}$−dimensional transport code by H. GRAD since it alternates between a 2−dimensional solution of the equilibrium equations and a 1−dimensional solution of the transport equations. A survey of methods of numerical solution of this type of problem describing the evolution of the axisymmetric equilibrium of the plasma in a Tokamak on the time−scale of diffusion phenomena within the plasma is given by J. BLUM − J. LE FOLL [1].

VI.3 APPLICATION TO THE TOKAMAK TORE SUPRA

VI.3.1 The Poloidal Field System of TORE Supra

The principal data for the Tokamak TORE Supra were given in Sections I.5.1 and I.5.5. Its poloidal field system consists of nine independent circuits in parallel, each one consisting of a certain number of turns of the corresponding coil and powered by its own generator. One of these nine circuits, which consists of windings around the iron core, is used mainly to induce the plasma current I_p and is called the ohmic heating circuit. The eight other circuits control the shape and position of the plasma.

Since we are considering here only configurations with up/down symmetry, the poloidal field system can be represented by the simplified diagram in Fig. VI.1. It consists of five circuits in parallel, the first being the ohmic heating circuit and the other four controlling the radial position and the shape of the plasma. Figure VI.2 represents the meridian section of the poloidal field coils B_i ($i \in \{1,\ldots,5\}$). The ohmic heating circuit consists of n_1 turns of coil B_1 with current I_1. It is controlled by a generator G_1 to which is applied a voltage V_1. From the diagram in Fig.VI.1 the equation for this circuit is

(VI.87)
$$V_1 = R_e \sum_{j=1}^{5} I_j + n_1 R_1 I_1 + \frac{n_1}{S_1} \int_{B_1} \dot{\psi} \, dS$$

where R_1 is the resistance per turn of the coil B_1 and R_e is an external resistance.

The equations for the four other circuits are

(VI.88)
$$V_1 + V_i = R_e \sum_{j=1}^{5} I_j + n_i R_i I_i + \frac{n_i}{S_i} \int_{B_i} \dot{\psi} \, dS, \quad i \in \{2,\ldots,5\}$$

where n_i is the number of turns of the coil B_i in the ith circuit, R_i is the resistance per turn of the coil B_i, V_i is the voltage applied to the generator G_i and I_i is the current in the ith circuit. The current I_i is related in the current density j_i in each coil B_i by the relation

(VI.89) $$I_i = S_i j_i \quad i \in \{1,\dots,5\} .$$

For the poloidal field system of TORE Supra, equations (VI.87)–(VI.89) replace the equation (VI.5) for the general model.

The voltage V_i applied to the ith circuit is calculated as being the sum of a pre-programmed voltage $V_i^{PR}(t)$ and a feedback voltage V_i^F:

(VI.90) $$V_i = V_i^{PR}(t) + V_i^F \ , \quad i \in \{1,\dots,5\}$$

provided always that V_i does not exceed a stipulated value V_i^{max}.

The purpose of the feedback system is to control the total plasma current I_p and the plasma boundary. The plasma shape is to be as close as possible to a desired shape such as a circle, for example. Since there is only a finite number of generators and hence of control parameters, we attempt to make the plasma pass through a finite number of points, namely the points P_1, P_2, P_3 and P_4 shown in Fig.VI.3. The voltages V_i^F of the proportional derivative feedback are as follows:

(VI.91)
$$\begin{aligned} V_i^F = {} & \alpha_{i,1}(I_p - I_p^{ref}) + \beta_{i,1}\frac{d}{dt}(I_p - I_p^{ref}) \\ & + \sum_{j=2}^{4} \alpha_{i,j}[\psi(P_j) - \psi(P_1)] + \sum_{j=2}^{4} \beta_{i,j}\frac{d}{dt}[\psi(P_j) - \psi(P_1)] \\ & + \alpha_{i,5}(I_4 - I_5) + \beta_{i,5}\frac{d}{dt}(I_4 - I_5), \quad i \in \{1,\dots,5\} \end{aligned}$$

where I_p^{ref} denotes the reference value of the plasma current and where the last two terms in V_i^F are intended to guarantee an equal division of the current between the coils B_4 and B_5. The quantities $\psi(P_j)$ are calculated from magnetic measurements of ψ and $\partial\psi/\partial n$ on the vacuum vessel (cf. Fig.VI.3) by the method presented in Section V.3.1 and applied to the case of TORE Supra in Section V.3.2.

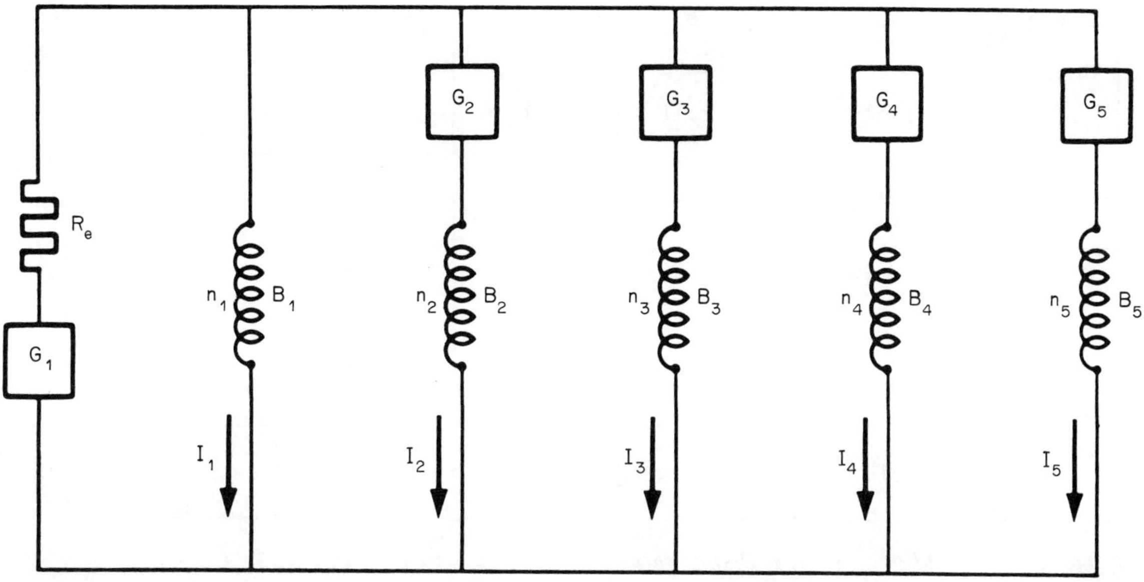

Fig.VI.1 Simplified diagram of the poloidal field system of TORE Supra (configurations with up/ down symmetry)

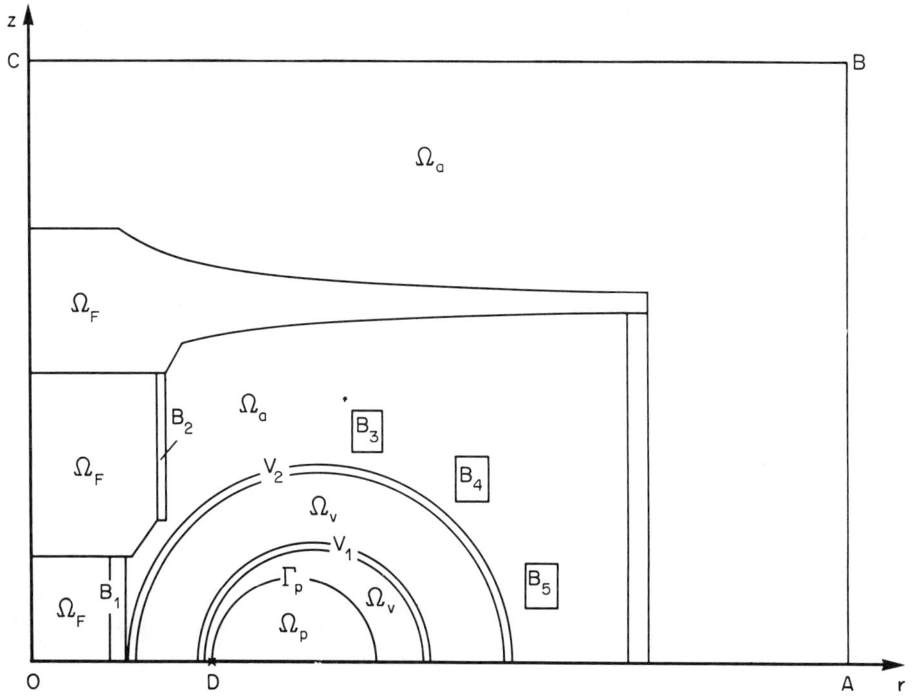

Fig.VI.2 Meridian section of TORE Supra (upper half−plane)

The matrices of gains α_{ij} and β_{ij} are given by Table VI.1 below:

Gains	$\alpha_{i,1}$	$\beta_{i,1}$	$\alpha_{i,2}$	$\beta_{i,2}$	$\alpha_{i,3}$	$\beta_{i,3}$	$\alpha_{i,4}$	$\beta_{i,4}$	$\alpha_{i,5}$	$\beta_{i,5}$
V_1^F	6×10^{-3}	2×10^{-5}	0	0	0	0	0	0	0	0
V_2^F	-6×10^{-4}	-2×10^{-6}	-1800	-6	-900	-3	-2250	-7.5	0	0
V_3^F	-3×10^{-3}	-10^{-5}	-900	-3	-1800	-6	-9000	-30	0	0
V_4^F	-3×10^{-3}	-10^{-5}	0	0	0	0	-15000	-50	0.9	3×10^{-3}
V_5^F	-3×10^{-3}	-10^{-5}	0	0	0	0	-15000	-50	-0.9	-3×10^{-3}

Table VI.1 *Matrix of feedback gains* (V *in volts,* I *in amperes,* ψ *in* Wb, t *in seconds).*

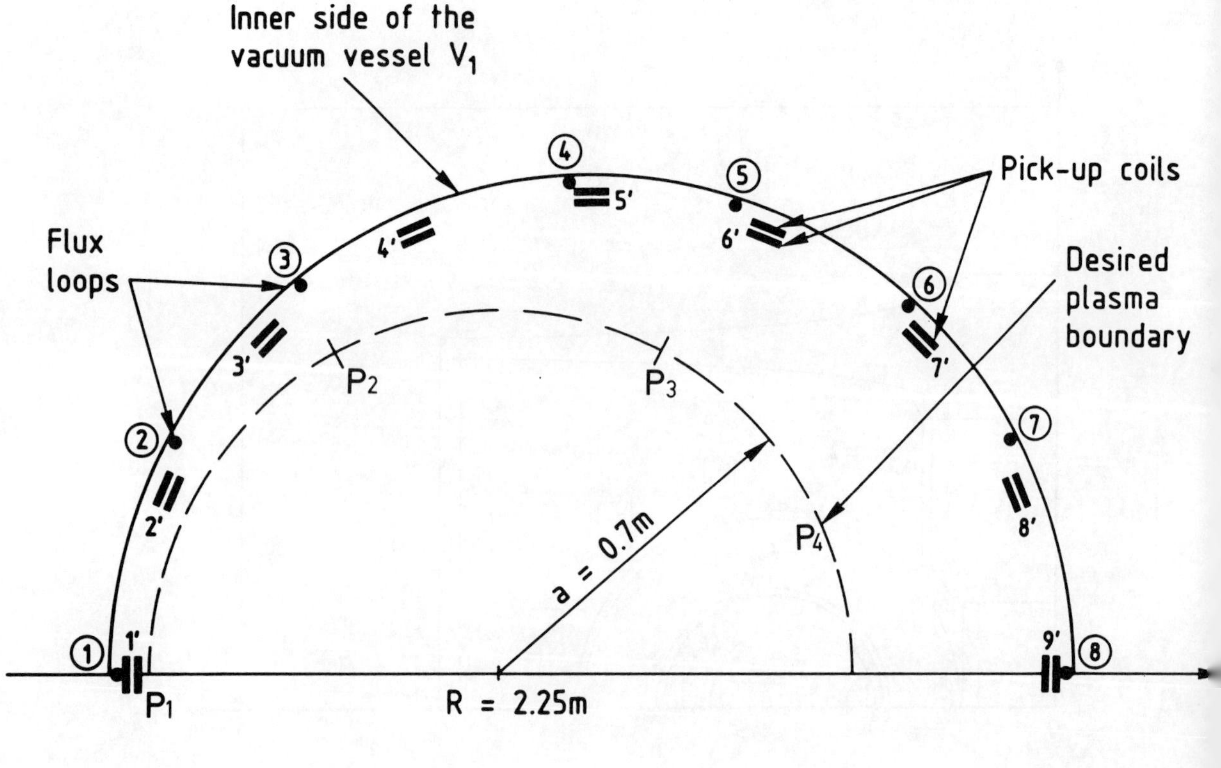

Fig.VI.3 Control of plasma shape

Another characteristic of TORE Supra is the existence of two vacuum vessels (cf. Fig.VI.2), namely an internal discharge chamber V_1 and a cryogenic vessel V_2, with the superconducting toroidal field coils between them.

In each of these vessels the equation for ψ is

$$\frac{\sigma_i}{r}\dot{\psi} + L\psi = 0 \quad \text{in} \quad V_i,\ i \in \{1,2\} \tag{VI.92}$$

These two equations replace equation (VI.9) for the general model.

The other equations (VI.3), (VI.4), (VI.10)–(VI.14) for the general model remain unchanged. In addition there is the symmetry condition.

$$\frac{\partial\psi}{\partial z} = 0 \quad \text{on} \quad Or \tag{VI.93}$$

Using a weak formulation of the type (VI.15) and the numerical methods of Section VI.2.1, an equation analogous to (VI.76) can be written down for solving this system at each time–step.

Numerical data for a simulation of the poloidal field system of TORE Supra are set out in Table VI.2.

Number of turns in each circuit: n_1, n_2, n_3, n_4, n_5	195, 176, 95, 96, 96
External resistance Re	0.1(Ω)
Resistance per turn R_1, R_2, R_3, R_4, R_5	0.114, 0.288, 0.399, 0.505, 0.583 (mΩ)
Conductivities σ_1,σ_2 of the vacuum vessels	54.5, 22.4 $(m\Omega \times m)^{-1}$
Maximum voltages: V_1^{max}, V_2^{max}, V_3^{max}, V_4^{max}, V_5^{max}	1400, 1500, 2200, 3000, 3000 (V)

Table VI.2 Specifications for the poloidal field system of TORE Supra

VI.3.2 The Transport Model

VI.3.2.1 Transport Coefficients

We adopt the empirical model of Section VI.1.2.7 with the following diffusion coefficients for electrons (Intor scaling):

$$D = 1.25 \times 10^{19} n_e^{-1}$$

$$(D \text{ in } m^2/s \text{ with } n_e \text{ in } m^{-3})$$

$$K_e = 5 \times 10^{19} (m \times s)^{-1}$$

The thermal conductivity K_i of the ions is assumed to be neo-classical and is given by the expressions in F.L. HINTON – R.D. HAZELTINE. The resistivity $\eta_{\shortparallel}$ of the plasma is given by the formula of L. SPITZER with neo-classical corrections of S.P. HIRSHMAN.

VI.3.2.2 Additional Heating and Current Drive:

The ohmic heating due to the plasma current I_p itself is insufficient to bring the plasma up to temperatures of the order of 10 KeV that are necessary to satisfy Lawson's criterion defined in Section I.1, especially as the plasma resistivity $\eta_{\shortparallel}$ varies as $T_e^{-3/2}$. Additional heat sources are therefore provided for heating the plasma. The main additional heat sources used in present Tokamaks are (cf. J. TACHON):

. heating by injection of neutral atoms: a beam of highly accelerated neutral atoms is injected into the plasma, and these travel through the magnetic configuration until they are ionized by electrons or ions. The injected atoms are thus transformed into confined ions which transfer their energy to the ions and electrons of the plasma.

. radio frequency heating: by means of antennae a wave is generated that at certain frequencies resonates with the natural frequencies of the plasma. These eigenmodes are excited in the cavity that the vacuum vessel filled with plasma constitutes, and their damping in the plasma contributes to heating it up. The frequencies that appear to be the most adapted to heating are the cyclotronic electron frequency, the cyclotronic ion frequency and the lower hybrid frequency (so-called because it involves the two kinds).

Moreover, in order to maintain constancy of the current I_p in the plasma in the continuous regime, which has to be the case in a fusion reactor, it is necessary to generate current within the plasma. The lower hybrid frequency heating is particularly well adapted for generating plasma current.

The sources due to additional heating enter into the terms s_2 and s_3 in equations (VI.69) and (VI.70) for the energy of the electrons and ions.

Certain equations in Section VI.1.2 have to be modified in the presence of current drive. Ohm's law (VI.49) becomes

(VI.94) $$E + u \times B = \eta(j - j_{RF})$$

where j_{RF} is the current density generated by the radio frequency heating.

Equation (VI.45) in Proposition (VI.6) then becomes:

(VI.95) $$\frac{\partial\psi}{\partial t} - \frac{\eta_{\parallel}\rho}{\mu_0 C_3^2}\frac{\partial}{\partial\rho}\left(\frac{C_2C_3}{\rho}\frac{\partial\psi}{\partial\rho}\right) = \eta_{\parallel} R_0 j_{RF}$$

and (VI.59) is

$$\frac{\partial\psi'}{\partial t} - \frac{1}{\mu_0}\frac{\partial}{\partial\rho}\left[\frac{\eta_{\parallel}\rho}{C_3^2}\frac{\partial}{\partial\rho}\left(\frac{C_2C_3}{\rho}\psi'\right)\right] = R_0\frac{\partial}{\partial\rho}(\eta_{\parallel} j_{RF})$$

The Joule effect term on the right hand side of (VI.46) or (VI.69) then becomes equal to:

$$\frac{\eta_{\parallel}}{\mu_0}\frac{\partial}{\partial\rho}(C_2\psi')\left[\frac{\rho}{\mu_0 C_3^2}\frac{\partial}{\partial\rho}\left(\frac{C_2C_3}{\rho}\psi'\right) + R_0 j_{RF}\right]$$

and the term s_2 is equal to the power source s_{RF} generated by the high frequency heating.

From now on we shall assume that the density of current generated in the plasma has the form

$$j_{RF} = j_0\,(1-\rho^2/\rho^2_{max})$$

and deduce s_{RF} from the ratio s_{RF}/j_{RF} given by the theory of N.J. FISCH.

VI.3.2.3 Radiation Losses

Here we shall consider two impurities: one heavy (nickel) and one light (carbon). They are defined proportionally by their ratio to the electron density n_e:

$$\text{(VI.96)} \qquad \begin{cases} n_{Ni} = \gamma_{Ni} n_e \\ n_C = \gamma_C n_e \end{cases}$$

The hydrogen ion density is then given by the neutrality relation:

$$\text{(VI.97)} \qquad n_e = n_H + n_{N_i} \bar{Z}_{N_i} + n_C \bar{Z}_C$$

where $\bar{Z}_{Ni}$ and $\bar{Z}_C$ denote the mean charge of the nickel and the carbon.

The total density n_i of the ions is

$$n_i = n_H + n_{N_i} + n_C$$

$$= n_e[1+\gamma_{N_i}(1-\bar{Z}_{N_i}) + \gamma_C(1-\bar{Z}_C)]$$

The effective charge Z_{eff} of the plasma is defined by

$$\text{(VI.98)} \qquad Z_{eff} = \frac{n_H + n_{N_i} \bar{Z^2}_{N_i} + n_C \bar{Z^2}_C}{n_e}$$

where $\bar{Z^2}_{Ni}$ and $\bar{Z^2}_C$ are the mean squares of the charges. We have then

$$Z_{eff} = 1 - \gamma_{N_i}[\bar{Z}_{N_i} - \bar{Z^2}_{N_i}] - \gamma_C[\bar{Z}_C - \bar{Z^2}_C]$$

The radiation losses are

$$\text{(VI.99)} \qquad \begin{cases} s_{N_i} = n_e n_{N_i} L_{N_i} \\ s_C = n_e n_C L_C \end{cases}$$

where L_{Ni} and L_C are the rates of radiation loss.

Under the coronal assumption the quantities $\bar{Z}$, $\bar{Z^2}$, L_{Ni}, L_C are given as functions of T_e by the curves taken from D.E. POST et al. The terms s_{Ni} and s_C feature in the term s_3 of the ion energy equation (VI.70).

VI.3.3 Simulation of Discharge Types in TORE Supra

We intend to simulate discharge with ohmic heating alone and with additional heating and current drive. These simulations have been presented in J. BLUM – J.LE FOLL – C. LELOUP.

The initial data for all these simulations corresponds to an equilibrium configuration with plasma having a peaked current density and a total current I_p of 300 kA, the flux in the iron core being 8.35 Wb because of premagnetization (it is hard to get the simulation started with a plasma current below 300 kA). The profile of electron density n_e is assumed parabolic in ρ at the initial instant, and the mean density is equal to 10^{19} m^{-3}. Each discharge is divided into three phases:

. the phase of rapid increase in plasma current where the voltage V_1 in the ohmic heating circuit is taken equal to zero and where R_e is 0.1 Ω. To avoid the creation of skin effects in the plasma current density profile, the diffusion coefficient and thermal conductivity of electrons are multiplied by a certain factor as soon as $\partial\langle j_T \rangle/\partial\rho$ becomes positive. The end of this phase is reached when $R_e\Sigma\ I_j$ reaches -1000 V.

. the phase of slow increase in plasma current where $V_1 = -1000$ V and $R_e = 0$. The end of this phase is reached when I_p attains 1.7 MA.

. the plateau phase where I_p^{ref} is taken equal to 1.7 MA in (VI.91).

Three types of discharge are then simulated:

. discharge with **ohmic heating alone**: the pre–programmed voltage V_1^{PR} is equal to -250V during the plateau phase. Figure VI.4 represents the equilibrium configuration obtained at t = 2s, during the plateau phase.

. discharge with **RF heating and current drive**: the lower hybrid frequency heating of electrons begins at the end of the slow current rise phase, i.e. at t = 800 ms. The injected power P_{RF} is equal to 6 MW. The total current I_{RF} generated is deduced from the ratio I_{RF}/P_{RF} calculated in G. TONON – D. MOULIN for the heating system characteristics of TORE Supra. Since the mean electron density $\bar{n}_e$ is equal to 6×10^{19} m^{-3} during the plateau phase, we obtain I_{RF}/P_{RF}= 0.1 A/W and so I_{RF} = 600 kA. The pre–programmed voltage V_1^{PR} is equal to zero during this phase.

. discharge with **neutral beam heating and RF heating.** The injection of particles is begun at the end of the phase of rapid current increase i.e. at t = 200 ms; a power of 6 MW is injected into the plasma in this way, and the distribution between the sources s_2 and s_3 is calculated from J.P. ROUBIN. At t = 800 ms the lower hybrid frequency heating is started, corresponding to a generation of 6 MW of power and 600 kA of current.

In all these discharges the plasma boundary is maintained circular, by means of the feedback defined by (VI.91) where the points P_j belong to a circle of radius 70 cm.

Figure VI.5 represents the evolution of the total plasma current I_p and the loop voltage V_T at the plasma boundary (V_T is equal to $d\psi_p/dt$ where ψ_p is the value of ψ on Γ_p) for each of the three discharges. The voltage V_T , which is equal to 1.1 V at t = 2s in the ohmic heating scenario, has a value of 0.156 V at the same instant in the second scenario (RF heating) and 0.06 V in the third with injection of neutral particles.

This low voltage obtained as a result of current generation should allow long discharges of 30s to be obtained in TORE Supra.

Figure VI.6 represents the temporal evolution of the parameters β_p (poloidal beta) and self- inductance ℓ_i for the three discharges.
At time t = 2s the mean ionic temperature $\bar{T}_i$ is equal to 570 eV in the ohmic heating scenario, 1.8 keV with RF heating, and 3.6 keV in the third scenario with neutral beam injection.

Simulations of the same type realized with the code SCED for the Tokamak JET have been presented in J. BLUM – J. LE FOLL [2].

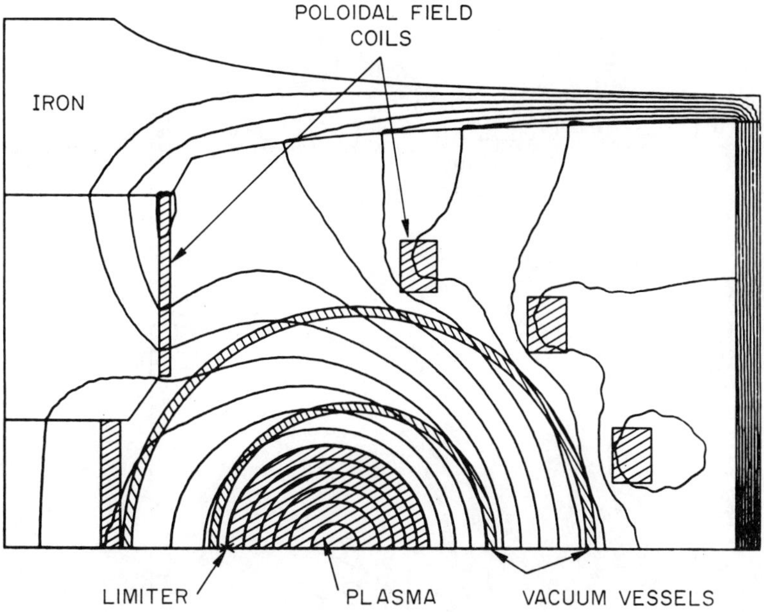

Fig.VI.4 : Equilibrium configuration during the plateau phase (ohmic heating)

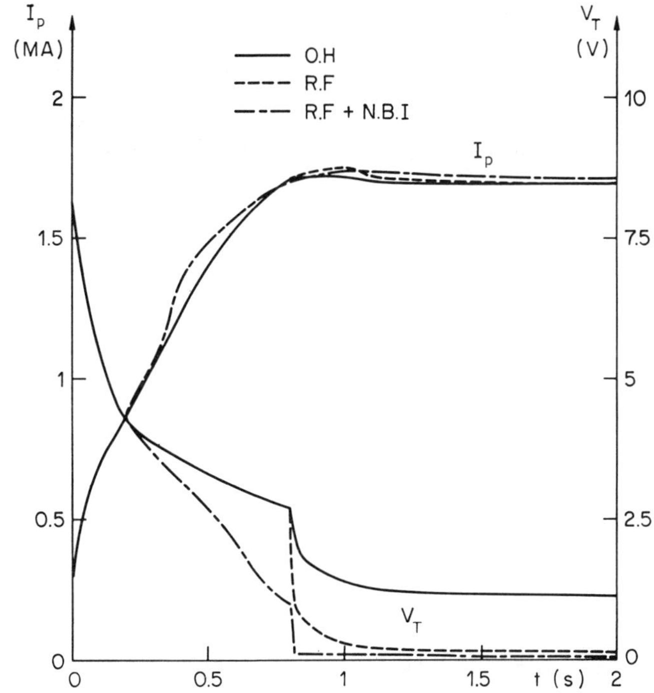

Fig.VI.5 : Evolution of the total plasma current I_p *and the loop voltage* V_T *in the three discharges*

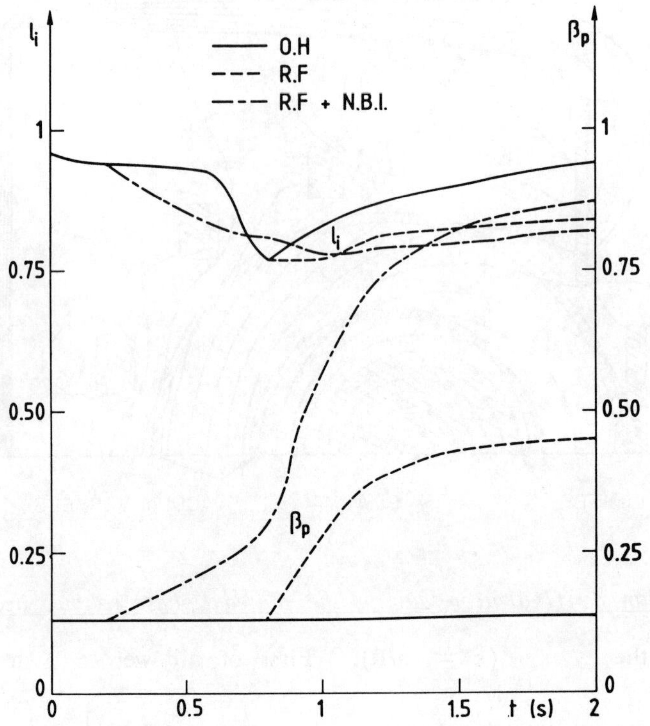

Fig.VI.6 Evolution of the parameters β_p and ℓ_i in the three discharges

7. Evolution of the equilibrium of a high aspect-ratio circular plasma; stability and control of the horizontal displacement of the plasma

In Chapter VI we studied the evolution of the equilibrium of a plasma of arbitrary section; in this chapter we shall consider the particular case of a plasma of circular section, for which a simplified theory can be developed.

In the first part we give a new presentation of the Shafranov analytic theory of equilibrium; this theory is a first order expansion in ε of the equilibrium equations, where ε is the inverse of the aspect ratio, i.e. the ratio of the minor radius a to the major radius R of the plasma ($\varepsilon = a/R$). First of all we establish the equilibrium integral equations on each magnetic surface (virial theorem) and deduce from them a first order expansion in ε for the poloidal field in the interior and the exterior of the plasma, as well as an expression for the plasma current density. We show that if the plasma has circular section then the magnetic surfaces in the interior of the plasma, or outside but in a neighbourhood of it, are likewise of circular section and displaced off–centre relative to the plasma by an amount to be calculated. Finally, we establish the expression for the magnetic field needed to maintain the plasma in equilibrium, and deduce from it the equation for the horizontal displacement of the plasma.

In the second part we model the ensemble of plasma and external circuits for the Tokamak TFR at Fontenay–aux–Roses, whose plasma can be assumed to be of circular section. First we define the flux pattern in the machine by solving the magnetostatic problem; we give algebraic laws exhibiting quasi–linear relationships between the fluxes and the currents in the coils and total plasma current, the coefficients depending on the degree of saturation of the magnetic circuit and on the radial position of the plasma. The currents in the various circuits (primary, pre–programming, feedback), the total plasma current and its radial position are then modelled by a system of differential equations; this system is coupled to a transport code describing the internal dynamics of the plasma. Finally, we compare such a simulation with the experimental results of a typical discharge of TFR.

In the third part we are concerned with the control of horizontal displacements of the plasma. We first compute the pre-programming necessary for realizing the equilibrium field of the plasma. This pre-programming splits into three stages : optimization of the distribution of windings in the primary circuit, voltage to be applied to the pre-programming circuit, learning method that improves the pre-programming from discharge to discharge. Next we study the stability of horizontal displacements of the plasma in TFR 600; using a simplified model we compute the time constant of the phenomenon from indices of various magnetic fields, and compare it with that observed experimentally. Finally we calculate the feedback gains that enable the plasma to be brought to a certain reference configuration and thus to stabilize the horizontal displacement instability of the plasma observed in TFR 600.

VII.1 THE SHAFRANOV THEORY OF EQUILIBRIUM

(cf. V.D. SHAFRANOV [3] and [4], V.S. MUKHOVATOV – V.D. SHAFRANOV.)

We consider the equilibrium of a plasma whose boundary is assumed circular with minor radius a and major radius R. We call the quantity R/a the aspect ratio of the Tokamak, and denote its inverse by ε; thus $\varepsilon = a/R$.

The Shafranov theory enables an analytic expansion to first order in ε to be given for the principal quantities that characterize the equilibrium of the plasma (poloidal flux and field, current density, position of magnetic surfaces).

In this chapter we are concerned exclusively with configurations that are symmetric with respect to the equatorial plane. The proofs of the following propositions are formal.

VII.1.1 *Integral Expression of the Equilibrium Equations*

VII.1.1.1 *The Virial Theorem*

Proposition VII.1

The following two integral forms of the equilibrium equation (I.4) are satisfied on each magnetic surface S in the interior of the plasma:

$$\text{(VII.1)} \qquad \int_V \left(3p + \frac{B^2}{2\mu_0}\right) dV = \int_S \left(p + \frac{B^2}{2\mu_0}\right)(\mathbf{OM}.\mathbf{dS})$$

$$\text{(VII.2)} \qquad \int_V \left[\left(p + \frac{B^2}{2\mu_0}\right)\frac{1}{r} - \frac{B_T^2}{\mu_0 r}\right] dV = \int_S \left(p + \frac{B^2}{2\mu_0}\right)(\mathbf{e_r}.\mathbf{dS})$$

where V denotes the volume enclosed by the surface S, where **dS** is the vector of modulus dS normal to S in M, and where $\mathbf{e_r}$ is the unit vector along the r− axis. □

Remark VII.1

The equality (VII.1) is known as the "virial theorem" (cf. V.D. SHAFRANOV [4]).□

Proof

We use the approach developed by L.E. ZAKHAROV − V.D. SHAFRANOV in order to define the method of moments. Multiply the equilibrium equation (I.4) by an arbitrary vector Q and use Maxwell's equations (I.2) and (I.3) : we obtain

$$\text{(VII.3)} \qquad Q.\nabla p = \frac{Q}{\mu_0}\,[(\nabla\times B)\times B] \ .$$

Now we have the following vectorial identity:

$$\text{(VII.4)} \qquad Q.[(\nabla\times B)\times B] = \nabla.[(Q.B)B - \frac{1}{2}B^2 Q] + \frac{1}{2}B^2\nabla.Q - B.(B.\nabla)Q \ .$$

Integrating (VII.3) over the volume V contained within a magnetic surface S and using (VII.4) and Gauss Theorem we obtain:

(VII.5) $$\int_V Q.\nabla p \, dV = \mu_o^{-1}\{\int_S (Q.B)(B.dS) - \int_S \frac{1}{2} B^2(Q.dS)$$

$$+ \int_V \frac{1}{2} B^2\nabla.Q \, dV - \int_V B.(B.\nabla)Q \, dV\} \; .$$

By definition, B.dS = 0 on a magnetic surface S and moreover

(VII.6) $$\int_V Q.\nabla p dV = \int_S pQ.dS - \int_V p \, \nabla.Q \, dV \; .$$

Equation (VII.5) can then be written

(VII.7) $$\int_V [(p + \frac{B^2}{2\mu_o})\nabla.Q - \frac{B.(B.\nabla)Q}{\mu_o}] \, dV = \int_S (p + \frac{B^2}{2\mu_o})(Q.dS) \; .$$

If we choose as Q the vector OM = $re_r + ze_z$ where e_r, e_z are the unit vectors along the r− axis and z− axis respectively, we have

(VII.8) $$\begin{cases} \nabla.OM = 3 \\ (B.\nabla)OM = B \end{cases}$$

and equation (VII.7) takes the form of the "virial theorem" (VII.1).

If we now take as Q the vector e_r we have

(VII.9) $$\begin{cases} \nabla.e_r = \frac{1}{r} \\ (B.\nabla)e_r = \frac{1}{r} B_T \end{cases}$$

where we recall that B_T is the toroidal component of the magnetic induction **B**. Equation (VII.7) then takes the form (VII.2). □

VII.1.1.2 Equilibrium Equation for the Pressure

To order zero in ε (cylindrical approximation) the magnetic surfaces are cylinders of circular section, concentric relative to the plasma section. To order 1 in ε (first toroidal approximation) the magnetic surfaces are, as we shall see later, tori of eccentric circular section relative to the plasma section. Consider one of the magnetic surfaces S lying inside the plasma, having minor radius ρ_1 and major radius R_1. Define polar coordinates (ρ, θ) in the meridian plane of the torus such that the origin O' is the centre of the section C of the magnetic surface S (see Fig.VII.1).

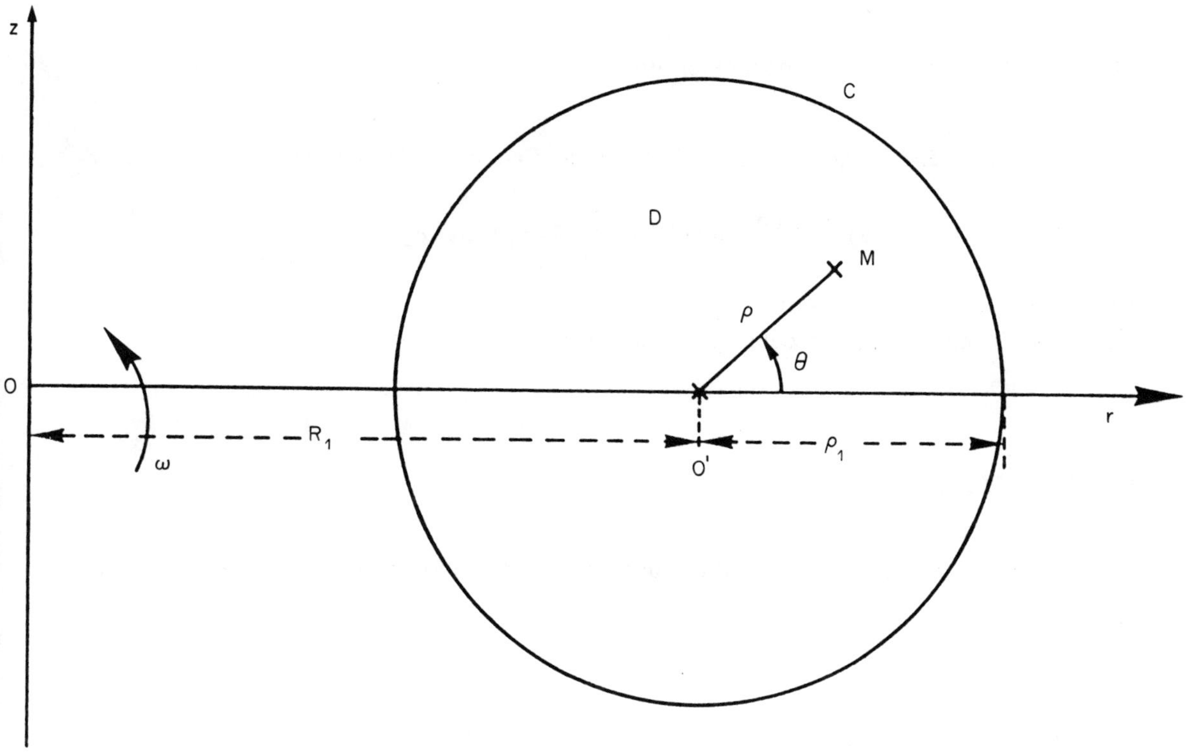

Fig.VII.1 : Coordinate system for the magnetic surface S.

We have:

$$\text{(VII.10)} \qquad \begin{cases} r = R_1 + \rho \cos\theta \\ z = \rho \sin \theta \ . \end{cases}$$

Define the average $\bar{A}$ of an arbitrary quantity A in the volume V contained inside the surface S as follows:

$$\text{(VII.11)} \qquad \bar{A} = \frac{\int_V A \, dV}{\int_V dV} = \frac{\int_0^{2\pi} \int_0^{\rho_1} A(\rho, \theta) r\rho \, d\rho \, d\theta}{\pi R_1 \rho_1^2} .$$

By (I.11) and (I.15) the toroidal component $\mathbf{B}_T$ of the magnetic field varies as $1/r$ on the line of flux C; to first order in ε it can be written

$$\text{(VII.12)} \qquad B_T(\rho_1, \theta) = B_T^{(o)}(\rho_1)(1 - \frac{\rho_1}{R_1} \cos \theta)$$

where $B_T^{(o)}$ denotes the component of order 0.

The orthoradial component $\mathbf{B}_\theta$ of $\mathbf{B}$ can to first order be written

$$\text{(VII.13)} \qquad B_\theta(\rho_1, \theta) = B_\theta^{(o)}(\rho_1)[1 + \frac{\rho_1}{R_1} \xi(\rho_1) \cos \theta]$$

where the function $\xi(\rho_1)$ is determined in the following proposition:

<u>Proposition VII.2:</u>

On each magnetic surface the equation of equilibrium for the pressure can be expressed as

$$\text{(VII.14)} \qquad \overline{p^{(o)}}(\rho_1) - p^{(o)}(\rho_1) = \frac{[B_\theta^{(o)}(\rho_1)]^2}{2\mu_o} + \frac{[B_T^{(o)}(\rho_1)]^2 - \overline{[B_T^{(o)}]^2}(\rho_1)}{2\mu_o} .$$

The coefficient $\xi(\rho_1)$ characterizing the first order term of B_θ in (VII.13) may be written

$$\text{(VII.15)} \qquad \xi(\rho_1) = \beta_p(\rho_1) + \frac{\ell_i(\rho_1)}{2} - 1$$

with

$$\beta_p(\rho_1) = \frac{2\mu_o[\overline{p^{(o)}(\rho_1)} - p^{(o)}(\rho_1)]}{[B_\theta^{(o)}(\rho_1)]^2}$$

$$\ell_i(\rho_1) = \frac{\overline{[B_\theta^{(o)}]^2(\rho_1)}}{[B_\theta^{(o)}(\rho_1)]^2} .$$

□

Proof

We shall rewrite the two integral equations (VII.1) and (VII.2) in the coordinate system $(0',\rho,\theta)$ of Fig. (VII.1).

Neglecting the radial component $\mathbf{B}_\rho$ of $\mathbf{B}$, which is legitimate up to first order in ε (see below), we have on C, by (VII.12) and (VII.13):

(VII.16)

$$B^2(\rho_1,\theta) = [B^{(o)}(\rho_1)]^2 + \frac{2\rho_1 \cos\theta}{R_1} \{\xi(\rho_1)[B_\theta^{(o)}(\rho_1)]^2 - [B_T^{(o)}(\rho_1)]^2\}$$

with

$$[B^{(o)}(\rho_1)]^2 = [B_\theta^{(o)}(\rho_1)]^2 + [B_T^{(o)}(\rho_1)]^2$$

Moreover, on C we have

$$\text{(VII.17)} \quad \begin{cases} \mathrm{OM}.\mathrm{dS} = (R_1+\rho_1 \cos\theta)(R_1 \cos\theta + \rho_1)\ \rho\ d\theta\ d\omega \\ \mathrm{e}_r.\mathrm{dS} = (R_1 + \rho_1 \cos\theta)\ \rho_1 \cos\theta\ d\theta\ d\omega\ . \end{cases}$$

Using (VII.11), (VII.16) and (VII.17), the virial theorem (VII.1) can to order zero be written as follows:

$$3[\overline{p^{(o)}(\rho_1)} - p^{(o)}(\rho_1)] = \frac{[B_\theta^{(o)}(\rho_1)]^2[3+2\xi(\rho_1)] - \overline{[B_\theta^{(o)}]^2(\rho_1)}}{2\mu_o}$$

(VII.18)

$$+ \frac{[B_T^{(o)}(\rho_1)]^2 - \overline{[B_T^{(o)}]^2(\rho_1)}}{2\mu_o} .$$

Note that to order zero the notion of average over V, defined by (VII.11), can equally well be expressed as the average over the meridian section of V (see Fig.VII.1), namely

$$\text{(VII.19)} \qquad \overline{A^{(o)}} = \frac{\int_D A^{(o)} dS}{\int_D dS} = \frac{2\int_o^{\rho_1} A^{(o)}(\rho)\ \rho\ d\rho}{\rho_1^2} .$$

Using (VII.16), (VII.17) and (VII.19), the integral equation (VII.2) can be written to order zero as

$$\overline{p^{(o)}(\rho_1)} - p^{(o)}(\rho_1) = \frac{[B_\theta^{(o)}(\rho_1)]^2[1+2\xi(\rho_1)] - \overline{[B_\theta^{(o)}]^2(\rho_1)}}{2\mu_o}$$

(VII.20)

$$+ \frac{\overline{[B_T^{(o)}]^2(\rho_1)} - [B_T^{(o)}(\rho_1)]^2}{2\mu_o} .$$

By subtraction and addition of (VII.18) and (VII.20) we obtain the relations (VII.14) and (VII.15). □

For ease of notation in what follows we shall denote the values of $\beta_p(a)$, $\ell_i(a)$ and $\xi(a)$ by β_p, ℓ_i and ξ.

The coefficient β_p, already defined by (I.95), is the ratio of the kinetic pressure $\bar{p}(a)$ of the plasma to the magnetic pressure due to the poloidal field $B_\theta^{(o)}(a)$. In the circular case Ampère's theorem gives

$$B_\theta^{(o)}(a) = -\frac{\mu_o I_p}{2\pi a} . \tag{VII.21}$$

As the pressure p vanishes on the plasma boundary, formula (I.95) becomes identical to the expression (VII.15) for β_p for $\rho_1 = a$.

The coefficient ℓ_i, already defined by (I.96), is the internal coefficient of self-inductance per unit length of the plasma.

The expression (I.96) is the same as the formula (VII.15) for ℓ_i when $\rho_1 = a$, in the circular case and at order 0.

The coefficient ξ, still called the asymmetry coefficient, characterizes via (VII.13) the distribution of the field B_θ on the plasma boundary.

VII.1.2 Expressions for the Magnetic Field and Current Density to Order 1; Relative Position of the Magnetic Surfaces

Let us now take polar coordinates (ρ, θ) having the centre of the plasma boundary as origin. Let R be the major radius of the plasma.

Denote by $\Delta(\rho_1)$ the eccentricity relative to the plasma of the line of flux of minor radius ρ_1, i.e. the algebraic measure (oriented by the positive r-axis) of the distance from the centre of the plasma to the centre of the line of flux. We suppose $\Delta(\rho_1)$ to be small compared with ρ_1. In the plasma coordinate system the equation of this line of flux is

$$\rho = \rho_1 + \Delta(\rho_1) \cos \theta . \tag{VII.22}$$

We then have the following proposition:

Proposition VII.3

In the interior of the plasma, the components of order 1 in B_θ, ψ and j_T are as follows:

$$B_\theta^{(1)}(\rho,\theta) = [\frac{\rho}{R} B_\theta^{(0)}(\rho)\xi(\rho) - \Delta(\rho) \frac{\partial B_\theta^{(o)}}{\partial \rho}(\rho)] \cos\theta \tag{VII.23}$$

$$\psi^{(1)}(\rho,\theta) = -\Delta(\rho) \frac{\partial \psi^{(o)}}{\partial \rho}(\rho) \cos\theta \tag{VII.24}$$

$$j_T^{(1)}(\rho,\theta) = \{[\frac{2}{B_\theta^{(0)}(\rho)} \frac{\partial p^{(0)}}{\partial \rho}(\rho) - j_T^{(0)}(\rho)]\frac{\rho}{R} - \Delta(\rho) \frac{\partial j_T^{(o)}}{\partial \rho}(\rho)\}\cos\theta \tag{VII.25}$$

where

$$\Delta(\rho) = \int_\rho^a \frac{\rho'}{R} [\xi(\rho') + 1]d\rho' \tag{VII.26}$$

with $\xi(\rho)$ given by (VII.15). □

Proof

If A denotes any one of the variables B_θ, ψ and j_T we have by (VII.22)

$$A^{(0)}(\rho,\theta) = A^{(0)}(\rho_1) + \Delta(\rho_1) \cos\theta \frac{\partial A^{(0)}}{\partial \rho}(\rho_1) \tag{VII.27}$$

on each line of flux.

Using the first-order expression (VII.13) for B_θ on the magnetic surface with minor radius ρ_1 we can write

$$B_\theta^{(0)}(\rho,\theta)+B_\theta^{(1)}(\rho,\theta) = B_\theta^{(0)}(\rho_1)[1 + \frac{\rho_1}{R} \xi(\rho_1) \cos\theta] . \tag{VII.28}$$

If we replace A by B_θ in (VII.27) and substitute this expression into (VII.28) we obtain (VII.23) for the first order component of B_θ.

If we replace A by ψ in (VII.27) we can write the poloidal flux ψ in the flux line of radius ρ_1 as:

(VII.29) $$\psi(\rho,\theta) = \psi^{(o)}(\rho_1) + \Delta(\rho_1)\cos\theta \frac{\partial\psi^{(o)}}{\partial\rho}(\rho_1) + \psi^{(1)}(\rho,\theta) .$$

Now by definition ψ is constant and equal to $\psi^{(o)}(\rho_1)$ on this line of flux, and from this we deduce the expression (VII.24) for $\psi^{(1)}(\rho,\theta)$.

By (I.11) the radial and orthoradial components B_ρ and B_θ of **B** can be expressed as functions of the flux ψ by:

(VII.30) $$\begin{cases} B_\rho = -\dfrac{1}{\rho r}\dfrac{\partial\psi}{\partial\theta} \\ B_\theta = \dfrac{1}{r}\dfrac{\partial\psi}{\partial\rho} \end{cases}$$

where

$$r = R + \rho\cos\theta .$$

The component $B_\theta^{(1)}$ of B_θ is then

(VII.31) $$B_\theta^{(1)}(\rho,\theta) = -\frac{\rho}{R^2}\frac{\partial\psi^{(o)}}{\partial\rho}(\rho)\cos\theta + \frac{1}{R}\frac{\partial\psi^{(1)}}{\partial\rho}(\rho,\theta) .$$

Using (VII.24) we obtain

(VII.32) $$B_\theta^{(1)}(\rho,\theta) = [-\frac{\rho}{R}B_\theta^{(o)}(\rho) - \frac{\partial}{\partial\rho}(B_\theta^{(o)}\Delta)]\cos\theta .$$

If we identify the expressions (VII.23) and (VII.32) for $B_\theta^{(1)}$ we obtain

(VII.33) $$\frac{d\Delta}{d\rho}(\rho) = -\frac{\rho}{R}[\xi(\rho) + 1] .$$

which gives the expression (VII.26) for the eccentricity $\Delta(\rho)$ of the line of flux of radius ρ relative to the plasma.

By the Grad–Shafranov equation (I.17) the plasma current density j_T can be written

(VII.34) $$j_T = r\frac{\partial p}{\partial\psi} + \frac{1}{2\mu_0 r}\frac{\partial f^2}{\partial\psi} .$$

If we put

$$\text{(VII.35)} \qquad j_T^{(o)}(\rho_1) = R\,\frac{\partial p}{\partial \psi}(\rho_1) + \frac{1}{2\mu_o R}\,\frac{\partial f^2}{\partial \psi}(\rho_1)$$

we have to first order in ε:

$$\text{(VII.36)} \qquad j_T(\rho,\theta) = j_T^{(o)}(\rho_1)\left(1 - \frac{\rho_1}{R}\cos\theta\right) + \frac{2\rho_1\cos\theta}{R\,B_\theta^{(o)}(\rho_1)}\,\frac{\partial p^{(o)}}{\partial \rho}(\rho_1)$$

on the flux line of radius ρ_1.

If we replace A by j_T in (VII.27) we can likewise write

$$\text{(VII.37)} \qquad j_T(\rho,\theta) = j_T^{(o)}(\rho_1) + \Delta(\rho_1)\,\frac{\partial j_T^{(o)}}{\partial \rho}(\rho_1)\cos\theta + j_T^{(1)}(\rho,\theta)$$

to first order in ε, and by identifying (VII.36) and (VII.37) we obtain the expression (VII.25) for $j_T^{(1)}(\rho,\theta)$. □

Remark VII.2

The relations (VII.23) and (VII.25) enable the first order components of B_θ and of j_T to be calculated from the zero order components of j_T and of p. Indeed, $B_\theta^{(o)}$ can be deduced from $j_T^{(o)}(\rho)$ by applying Ampère's theorem:

$$\text{(VII.38)} \qquad B_\theta^{(o)}(\rho) = -\frac{\mu_o}{\rho}\int_o^\rho j_T^{(o)}(\rho')\rho'\,d\rho'$$

and the functions $\xi(\rho)$ and $\Delta(\rho)$ can likewise be deduced from $j_T^{(o)}$ and $p^{(o)}$ by applying (VII.15) and (VII.26).

Note further that the magnetic axis is displaced outwards relative to the centre of the plasma by the following quantity, which can be deduced from (VII.26):

$$\text{(VII.39)} \qquad \Delta(0) = \int_o^a \frac{\rho}{R}\,[\xi(\rho) + 1]\,d\rho\ .$$

□

VII.1.3 Expressions for the Flux and the Poloidal Field Outside the Plasma

The equation (I.20) for ψ in the air or vacuum surrounding the plasma is

$$\nabla.(\frac{1}{r}\nabla\psi) = 0 \text{ . *} \qquad \text{(VII.40)}$$

In terms of the polar coordinates (ρ,θ) having the centre of the plasma as origin, this equation becomes

$$\frac{1}{\rho}\frac{\partial}{\partial\rho}\left(\frac{\rho}{R+\rho\cos\theta}\frac{\partial\psi}{\partial\rho}\right) + \frac{1}{\rho^2}\frac{\partial}{\partial\theta}\left(\frac{1}{R+\rho\cos\theta}\frac{\partial\psi}{\partial\theta}\right) = 0 \text{ .} \qquad \text{(VII.41)}$$

We then have the following proposition:

Proposition VII.4

To order 1 in ε, the flux and the poloidal field outside and in a neighbourhood of the plasma (i.e. at a distance ρ from the centre of the plasma that is greater than a and small compared with R) can be expressed as:

$$\psi(\rho,\theta) = \frac{\mu_o R I_p}{2\pi}\{\text{Log}\frac{8R}{\rho} - 2 - \frac{\rho\cos\theta}{2R}[\text{Log}\frac{\rho}{a} + (1 - \frac{a^2}{\rho^2})(\xi + \frac{1}{2})]\} \qquad \text{(VII.42)}$$

$$B_\theta(\rho,\theta) = -\frac{\mu_0 I_p}{2\pi\rho}\{1 + \frac{\rho\cos\theta}{2R}[\text{Log}\frac{\rho}{a} - 1 + (1 + \frac{a^2}{\rho^2})(\xi + \frac{1}{2})]\} \qquad \text{(VII.43)}$$

$$B_\rho(\rho,\theta) = -\frac{\mu_0 I_p \sin\theta}{4\pi R}[\text{Log}\frac{\rho}{a} + (1 - \frac{a^2}{\rho^2})(\xi + \frac{1}{2})] \text{ .} \qquad \text{(VII.44)}$$

---

* Footnote The operators $\nabla\cdot$ and ∇ are here the 2-dimensional operators in the (r,z)-plane, while (VI.41) involved the 3-dimensional operators $\nabla\cdot$ and ∇ in (r,z,ω)-space.

The lines of flux outside the plasma are then circles displaced eccentrically relative to the plasma by an amount $\Delta(\rho')$ given by:

$$\text{(VII.45)} \qquad \Delta(\rho') = -\frac{\rho'^2}{2R} \operatorname{Log} \frac{\rho'}{a} - \frac{\rho'^2 - a^2}{2R}(\xi + \frac{1}{2})$$

where ρ' denotes the radius of the circle. □

<u>*Remark VII.3*</u>

By (VII.42), (VII.43) and (VII.44) the flux and the poloidal field outside the plasma depend, to first order in ε, on the internal current density distribution and on the plasma pressure by the single coefficient $\xi = \xi(a)$ determined by (VII.15) and which, by (VII.13), characterizes the variation of the field B_θ on the plasma boundary. □

<u>Proof of Proposition VII.4</u>

To order 0 in ε, equation (VII.41) is

$$\text{(VII.46)} \qquad \frac{\partial}{\partial\rho}(\rho \frac{\partial\psi}{\partial\rho}) + \frac{1}{\rho}\frac{\partial^2\psi}{\partial\theta^2} = 0 \ .$$

Since to order 0 the magnetic surfaces are cylindrical of circular section, $\psi^{(0)}$ depends only on ρ and takes the form

$$\text{(VII.47)} \qquad \psi^{(0)}(\rho) = \alpha \operatorname{Log} \rho + b \ .$$

More precisely, $\psi^{(0)}$ is the flux due to a toroidal ring of current I_p and major radius R:

$$\text{(VII.48)} \qquad \psi^{(0)}(\rho) = \frac{\mu_0 R I_p}{2\pi}(\operatorname{Log} \frac{8R}{\rho} - 2) \ .$$

From (VII.41) we then deduce the equation for $\psi^{(1)}(\rho,\theta)$:

$$\text{(VII.49)} \qquad \frac{\partial}{\partial\rho}(\rho \frac{\partial\psi^{(1)}}{\partial\rho}) + \frac{1}{\rho}\frac{\partial^2\psi^{(1)}}{\partial\theta^2} = \frac{\partial}{\partial\rho}(\frac{\rho^2}{R}\frac{\partial\psi^{(0)}}{\partial\rho}) \cos\theta \ .$$

The solution to (VII.49) is

(VII.50) $$\psi^{(1)}(\rho,\theta) = -\frac{\mu_0 I_p}{4\pi}\rho \text{ Log } \rho \cos\theta + C\rho\cos\theta + \frac{C'\cos\theta}{\rho}$$

where C and C' are two arbitrary constants.

By (VII.48) and (VII.50) the poloidal flux ψ is to order 1 in ε:

(VII.51) $$\psi(\rho,\theta) = \frac{\mu_0 R I_p}{2\pi}\left(\text{Log }\frac{8R}{\rho} - 2\right) + \left[\frac{\mu_0 I_p}{4\pi}\rho\left(\text{Log }\frac{8R}{\rho} - 1\right) + \frac{C_1}{\rho} + C_2\,\rho\right]\cos\theta$$

where C_1 and C_2 are two constants to be determined later.

From (VII.30) and (VII.51) we can deduce the expression for the radial and orthoradial components of **B** to order 1:

(VII.52) $$\begin{cases} B_\rho(\rho,\theta) = \left[\dfrac{\mu_0 I_p}{4\pi}\left(\text{Log }\dfrac{8R}{\rho} - 1\right) + \dfrac{C_1}{\rho^2} + C_2\right]\dfrac{\sin\theta}{R} \\[2ex] B_\theta(\rho,\theta) = -\dfrac{\mu_0 I_p}{2\pi\,\rho} + \left[\dfrac{\mu_0 I_p}{4\pi}\text{ Log }\dfrac{8R}{\rho} - \dfrac{C_1}{\rho^2} + C_2\right]\dfrac{\cos\theta}{R}\ . \end{cases}$$

The plasma boundary is the circle of radius a; it is a line of flux and, moreover, by (VII.13) and (VII.21) we have:

(VII.53) $$\begin{cases} B_\rho(a,\theta) = 0 \\[2ex] B_\theta(a,\theta) = -\dfrac{\mu_0 I_p}{2\pi\, a}\left(1 + \dfrac{a}{R}\,\xi\cos\theta\right)\ . \end{cases}$$

Putting $\rho = a$ in (VII.52) and identifying those equations with (VII.53) we determine the two constants C_1 and C_2:

(VII.54) $$\begin{cases} C_1 = \dfrac{\mu_0\, I_p\, a^2}{4\pi}\left(\xi + \dfrac{1}{2}\right) \\[2ex] C_2 = -\dfrac{\mu_0\, I_p}{4\pi}\left(\text{Log }\dfrac{8R}{a} + \xi - \dfrac{1}{2}\right) \end{cases}$$

The expressions (VII.51), (VII.52) for ψ and for $\mathbf{B}_p$ then have the form (VII.42), (VII.43) and (VII.44).

It can easily be shown from (VII.42) that the lines of flux outside but in a neighbourhood of the plasma are circles of $C_{\rho'}$ of radius ρ' (which we assume small relative to R) displaced eccentrically with respect to the plasma by an amount $\Delta(\rho')$. Writing down that ψ is constant on $C_{\rho'}$ whose equation is

$$\rho - \Delta(\rho')\cos\theta = \rho'$$

we obtain the expression (VII.45) for $\Delta(\rho')$. □

From V.D. SHAFRANOV [3], the poloidal flux ψ is the sum of the flux ψ_1 due to the plasma current and the flux ψ_2 representing the external contribution that is equal to $C_2\rho\cos\theta$. Thus we have:

$$\text{(VII.55)} \quad \begin{cases} \psi = \psi_1 + \psi_2 \\[2ex] \psi_1 = \dfrac{\mu_o R I_p}{2\pi} \{\text{Log}\, \dfrac{8R}{\rho} - 2 + \dfrac{\rho\cos\theta}{2R}[\text{Log}\, \dfrac{8R}{\rho} - 1 + \dfrac{a^2}{\rho^2}(\xi + \dfrac{1}{2})]\} \\[2ex] \psi_2 = -\dfrac{\mu_o I_p \rho\cos\theta}{4\pi}(\text{Log}\, \dfrac{8R}{a} + \xi - \dfrac{1}{2}) . \end{cases}$$

The flux ψ_2 is proportional to (r− R), and by (I.8) it is the flux created by a homogeneous vertical field B_{eq} equal to C_2/R :

$$\text{(VII.56)} \qquad B_{eq} = -\frac{\mu_o I_p}{4\pi R}(\text{Log}\, \frac{8R}{a} + \xi - \frac{1}{2}) .$$

The plasma is thus in equilibrium under the action of its own field (creating the flux ψ_1) and under the action of a homogeneous vertical external field B_{eq} (creating the flux ψ_2):

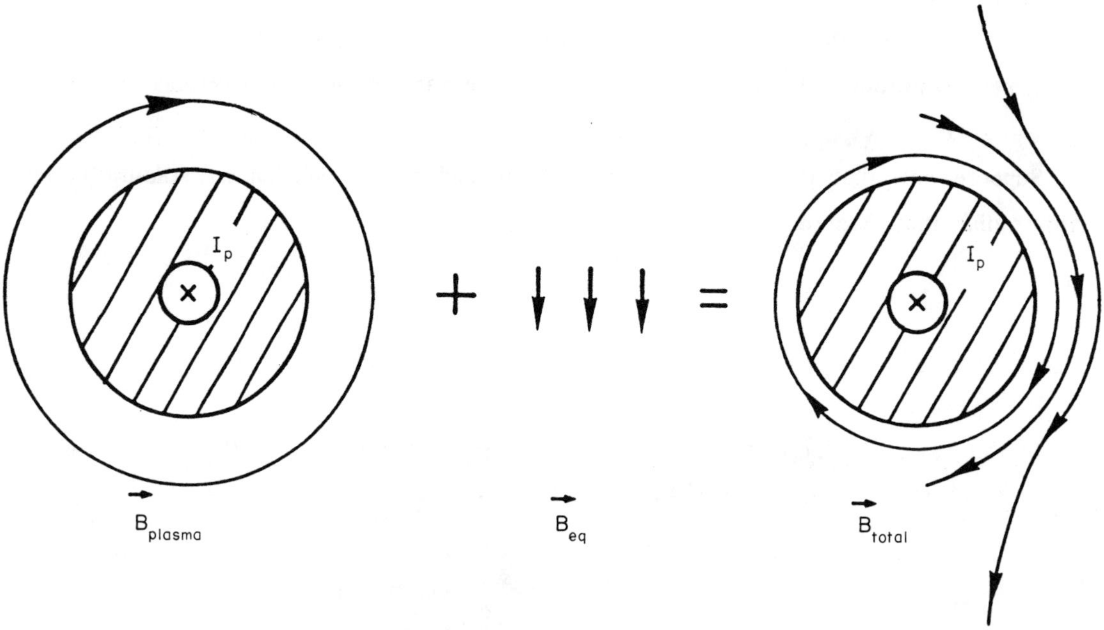

VII.1.4 Equation for the Horizontal Displacement Δ_H of the Plasma:

We must distinguish two cases: that of the perfect superconducting thick shell and that of the thin shell (vacuum vessel or liner).

The time for the magnetic field to penetrate into the thickness d_V of a vessel V of conductivity σ_V is equal to

(VII.57) $$\tau = \mu_0 \sigma_V d_V^2 .$$

The vessel V can be considered as a perfect shell if

(VII.58) $$T \ll \tau$$

where T is the duration of the discharge. In this case the eddy currents induced in the shell have no time to penetrate it, and it is the image currents that create the vertical equilibrium field B_{eq} given by (VII.56). The shell is then a magnetic surface.

Now consider the contrary hypothesis to (VII.58), namely

(VII.59) $\tau \ll T$.

In this case the currents have diffused in the shell and are distributed homogeneously throughout the thickness. This is the case in present Tokamaks.

Suppose now that the shell is a torus of circular section with minor radius b_o and major radius R_o. We then have the following proposition:

<u>Proposition VII.5</u>

In the case of a perfect shell the position Δ_H of the centre of the plasma relative to the centre of the shell is given by

(VII.60) $$\Delta_H = \Delta_o = \frac{b_o^2}{2R_o} \left[\text{Log} \frac{b_o}{a} + \left(1 - \frac{a^2}{b_o^2}\right)\left(\xi + \frac{1}{2}\right)\right] .$$

In the case of a thin vessel the equation for Δ_H is the following differential equation

(VII.61) $$\frac{d}{dt} \left[I_p(\Delta_H - \Delta_o)\right] = \frac{2\pi b_o^2}{\mu_o \tau_v} (B_v^3 - B_{eq})$$

where

$$\tau_v = \frac{\mu_o \, \sigma_v \, d_v \, b_o}{2}$$

is the attenuation time for the magnetic field in the shell and where B_v^3 is the vertical field imposed outside the vessel. □

<u>Proof</u>

In the case of a perfect shell, which is a magnetic surface, the position of its centre relative to the centre of the plasma is given by (VII.45) with $\rho' = b_o$. The position Δ_H of the plasma relative to the shell is nothing other than $-\Delta(b_o)$, which we denote by Δ_o and which is given by (VII.60).

In the case of a thin shell, by Ampere's theorem the orthoradial component $\mathbf{B}_\theta$ of **B** has a jump discontinuity across the vessel and, by the law of conservation of magnetic induction, the radial component $\mathbf{B}_\rho$ of **B** is continuous across the vessel. Thus for the components of order 1 of B_θ and B_ρ we have:

$$\text{(VII.62)} \qquad \begin{cases} (B_\theta^{(1)})_i - (B_\theta^{(1)})_e = \mu_o \, d_v \, j_v^{(1)} \\ \\ (B_\rho^{(1)})_i - (B_\rho^{(1)})_e = 0 \end{cases}$$

where the suffices i and e denote quantities taken in the interior and the exterior of the vessel respectively, and where $j_v^{(1)}$ denotes the dipolar component (in cos θ) of the current density in the vessel.

According to Faraday's Law and Ohm's Law, and using (VII.30), the transmission conditions (VII.62) can be written:

$$\text{(VII.63)} \qquad \begin{cases} \left(\dfrac{\partial\psi^{(1)}}{\partial\rho}\right)_i - \left(\dfrac{\partial\psi^{(1)}}{\partial\rho}\right)_e = - \mu_o \, d_v \, \sigma_v \, \dfrac{\partial\psi^{(1)}}{\partial t} \\ \\ (\psi^{(1)})_i - (\psi^{(1)})_e = 0 \, . \end{cases}$$

Using the expression (VII.51) for ψ we obtain

$$\text{(VII.64)} \qquad \begin{cases} \left[(C_2)_i - (C_2)_e - \dfrac{(C_1)_i - (C_1)_e}{b_o^2}\right] \cos\theta = - \mu_o \, d_v \, \sigma_v \, \dfrac{\partial\psi^{(1)}}{\partial t} \\ \\ (C_2)_i - (C_2)_e + \dfrac{(C_1)_i - (C_1)_e}{b_o^2} = 0 \, . \end{cases}$$

The equation for the shell in the plasma coordinate system is

$$\text{(VII.65)} \qquad \rho = b_o - \Delta_H \cos\theta \quad .$$

On the vessel, the term of order 1 in ψ given by (VII.42) is

$$\text{(VII.66)} \qquad \psi^{(1)} = \frac{\mu_o \, R_o \, I_p}{2\pi \, b_o} (\Delta_H - \Delta_o) \cos\theta$$

where Δ_o is given by (VII.60).

With the term $C_2\rho \cos\theta$ representing the contribution of the vertical field outside the plasma, we have

$$\text{(VII.67)} \quad \begin{cases} (C_2)_e = R_o\, B_V^3 \\ \\ (C_2)_i = R_o\, B_{eq} \,. \end{cases}$$

By eliminating C_1 between the two equations of (VII.64) and using (VII.66) and (VII.67) we obtain the differential equation (VII.61) for Δ_H. □

VII.2 MODELLING ENSEMBLE OF THE PLASMA AND EXTERNAL CIRCUITS CONFIGURATION IN TFR:

Our aim is to write down the evolution equations for the currents in the various circuits of TFR as well as for the total plasma current I_p. To do this, we need first of all to study the flux pattern in the machine, solving a magnetostatic problem.

VII 2.1 The Magnetostatic Problem

We have to determine the flux function $\psi(r,z)$ that satisfies

$$\text{(VII.68)} \qquad L\psi = j_T$$

where j_T denotes the toroidal current density in the coils and the plasma, and where the operator L is defined by (I.13). This operator is nonlinear in the iron and linear elsewhere (air, coils, plasma). The current density j_T is given and assumed homogeneous in each of the coils B_i. The boundary condition for ψ is the condition (I.25), while the transmission conditions on the air– iron interface are the conditions (I.26).

VII.2.1.1 Expression for the Plasma Current Density

The plasma boundary is here fixed and circular. It remains to define the toroidal current density in the plasma as a function of r and z, which cannot be arbitrary for it must satisfy the equilibrium equation (I.4). To do this we shall use the expression

(VII.25) for j_T to first order in ε. The expressions for j_T and p to order 0 are chosen a priori as follows:

$$\text{(VII.69)} \quad \begin{cases} j_T^{(o)}(\rho) = j_o(1 - \frac{\rho^2}{a^2})^{q_o} \\ p^{(o)}(\rho) = p_o(1 - \frac{\rho^2}{a^2}) \end{cases}$$

where the parameter q_o is free. By (VII.38) and (VII.19) we then have

$$\text{(VII.70)} \quad \begin{cases} B_\theta^{(o)}(\rho) = \frac{\mu_o\ a^2\ j_o}{2(q_o+1)\rho}\ [(1 - \frac{\rho^2}{a^2})^{q_o+1} - 1] \\ \bar{p}^{(o)}(\rho) = p_o(1 - \frac{\rho^2}{2a^2})\ . \end{cases}$$

With the help of (VII.15) and (VII.26) we deduce $\xi(\rho)$ and $\Delta(\rho)$ from this, and then these are used to calculate $j_T^{(1)}$ via (VII.25).

Let us examine some particular cases:

. flat current : $q_o = 0$

In this case we have:

$$\text{(VII.71)} \quad \begin{cases} B_\theta^{(o)}(\rho) = - \frac{\mu_o\ j_o\ \rho}{2} \\ \ell_i(\rho) = \frac{1}{2} \\ \beta_p(\rho) = \beta_p = \frac{4p_o}{\mu_o\ a^2\ j_o} \\ \xi(\rho) = \beta_p - 0.75 \\ \Delta(\rho) = \frac{\beta_p + 0.25}{2R_o}\ (a^2 - \rho^2) \\ j_T^{(1)}(\rho,\theta) = (2\beta_p - 1)\ \frac{j_o\ \rho\ \cos\ \theta}{R_o}\ . \end{cases}$$

. parabolic current : $q_0 = 1$.

If we put $X = \frac{\rho^2}{2a^2}$ we have

$$
(VII.72)\quad
\begin{cases}
B_\theta^{(0)}(\rho) = -\dfrac{\mu_0\, j_0\, \rho}{2}\,(1 - X) \\[2ex]
\ell_i(\rho) = \dfrac{1 - 4X/3 + X^2/2}{2(1 - X)^2} \\[2ex]
\beta_p(\rho) = \dfrac{\beta_p}{4(1-X)^2} \quad \text{with} \quad \beta_p = \dfrac{16 p_0}{\mu_0\, a^2\, j_0^2} \\[2ex]
\xi(\rho) = \dfrac{\beta_p - 3 + 20X/3 - 7X^2/2}{4(1-X)^2} \\[2ex]
\Delta(\rho) = \dfrac{a^2}{R_0}\left[\dfrac{\beta_p}{2} + \dfrac{7}{48} - \dfrac{X}{8} - \dfrac{6\beta_p + 1}{24(1-X)} + 1/12\ \mathrm{Log}(2-2X)\right] \\[2ex]
j_T^{(1)}(\rho,\theta) = \left[\beta_p - \dfrac{17}{24} + \dfrac{7X}{4} - \dfrac{1}{12(1-X)} + \dfrac{1}{6}\,\mathrm{Log}(2-2X)\right]\dfrac{\rho\, j_0 \cos\theta}{R_0}.
\end{cases}
$$

The plasma current density is the more "peaked", the greater the value of q_0. This "peaking" is in general characterized by the coefficient 1_i which is 0.5 in the case of a flat current and 11/12 in the case of parabolic current. We have similarly studied cases where the current is even more peaked: $q_0 = 2$ (corresponding to $1_i = 1.22$) and $q_0 = 3$ (corresponding to $1_i = 1.45$). When q_0 and β_p are fixed the plasma current density is a given function of ρ and θ, and the right hand side of (VII.68) is well– defined.

VII.2.1.2 Numerical Solution

The problem of determining the function $\psi(r, z)$ satisfying (VII.68) differs from the problem treated in Chapter I by the fact that the plasma boundary is fixed and the current density is a given function of ρ and θ. Therefore the only nonlinearity in the problem is the local magnetic permeability μ of the iron which is a given function of B_p^2.

This problem has been solved by means of the code MAGNETX (cf. J. LE FOLL – B. THOORIS). This code uses Newton iterations to resolve the nonlinearity connected with the function $\bar{\mu}(B_p^2)$ and conformal finite elements P_1 for the discretization in space. We are going to use this tool now to determine the flux pattern in TFR.

VII.2.2 The Flux Pattern in TFR

Recall (cf. Section I.5.3 and Fig. I.10) that the Tokamak TFR has only two types of coil : the internal coils (fixed against the iron core) and the external coils (fixed against the return arms). Systematic studies have been carried out with the magnetostatic code for the currents in the internal and external coils separately, then with plasma current exclusively, and finally with currents simultaneously in the coils and the plasma. Such simulations have been performed for various degrees of saturation of the iron, various positions of the plasma, and the different values of β_p and q_0 in the expression for the plasma current density. The vacuum vessel (Fig.VII.2) is a torus of

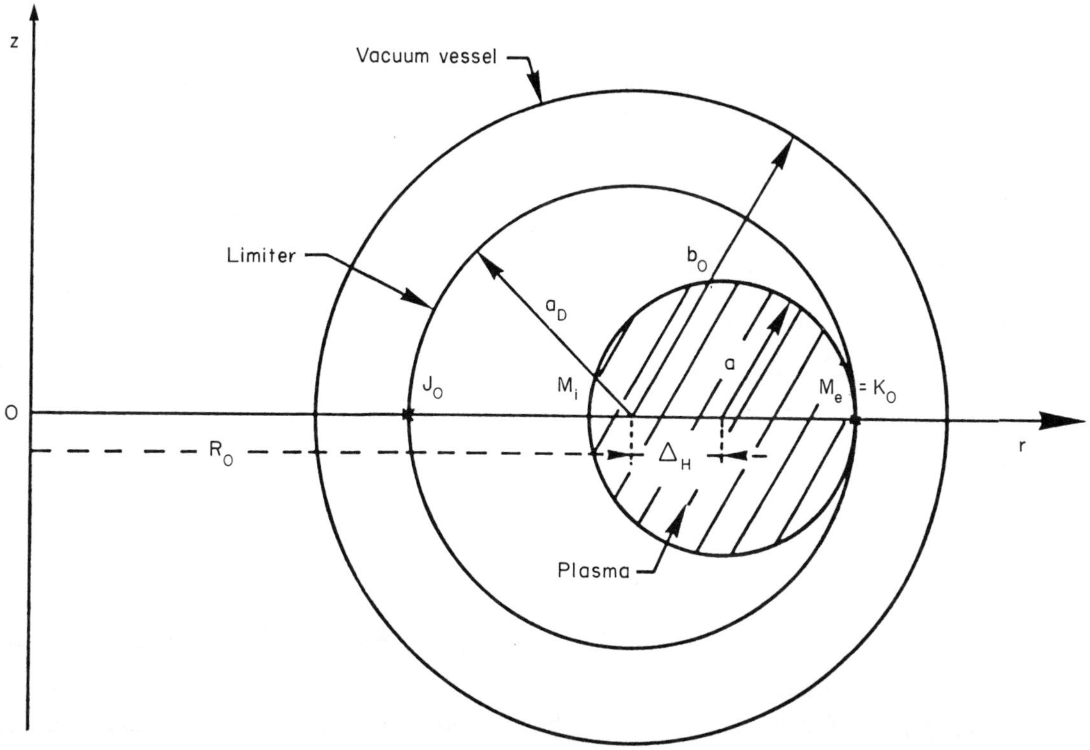

Fig.VII.2 : Position of the vacuum vessel, limiter and plasma

circular section with major radius R_o = 98 cm and minor radius b_o = 26.5 cm, the limiter is a circle of minor radius a_D = 24 cm concentric with the vacuum vessel. The plasma boundary is the circle tangent internally to the limiter D; its position is defined by the horizontal distance Δ_H by which the centre of the plasma is displaced relative to the centre of the vacuum vessel. If Δ_H is positive the point of contact between the plasma and the limiter is the outer point K_o; if Δ_H is negative it is the inner point J_o. The minor and major radius of the plasma are then given by the relations:

(VII.73) $$\begin{cases} a = a_D - |\Delta_H| \\ R = R_o + \Delta_H . \end{cases}$$

From simulations carried out using the code MAGNETX we obtain the following laws describing the flux pattern in TFR.

VII.2.2.1 *Expression for the Flux ψ_F in the Iron Core, as a Function of the Sum of Ampère–Turns*:

This flux is defined by

(VII.74) $$\psi_F = 2\pi\ \psi(F)$$

where F is the point of the equatorial plane which is the boundary between the iron core and the air (see Fig. VII.3).

The algebraic sum of Ampère– turns is defined by

(VII.75) $$\Sigma I = I_i + I_e + K(\Delta_H) I_p$$

with

$$K(\Delta_H) = k_o + k_1\ \Delta_H$$

$$k_o = 1.02\ ,\ k_1 = -0.5 \times 10^{-3}\ \mathrm{cm}^{-1}$$

where I_i and I_e are the total currents in the internal and external coils respectively. The sum of Ampere turns is weighted by the coefficient K because the plasma, which is in the equatorial plane, has a contribution slightly higher than that of the coils which are not.

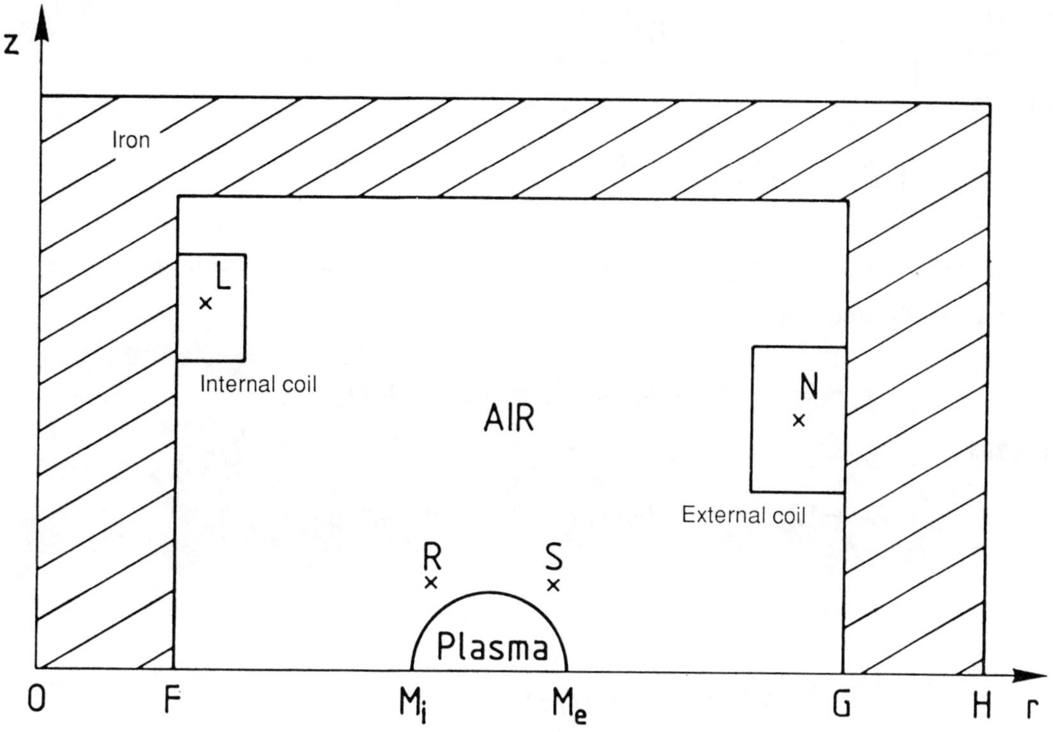

Fig.VII.3 : Characteristic points of a meridian section of TFR

From the simulation with the magnetostatic code, confirmed by experimental tests, we find that ψ_F depends only on Σ I:

(VII.76) $$\psi_F = G(\Sigma\ I)$$

where G is given by the graph in Fig. (VII.4). The quantity ψ_F characterizes the degree of saturation of the iron core.

VII.2.2.2 Expressions for the Leakage Fluxes ψ_i and ψ_e Between the Equatorial Plane and the Coils:

These fluxes are defined as follows (see Fig.VII.3):

$$\text{(VII.77)} \quad \begin{cases} \psi_i = 2\pi[\psi(L) - \psi(F)] \\ \\ \psi_e = 2\pi[\psi(N) - \psi(G)] \,. \end{cases}$$

They can be expressed quasi-linearly in terms of the currents, with the coefficients depending on ψ_F and Δ_H.

$$\text{(VII.78)} \quad \begin{cases} \psi_i = L_1(\psi_F) I_i + L_2(\psi_F) I_e + L_3(\psi_F, \Delta_H) I_p \\ \\ \psi_e = L_4(\psi_F) I_i + L_5(\psi_F) I_e + L_6(\psi_F, \Delta_H) I_p \end{cases}$$

with

$$L_3(\psi_F, \Delta_H) = L_3^o(\psi_F) + L_3' \, \Delta_H$$

$$L_6(\psi_F, \Delta_H) = L_6^o(\psi_F) + L_6' \, \Delta_H \,.$$

The functions L_1, L_2, L_3, L_4, L_5 and L_6 have been identified by simulations with the currents I_i, I_e and I_p separately, and the additivity has been verified with an accuracy of 10% when these currents exist simultaneously. These functions are shown in Fig.VII.5 as functions of ψ_F, while

$$L_3' = -L_6' = 0.22 \ \mu H/cm \,.$$

VI.2.2.3 Expressions for Leakage Fields in the Equatorial Plane

Consider the following three regions:

The first (j=1) region lying between the points F and G in the equatorial plane (see Fig.VII.3), so that B_v^1 is the mean field in the air:

(VII.79) $$B_V^1 = \frac{2\pi[\psi(G) - \psi(F)]}{\sigma_1}$$

with $\sigma_1 = \pi(r_G^2 - r_F^2)$ where r_G and r_F are the abscissae of the points G and F.

The second (j = 2) region lying between F and M_i, the innermost point of the plasma, so that B_V^2 is the mean field between the iron core and the plasma:

(VII.80) $$B_V^2 = \frac{2\pi[\psi(M_i) - \psi(F)]}{\sigma_2}$$

with $\sigma_2 = \pi(r_i^2 - r_F^2)$ where r_i is the abscissa of M_i.

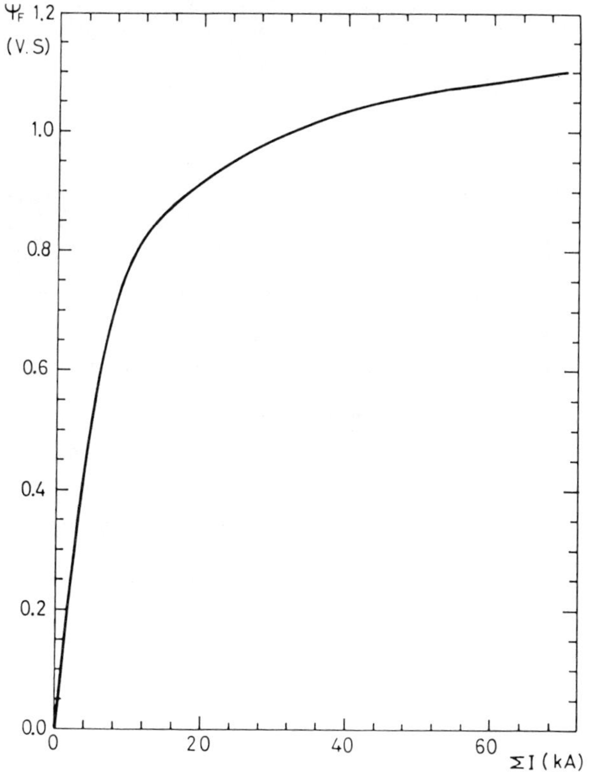

Fig.VII.4 : Saturation curve for the iron core

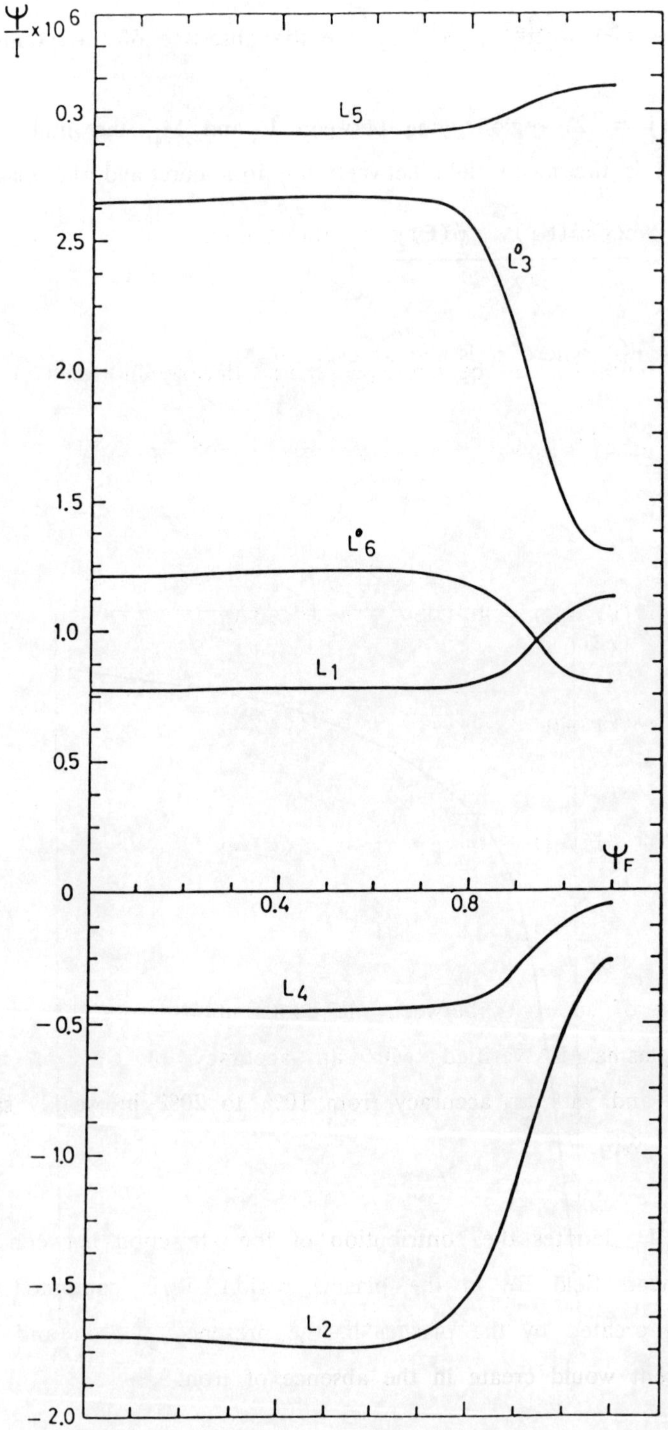

Fig. VII.5 : Leakage self−inductances between the equatorial plane and the coils.

The third (j = 3) region is that of the plasma between M_i and M_e, so that B_V^3 is the mean field in the plasma:

(VII.81) $$B_V^3 = \frac{2\pi[\psi(M_e) - \psi(M_i)]}{\sigma_3}$$

with $\sigma_3 = \pi(r_e^2 - r_i^2)$

where r_e is the abscissa of M_e. Clearly, the positions of M_i and M_e depend on Δ_H (see Fig.VII.2).

The leakage fields B_V^j can be expressed quasi-linearly in terms of the currents:

(VII.82) $$B_V^j = b_i^j(\psi_F,\Delta_H)I_i + b_e^j(\psi_F,\Delta_H)I_e + b_p^j(\psi_F,\Delta_H)I_p$$

with

$$\begin{cases} b_e^j(\psi_F,\Delta_H) = \overline{b_e^j}(\psi_F) + b_e'^j(\psi_F)\Delta_H \\ b_i^j(\psi_F,\Delta_H) = \overline{b_i^j}(\psi_F) + b_i'^j(\psi_F)\Delta_H \\ b_p^j(\psi_F,\Delta_H) = \overline{b_p^j}(\psi_F) + b_p'^j(\psi_F)\Delta_H \end{cases}$$

$$j \in \{1,2,3\} .$$

The principle of additivity between the contributions from the external and internal coils and the plasma is verified with an accuracy of 5% in nonsaturated cases ($|\psi_F| < 0.8$Wb) and with an accuracy from 10% to 20% in weakly saturated cases (ψ_F of the order of 1 Wb).

The term $b_p^3 I_p$ denotes the contribution of the attraction between the iron and the plasma to the mean field B_V^3 in the plasma region; it is computed as the difference between the field created by the plasma in the presence of iron and the field that the same plasma current would create in the absence of iron.

The functions $\overline{b_i^3}$, $\overline{b_e^3}$ and $\overline{b_p^3}$ are shown as functions of ψ_F in Figs.VII.6. The coefficients $b_e'^1$ and $b_i'^1$ are zero; the functions $b_i'^3$, $b_e'^3$ and $b_p'^3$ are shown as functions of ψ_F in Figs.VII.7.

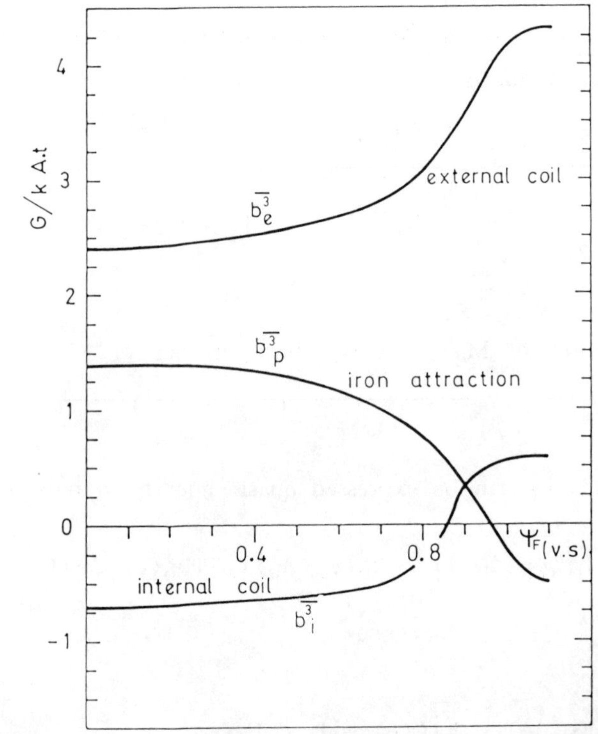

Fig. VII.6 : Contributions of the currents of the internal coils, external coils and plasma to the vertical leakage fields in the plasma region (coefficients $\overline{b_i^3}$, $\overline{b_e^3}$, $-\overline{b_p^3}$).

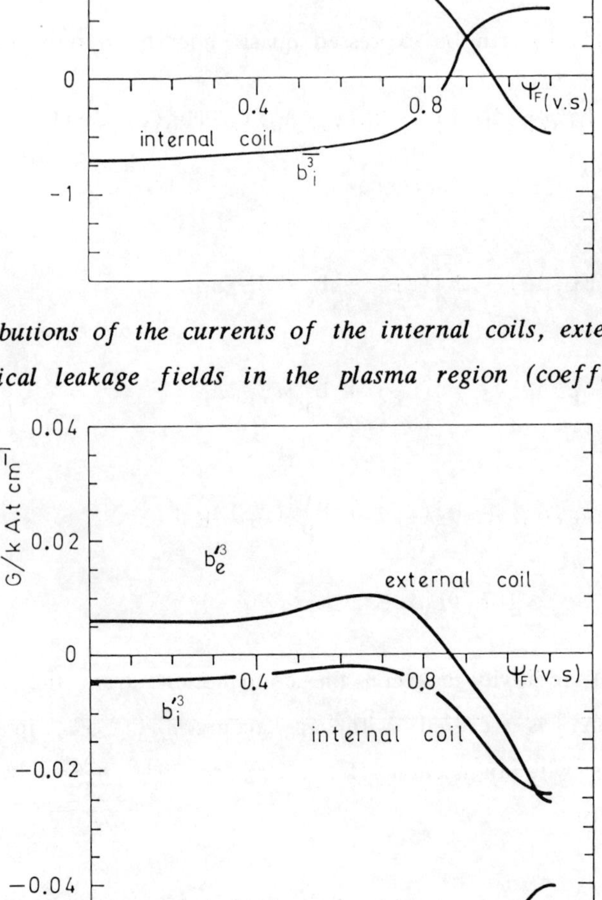

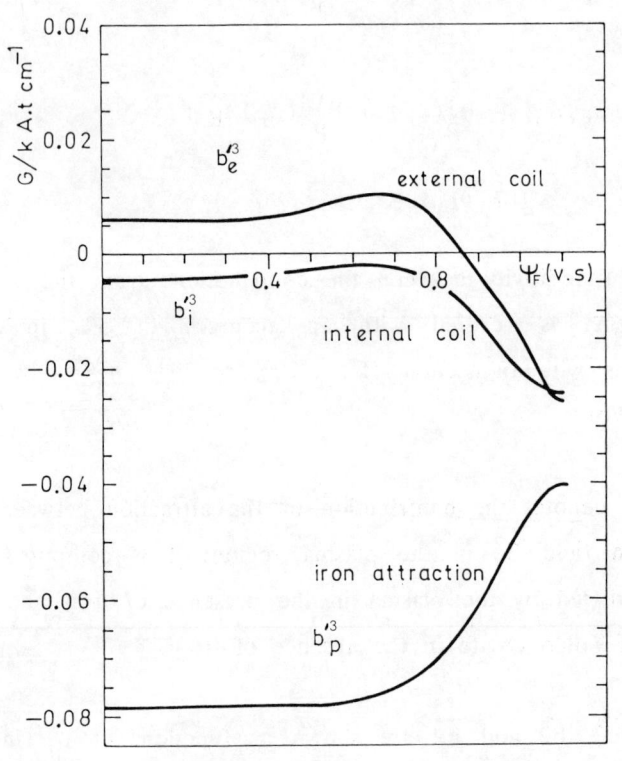

Fig. VII.7 : Gradients of these contributions with respect to the plasma displacement Δ_H *(coefficients* $b_i'^3$, $b_e'^3$, $-b_p'^3$).

The mean fields created by the internal and external coils were measured on the machine by flux loops between the points R and S (cf.Fig.VII.3) for various degrees of saturation of the iron, and also computed by the magnetostatic code. The comparison between measurements and computed values is shown in Fig.VII.8. Agreement between measurements and computations is realized to within 10%.

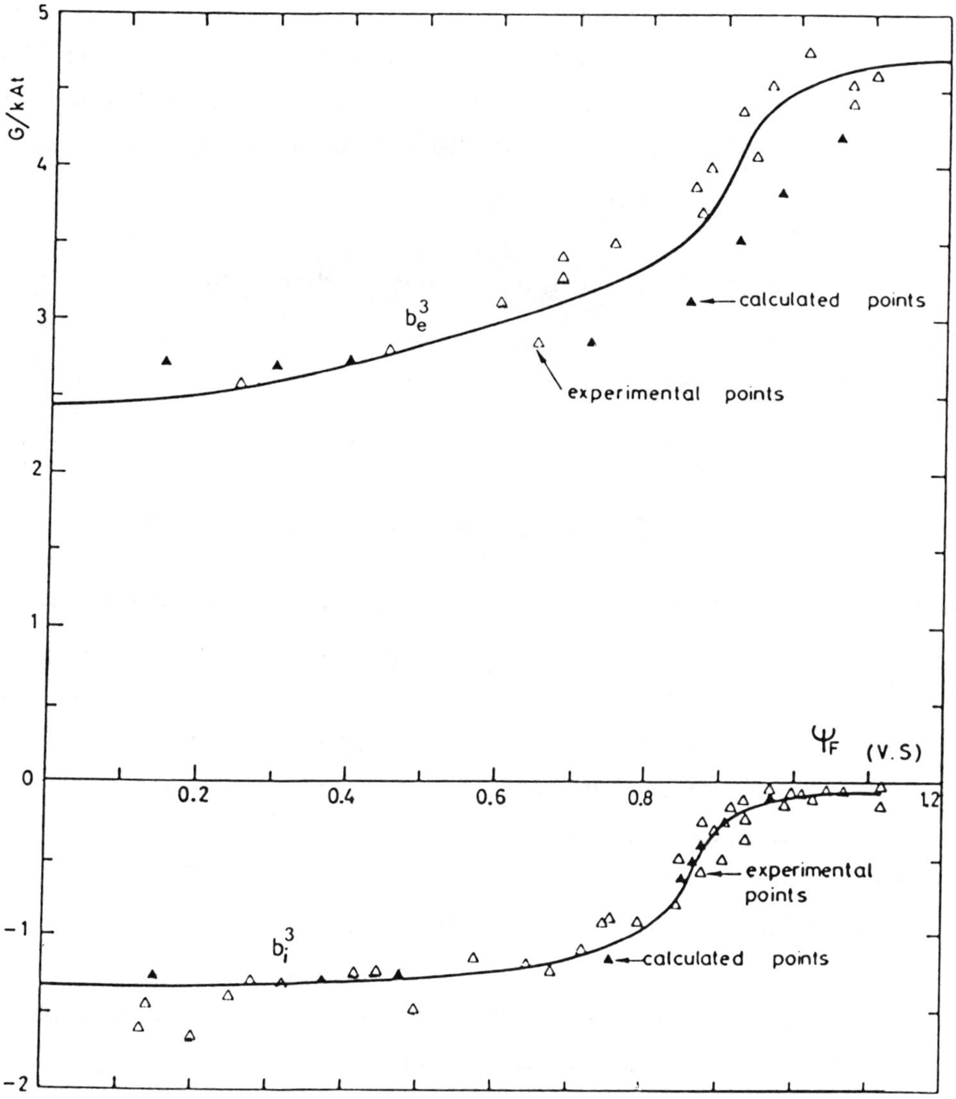

Fig.VII.8 : Comparison between measurements and computations for the contribution of the internal and external coils to the leakage field as a function of the degree of saturation of the iron.

VII.2.2.4 Determination of an Equilibrium Configuration

The field B_V^3 is nothing other than the vertical field in the plasma region imposed from outside. By (VII.61) the plasma is in equilibrium when

$$B_V^3 = B_{eq} \ . \tag{VII.83}$$

For the plasma its total current I_p, its position Δ_H, and its internal parameters β_p and q_0, are all given, as is the flux ψ_F in the iron core. The field B_{eq} given by (VII.56) is then calculated, as well as the sum Σ I of the Ampère–turns using (VII.76). From (VII.83), (VII.82) and (VII.76), in order to compute the currents I_i and I_e in the internal and external coils that realize the plasma equilibrium we have to solve the following linear system:

$$\begin{cases} b_i^3(\psi_F,\Delta_H)\,I_i + b_e^3(\psi_F,\Delta_H)\,I_e = B_{eq} - b_p^3(\psi_F,\Delta_H)\,I_p \\ I_i + I_e = G^{-1}(\psi_F) - K(\Delta_H)\,I_p \ . \end{cases} \tag{VII.84}$$

In this way for I_p = 200 kA, Δ_H = 0, ψ_F = 0.1 Wb, β_p = 0.5, q_0 = 3 we obtain

$$I_e = -\ 119 \text{ kA} \ , \quad I_i = -\ 85 \text{ kA} \ .$$

Using the magnetostatic code we can then compute the configurations obtained when these values of current are introduced into the plasma and the coils. This configuration is shown in Fig.VII.9. In computing this configuration the plasma boundary was assumed fixed and circular. Now, we observe that the flux lines in this region are not perfectly circular, so the fixed plasma boundary is therefore not a flux line. This outcome is not self–consistent; it distinguishes this type of method, using a magnetostatic code with fixed plasma boundary having current density given to 1st order in ε, from the self–consistent approach with free boundary defined in Chapter I.

The relations established in this section will enable us to formulate a simple model describing the set of external plasma circuits in TFR.

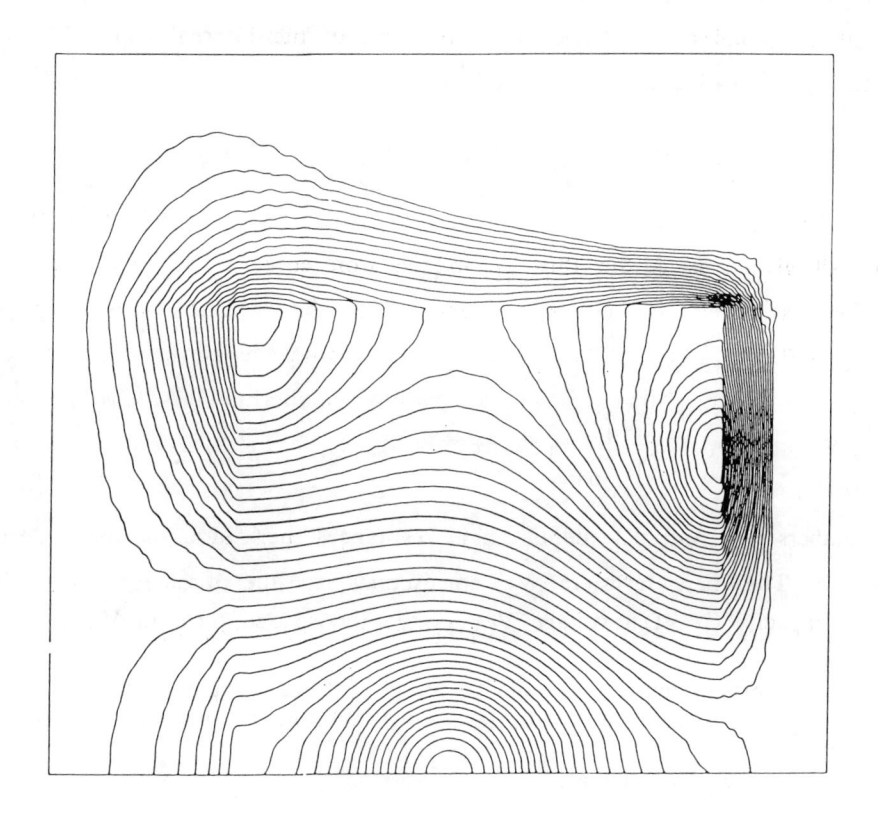

Fig. VII.9 : Example of an equilibrium configuration obtained by the magnetostatic code.

VII.2.3 The Equations for the Various Circuits in TFR

From the laws established in the previous section we are in a position to state the equations for the various circuits (primary, plasma, pre-programming, feedback) in TFR. The flux created by the internal or external coils or by the plasma can be deduced from ψ_F, ψ_i, ψ_e and the B_V^j. Each turn of the internal coil gives rise to the flux $\psi_F + \psi_i$, each turn of the external coil to the flux $\psi_F + \sigma_1 B_V^1 + \psi_e$, and the plasma ring to the flux $\psi_F + \sigma_2 B_V^2$.

VII.2.3.1 Equations of the Primary Circuit

The primary circuit, whose purpose is to induce current in the plasma, consists of n_1^e turns of the external coil in series with n_1^i turns of the internal coil. The equation of the current I_1 circulating in this circuit is

$$(n_1^e + n_1^i)\frac{d\psi_F}{dt} + n_1^e[\sigma_1 \frac{dB_V^1}{dt} + \frac{d\psi_e}{dt}] + n_1^i \frac{d\psi_i}{dt}$$

$$\text{(VII.85)} \qquad + (n_1^e R_1^e + n_1^i R_1^i)I_1 = V_1$$

with

$$V_1 = [V_{C_1} - \frac{1}{C_1}\int_0^t I_1(\tau)d\tau]^+ + G_1(I_p - I_p^{ref}) .$$

The real numbers R_1^e and R_1^i denote the resistances per turn of the external and internal coils. The first term in V_1 corresponds to the discharge of a bank of condensers of capacity C_1 of a voltage V_{C_1}, while the second term in V_1 is the voltage that maintains the plasma current around a certain reference value I_p^{ref}, with G_1 being the feedback gain.

VII.2.3.2 Equation for the Total Plasma Current

This may be written

$$\text{(VII.86)} \qquad \frac{d\psi_F}{dt} + \frac{d}{dt}(\sigma_2 B_V^2) = V_p$$

where V_p is the loop voltage at the plasma boundary.

VII.2.3.3 Equation for the Pre–Programming Circuit

This circuit, which is intended to bring about equilibrium of the plasma, consists of n_3 external windings and the same number of internal windings, mounted in opposition so that they induce no flux in the iron core. In this way a current dipole is created that generates the additional vertical field necessary to realize equilibrium of the plasma. The equation for the current I_3 that flows around this circuit at a voltage V_3 is then

$$n_3(\sigma_1 \frac{dB_v^1}{dt} + \frac{d\psi_e}{dt} - \frac{d\psi_i}{dt}) + L_3 \frac{dI_3}{dt}$$

(VII.87)

$$+ [n_3(R_1^e + R_1^i) + R_3]I_3 = V_3$$

where L_3 and R_3 are the self-inductance and the resistance introduced in order to decouple the pre-programming and feedback circuits.

VII.2.3.4 Equation for the Feedback Circuit

This circuit, aiming to stabilize the horizontal displacements of the plasma, consists of n_4 external windings and the same number of internal windings mounted in opposition. The equation for the current I_4 in this circuit at voltage V_4 is

(VII.88) $\quad n_4(\sigma_1 \dfrac{dB_v^1}{dt} + \dfrac{d\psi_e}{dt} - \dfrac{d\psi_1}{dt}) + n_4(R_1^e + R_1^i) I_4 = V_4$.

VII.2.3.5 The System of Differential Equations for the Circuit Currents

We disregard the equation for the homogeneous current in the vacuum vessel since this current is negligible compared with the plasma current.

The functions ψ_F, ψ_e, ψ_i and B_v^j appearing in equations (VII.85) — (VII.88) are given by (VII.76), (VII.78) and (VII.82) with

(VII.89) $\quad \begin{cases} I_e = n_1^e I_1 + n_3 I_3 + n_4 I_4 \\ I_i = n_1^i I_1 - n_3 I_3 - n_4 I_4 + I_{PR} \\ \Sigma I = (n_1^e + n_1^i)I_1 + I_p + I_{PR} \end{cases}$

where I_{PR} is a constant premagnetization current whose purpose is to set off the discharge with a flux in the iron core of the order of 1 Wb.

To the circuit equations (VII.85) — (VII.88) there must be added the equation (VII.61) for the horizontal displacement Δ_H of the plasma, where B_v^3 is given by (VII.82), B_{eq} by (VII.56), and a and R by (VII.73).

If I denotes the vector $\{I_1, I_p, I_3, I_4, \Delta_H\}$ then the system of equations (VII.85) – (VII.88) and (VII.61) can be written

$$M(I,\xi)\frac{dI}{dt} + R(I, \xi, \frac{d\xi}{dt})I = V(t,I) \tag{VII.90}$$

where M and R are 5 × 5 matrices depending on I, ξ and $d\xi/dt$, and where V is the vector $(V_1, V_p, V_3, V_4, 0)$ (recall that $\xi = \beta_p + 1/2\,\ell_i - 1$). The 4 × 4 submatrix of M relating to the circuit equations is the matrix of self–inductances and mutual inductances, and the 4 × 4 submatrix of R is that of resistances.

The initial condition for the system (VII.90) is

$$I(0) = I_o \,. \tag{VII.91}$$

<u>*Remark VII.4*</u> *:* <u>*Generalization for other Tokamaks:*</u>

The description of the flux pattern given in Section VII.2.2 is for a Tokamak with weakly saturated iron. In the case of a Tokamak with air transformer, since the operator L is linear the self–inductances and mutual inductances of the various circuits are constant, and can be deduced from the coefficient of mutual induction between two coaxial windings of radii a and b situated in two planes separated by a distance c (cf. E. DURAND):

$$M = \frac{\mu_o}{2\pi k}\sqrt{ab}\,[(1 - \frac{k^2}{2})\,\mathcal{F}_1(k) - \mathcal{F}_2(k)] \tag{VII.92}$$

with

$$k = \sqrt{\frac{4ab}{(a+b)^2+c^2}}$$

and where

$$\mathcal{F}_1(k) = \int_o^{\pi/2} \frac{d\theta}{\sqrt{1-k^2\sin^2\theta}}$$

$$\mathcal{F}_2(k) = \int_o^{\pi/2} \sqrt{1-k^2\sin^2\theta}\,d\theta$$

are complete elliptic integrals of the first and second kind.

In the case of a Tokamak with iron transformer which can be saturated (as in TORE Supra or JET) the additivity principles of Section VII.2.2 are no longer satisfied; the matrix M of self– and mutual inductances can nevertheless be calculated as follows.

The definition of the mutual inductance between two circuits i and j in the presence of a ferromagnetic substance is

(VII.93) $$M_{ij} = 2\pi \frac{\partial \bar{\psi}_i}{\partial I_j}$$

where $\bar{\psi}_i$ is the average of ψ over the coil B_i. By linearizing equation (VII.68), the variation $\partial\psi$ of ψ resulting from a variation δj in the current density may be calculated by

(VII.94) $$(L + L')\,\delta\psi = \delta j$$

where L' is given in Remark I.11. From this it is easy to deduce the matrix of the M_{ij} around a particular equilibrium configuration. To obtain the matrix M for a complete discharge it is necessary to consider a succession of equilibrium states, calculate this matrix for each of them and interpolate between two consecutive equilibria.

Even though the method for obtaining the system (VII.90) that has been presented in Sections VII.2.2 and VII.2.3 is characteristic of TFR, it is nevertheless the case that the expression for this system in the form (VII.90) is equally valid for air and saturated iron transformers. □

VII.2.4 Coupling with a Transport Code

The system (VII.90) requires the function $\xi(t)$ and the loop voltage $V_p(t)$ at the plasma boundary to be known. To simulate $\beta_p(t)$, $l_i(t)$ and $V_p(t)$ we use the system of transport equations from Section VI.1.2, taking account of the circular shape of the lines of flux and the high aspect ratio.

The toroidal field B_T is assumed to be constant inside the plasma, and equal to the external toroidal field B_o, so that the variable ρ defined by (VI.44) is identified with the minor radius of the magnetic surface being considered.

The volume V inside this surface is $2\ \pi^2\ R_o\ \rho^2$, so that

(VII.95) $$V' = 4\pi^2 R_o \rho .$$

The geometrical coefficients defined on the magnetic surfaces are then equal to:

(VII.96) $$\begin{cases} <|\nabla_\rho|^2> = 1 \\ <|\nabla_\rho|^2/r^2> = <1/r^2> = 1/R_o^2 . \end{cases}$$

Hence

(VII.97) $$C_1 = 4\pi^2 R_o \rho \qquad C_2 = C_3 = \frac{4\pi^2 \rho}{R_o} .$$

Thus we easily have the following proposition:

Proposition VII.6

To order 0 in ε (circular cylindrical approximation) the diffusion equations (VI.68), (VI.69), (VI.70) and (VI.45) take the form:

(VII.98) $$\frac{\partial n_e}{\partial t} - \frac{1}{\rho}\frac{\partial}{\partial \rho}(D\rho \frac{\partial n_e}{\partial \rho}) = s_1$$

(VII.99) $$\frac{3}{2}\frac{\partial p_e}{\partial t} - \frac{1}{\rho}\frac{\partial}{\partial \rho}(K_e\ \rho \frac{\partial T_e}{\partial \rho}) - \frac{5k}{2\rho}\frac{\partial}{\partial \rho}(DT_e\ \rho \frac{\partial n_e}{\partial \rho}) = \eta_{\parallel}\ j_T^2 + \frac{D}{n_e}\frac{\partial n_e}{\partial \rho}\frac{\partial p_i}{\partial \rho} - Q_\Delta + s_2$$

(VII.100) $$\frac{3}{2}\frac{\partial p_i}{\partial t} - \frac{1}{\rho}\frac{\partial}{\partial \rho}(K_i\ \rho \frac{\partial T_i}{\partial \rho}) - \frac{5k}{2\rho}\frac{\partial}{\partial \rho}(\frac{DT_i \rho}{Z}\frac{\partial n_e}{\partial \rho}) + \frac{D}{n_e}\frac{\partial n_e}{\partial \rho}\frac{\partial p_i}{\partial \rho} = Q_\Delta + s_3$$

(VII.101) $$\frac{\partial \psi}{\partial t} - \frac{\eta_{\parallel}}{\mu_o \rho}\frac{\partial}{\partial \rho}(\rho \frac{\partial \psi}{\partial \rho}) = 0 .$$ □

The proof is clear, using (VII.95), (VII.97) and the fact that in the cylindrical approximation

$$j_T = \frac{1}{\mu_0 R_o \rho} \frac{\partial}{\partial \rho} (\rho \frac{\partial \psi}{\partial \rho}) . \tag{VII.102}$$

Note that differentiating (VII.101) with respect to ρ and using (VII.30) gives the diffusion equation for B_θ:

$$\frac{\partial B_\theta}{\partial t} - \frac{1}{\mu_0} \frac{\partial}{\partial \rho}[\frac{\eta_{||}}{\rho} \frac{\partial}{\partial \rho} (\rho B_\theta)] = 0 . \tag{VII.103}$$

Differentiating (VII.103) with respect to ρ and using (VII.102) gives the diffusion equation for j_T:

$$\frac{\partial j_T}{\partial t} - \frac{1}{\mu_o \rho} \frac{\partial}{\partial \rho} [\rho \frac{\partial}{\partial \rho} (\eta_{||} j_T)] = 0 . \tag{VII.104}$$

The boundary condition at $\rho = a$ for (VII.104) is obtained by differentiating with respect to time the equation that defines the total plasma current I_p:

$$I_p = 2\pi \int_0^{a(t)} j_T \rho \, d\rho ,$$

and by using (VII.104) the following mixed condition is obtained:

$$\frac{dI_p}{dt} = 2\pi a [\frac{1}{\mu_0} \frac{\partial}{\partial \rho} (\eta_{||} j_T) + \frac{da}{dt} j_T]_{\rho = a} . \tag{VII.105}$$

The boundary conditions at $\rho = a$ for the equations (VII.98) – (VII.100) are the Dirichlet conditions (VI.71) which here take the form:

$$\begin{cases} n_e(a) = n_e^o \\ T_e(a) = T_e^o \\ T_i(a) = T_i^o . \end{cases} \tag{VII.106}$$

Finally, for $\rho = 0$ we have the following symmetry conditions:

$$\frac{\partial n_e}{\partial \rho}(0) = \frac{\partial T_e}{\partial \rho}(0) = \frac{\partial T_i}{\partial \rho}(0) = \frac{\partial j_T}{\partial \rho}(0) = 0 . \tag{VII.107}$$

Consider the system of equations (VII.98), (VII.99), (VII.100) and (VII.104). If u denotes the vector $\{n_e, T_e, T_i, j_T\}$ this system can be formulated as follows:

$$
\text{(VII.108)} \quad
\begin{cases}
\dfrac{\partial u}{\partial t} + \mathscr{A}_o(u) = s \\[2ex]
\dfrac{\partial u}{\partial \rho} = 0 \qquad \text{for} \qquad \rho = 0 \\[2ex]
\alpha_o(u)\, \dfrac{\partial u}{\partial \rho} + \beta_o(u, I)u = \gamma_o(I) \qquad \text{for} \qquad \rho = a \\[2ex]
u = u_o(\rho) \qquad \text{for} \qquad t = 0
\end{cases}
$$

where $\mathscr{A}_o$ is a 2nd order nonlinear operator (of parabolic type) and s is a source– vector.

The voltage per turn V_p and the parameters β_p and ℓ_i are obtained from u by the following relations deduced from (VII.101), (VII.15) and (VII.21):

$$
\text{(VII.109)} \quad
\begin{cases}
V_p = 2\pi\, R_o(\eta_{\parallel}\, j_T)\big|_{\rho=a} \\[2ex]
\beta_p = \dfrac{8\pi^2\, a^2\, k\, \overline{n_e\, (T_e + T_i)}}{\mu_o\, I_p^2} \\[2ex]
\ell_i = \dfrac{8\pi^2 \displaystyle\int_o^a B_\theta^2(\rho)\, \rho\, d\rho}{\mu_o^2\, I_p^2} \, .
\end{cases}
$$

The complete system (circuits + plasma) can then be represented as a coupling between the system (VII.90) of the differential equations for the circuits, and the system (VII.108) of the equations of transport, as follows:

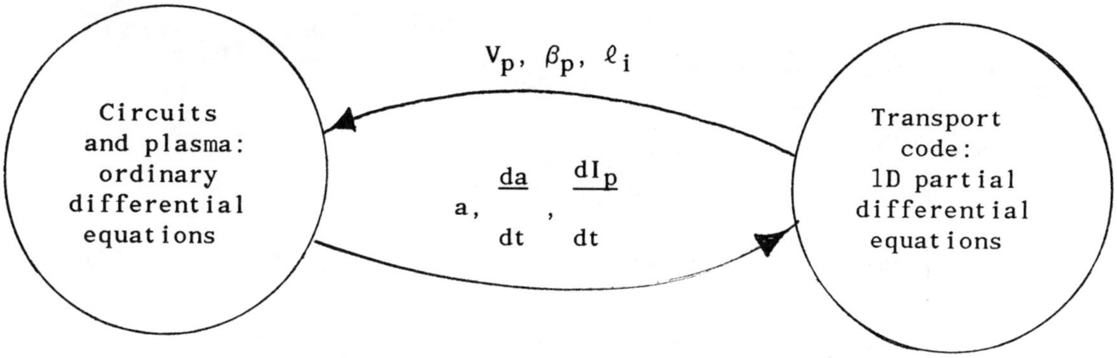

VII.2.5 *Numerical Solution*

At each time-step we first solve the system (VII.90) of differential equations by a modified Euler method. If I^n denotes I at the instant ($n\Delta t$) then I^{n+1} is computed from I^n by

$$\text{(VII.110)} \quad \begin{cases} M(I^n,\xi^n)\dfrac{\hat{I}^{n+1} - I^n}{\Delta t} + R(I^n,\xi^n,\dfrac{d\xi^n}{dt}) \dfrac{\hat{I}^{n+1} + I^n}{2} = V^n \\[2ex] M(\dfrac{I^n+\hat{I}^{n+1}}{2},\ \xi^n) \dfrac{I^{n+1} - I^n}{\Delta t} + R(\dfrac{I^n+\hat{I}^{n+1}}{2},\ \xi^n,\dfrac{d\xi^n}{dt}) \dfrac{I^n+I^{n+1}}{2} = V^n . \end{cases}$$

Next we carry out the same time-step for the system of diffusion equations (VII.108) using a semi-implicit time scheme:

$$\text{(VII.111)} \quad \frac{u^{n+1}-u^n}{\Delta t} + \mathscr{A}_o\,[\theta\, u^{n+1} + (1-\theta)u^n] = \theta\, s^{n+1} + (1-\theta)s^n$$

with the derivative terms appearing in the boundary condition (VII.105) calculated by

$$\text{(VII.112)} \quad \begin{cases} (\dfrac{dI_p}{dt})^n = \dfrac{I_p^{n+1} - I_p^n}{\Delta t} \\[2ex] (\dfrac{da}{dt})^n = -\ \text{sgn}(\Delta_H^n)\ \dfrac{\Delta_H^{n+1} - \Delta_H^n}{\Delta t} . \end{cases}$$

a)

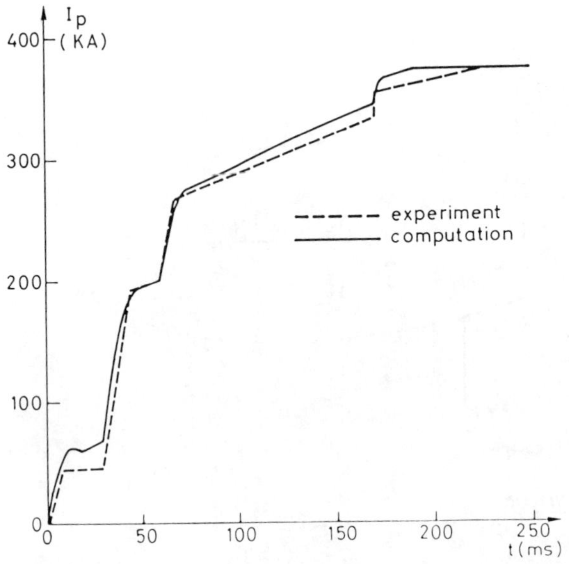

Fig.VII.10 : Comparison of simulation and experimental results for the total plasma current and the loop voltage

The space variable ρ belongs to the interval [0, a(t)] that varies with time; as in Section VI.2.2 we introduce the reduced variable $\bar{\rho} = \rho/a$ which belongs to the fixed interval [0, 1]. In the system (VII.108) we then have to replace the operator $\partial/\partial\rho$ by $(1/a)\partial/\partial\bar{\rho}$ and the operator $\partial/\partial t$ by $\partial/\partial t - (\bar{\rho}/a)(da/dt)\partial/\partial\bar{\rho}$. The operator $\mathscr{A}_0$ is approximated by a finite difference scheme and the four diffusion equations are solved successively by a fractional step method using the code MAKOKOT (cf. C. MERCIER – J.P. BOUJOT – F. WERKOFF).

The coupling between the two systems is therefore explicit.

Figure VII.10 shows the comparison between the result of a simulation and experimental results for a typical discharge of TFR 400 with four successive discharges of banks of condensers, corresponding to peaks in the voltage per turn. Figure VII.11 shows the time variation of the parameters β_p and 1_i obtained on the simulation of this discharge.

The simulation time for a complete discharge using this coupled system (circuits + plasma) is of the order of one minute on CDC 7600. It is therefore much smaller by an order of magnitude than the model developed in Chapter VI which needed half an hour

b)

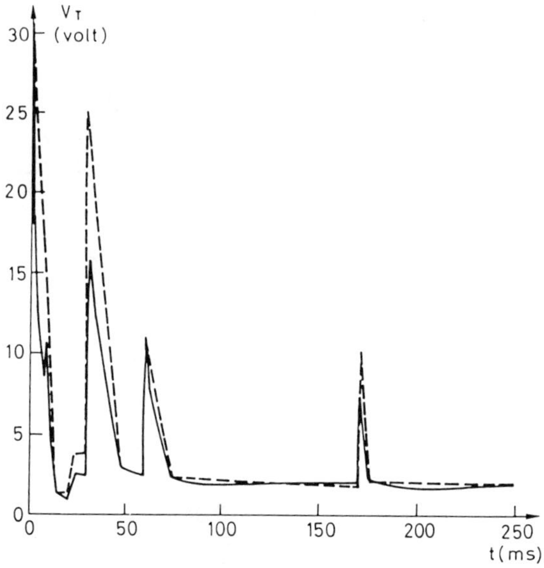

Fig.VII.10

on CRAY 1 for a simulation of this kind. This is explained easily by the fact that the system Eq of Chapter VI consisted of an evolutive 2–dimensional problem for $\psi(r,z,t)$. In Chapter VII, by working analytically to order 1 in ε, and then using preliminary numerical work to obtain a simplified description of the flux pattern in the machine, we reduced the description of the evolution of the equilibrium to a system of ordinary differential equations. The computation time which, in the self–consistent model of Chapter VI was essentially devoted to the 2–dimensional evolution of the equilibrium, is here mainly used up by the 1–dimensional simulation of diffusion in the plasma. Note, however, that the simplified model in this chapter is not self–consistent and applies only to plasmas of circular section in Tokamaks of high aspect ratio.

VII.3 STABILITY AND CONTROL OF HORIZONTAL DISPLACEMENTS OF THE PLASMA

The control system for horizontal displacements of the plasma consists of:

. a pre–programming (or open loop control) intended to create a field B_V^3 in the plasma region equal to the equilibrium field B_{eq}, for a certain reference position Δ_{ref};

. a feedback (or closed loop control) intended to stabilize the horizontal displacements of the plasma around the reference position.

a)

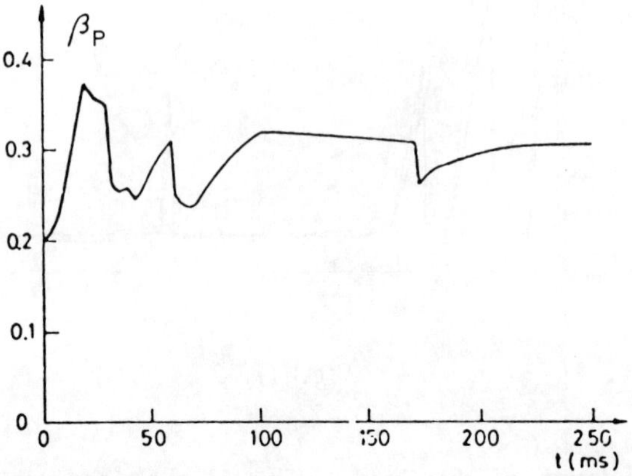

Fig.VII.11 : Time evolution of β_p and ℓ_i during a typical discharge.

VII.3.1 The Pre-Programming

This itself is divided into three stages:

. optimization of the configuration of the primary circuit and in particular the distribution Λ_i of numbers of internal and external windings:

(VII.113) $$\Lambda_i = \frac{n_1^i}{n_1^e + n_1^i}$$

. determination of the voltage V_3 to apply to the pre-programming circuit

. elaboration of a learning method that improves the pre-programming from discharge to discharge.

In the three stages, the object is to minimize the functional:

(VII.114) $$J = \frac{1}{2}\int_o^T [B_v^3(\Delta_{ref}) - B_{eq}(\Delta_{ref})]^2 \, dt .$$

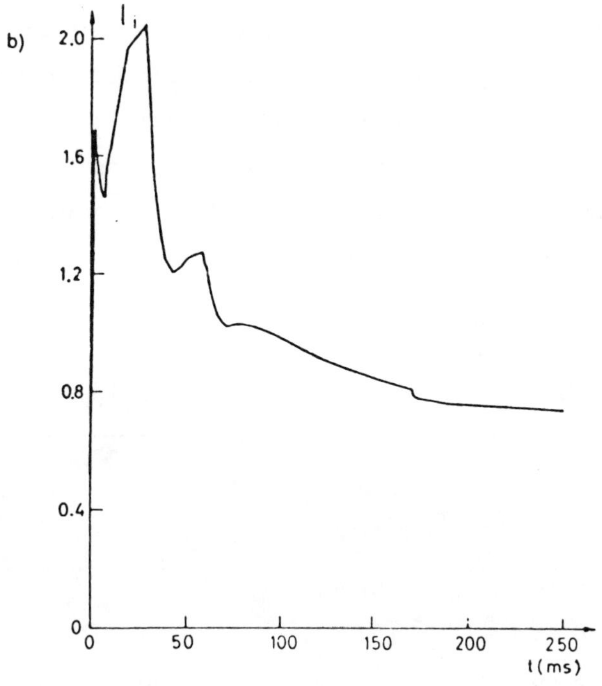

Fig. VII.11

The system of equations of state is the system of the ordinary differential equations (VII.90) for the currents in the various circuits and for the position Δ_H of the plasma, coupled to the system of transport equations (VII.108).

The optimization of J mainly concerns the circuit equations, with the transport code serving to simulate $V_p(t)$, $\beta_p(t)$, $l_i(t)$ which hardly depend on Λ_i and V_3. This is why, in the numerical solution of the problem, the optimization of J is carried out on the system (VII.90) of ordinary differential equations; the evolution of the internal characteristics V_p, β_p, l_i is next recomputed with the optimal values of the parameters Λ_i or V_3 obtained in the optimization, and then the optimization problem is solved again with the new functions $V_p(t)$, $\beta_p(t)$, $l_i(t)$, and so on.

The algorithm thus consists of internal iterations used to optimize the cost-function J, and external iterations that are used to recompute the parameters of the plasma and certain nonlinearities of the problem. Let us now look at the details of this algorithm in the three stages of the pre-programming.

VII.3.1.1 Optimization of the Configuration of the Primary

We restrict ourselves to the system consisting of the primary circuit and the plasma, modelled by equations (VII.85) and (VII.86). The total number of primary windings $n_1 = (n_1^e + n_1^i)$ is fixed. The currents I_1 and I_p depend only slightly on Λ_i defined by (VII.113), while the field B_v^3 depends on it heavily. Using (VII.56) and (VII.82) the functional J defined by (VII.114) can then be expressed as a trinomial function of the variable Λ_i, and its minimum is attained at

(VII.115)

$$\Lambda_i^{opt} = \frac{\int_0^T I_1(t)[b_e^3(\psi_F,\Delta_{ref})-b_i^3(\psi_F,\Delta_{ref})][n_1 I_1 b_e^3+b_i^3 I_{PR}+b_p^3 I_p-B_{eq}]dt}{n_1 \int_0^T I_1^2[b_e^3-b_i^3]^2\, dt}$$

where ψ_F is defined by (VII.76) and (VII.89).

In this case the external iterations have as much to do with the computation of $I_1(t)$ and $I_p(t)$ by the equations (VII.85) and (VII.86) as with that of $V_p(t)$, $\beta_p(t)$, $l_i(t)$ by (VII.108) and (VII.109). The optimization of J at each external iteration is limited to the computation of Λ_i^{opt} by (VII.115).

Three external iterations are sufficient to make the system converge and, for $\Delta_{ref} = 2$cm, $n_1^e + n_1^i = 240$, $I_{PR} = 6$ kA we obtain: $\Lambda_i^{opt} = 0.45$. This value of Λ_i may be compared with the determination of Λ in Chapter IV which, for $\beta_p = 0$ and $l_i = 1.5$, corresponds approximately to the value of 0.45. The distribution of windings that is as close as possible to this value of Λ_i^{opt} and experimentally possible is:

$$n_1^e = 128$$

$$n_1^i = 112 \; .$$

This is the distribution that is also confirmed to be optimal from the experimental point of view.

We have thus optimized the "average" equilibrium configuration during the discharge; in fact at certain instants B_v^3 will be less than B_{eq}, while at others it will be greater. We next try to realize equilibrium at each instant by means of the pre-programming circuit.

VII.3.1.2 Optimization of the Voltage V_3 Applied to the Pre–Programming Circuit

We introduce the pre–programming circuit modelled by (VII.87), and seek that voltage $V_3(t)$ which minimizes J. The matrix M for the system (VII.90), representing (VII.85) – (VII.87), is a 3 × 3 matrix computed at each external iteration. In the optimization of J we take it as a matrix depending only on time, so that at each external iteration the system (VII.90) can be regarded as linear. At the nth external iteration we thus have to solve an optimal control problem whose equations of state are

$$\text{(VII.116)} \quad \begin{cases} M(I^n, \xi^n)\,\dfrac{dI}{dt} + RI = V(I^n, t) \\[2ex] I(0) = I_o \end{cases}$$

with

$$V = (V_1(I^n, t)\;,\; V_p^n(t),\; V_3(t))$$

where I^n denotes the vector I at the end of the (n–1) external iteration. We seek $V_3^{n+1}(t)$ such that

$$\text{(VII.117)} \qquad J^n(V_3^{n+1}\;,\; I^{n+1}) = \inf_{V_3 \in U_{ad}} J^n(V_3, I)$$

with

$$J^n(V_3, I) = \frac{1}{2}\int_o^T [(b_e^3)^n\, I_e + (b_i^3)^n\, I_i + (b_p^3)^n\, I_p - B_{eq}]^2 dt$$

where $(b_e^3)^n$, $(b_i^3)^n$, $(b_p^3)^n$ are computed from $\psi_P^n(t)$ and Δ_{ref}, and where U_{ad} is the domain of admissible voltages.

The optimality system for this linear quadratic optimal control system is

(VII.118)
$$\begin{cases} M^n \dfrac{dI^{n+1}}{dt} + RI^{n+1} = (V_1^n, V_p^n, V_3^{n+1}) \\ I^{n+1}(0) = I_o \end{cases}$$

(VII.119)
$$\begin{cases} \dfrac{d}{dt}(M^n p^{n+1}) - R\, p^{n+1} = [(b_e^3)^n I_e^{n+1} \\ \qquad + (b_i^3)^n I_i^{n+1} + (b_p^3)^n I_p^{n+1} \\ \qquad + \dfrac{\mu_o}{4\pi R}\left(\text{Log}\, \dfrac{8R}{a} + \xi^n - \dfrac{1}{2}\right) I_p^{n+1}]S^n \\ p^{n+1}(T) = 0 \end{cases}$$

(VII.120)
$$\int_o^T p_3^{n+1}(V' - V_3^{n+1})dt \leqslant 0 \ , \ \forall\, V' \in U_{ad}$$

where

$$S^n = \begin{bmatrix} n_1^e (b_e^3)^n + n_1^i (b_i^3)^n \\ (b_p^3)^n + \dfrac{\mu_o}{4\pi R}\left(\text{Log}\, \dfrac{8R}{a} + \xi^n - \dfrac{1}{2}\right) \\ n_3[(b_e^3)^n - (b_i^3)^n] \end{bmatrix}$$

and p^{n+1} denotes the adjoint state.

If this problem is unconstrained the optimality condition (VII.120) becomes

(VII.121) $$p_3^{n+1}(t) = 0 \ , \ \forall\, t \in [0,T] \ .$$

If

$$U_{ad} = \{V(t) \ \text{ such that } \ V_{min} \leqslant V \leqslant V_{max} \ , \ \forall\, t \in [0,T]\}$$

then the optimality condition is the following alternative:

(VII.122)
$$\begin{cases} \text{either} \quad (p_3^{n+1} > 0) \quad \text{and} \quad V_3^{n+1} = V_{max} \\ \text{or} \quad (p_3^{n+1} < 0) \quad \text{and} \quad V_3^{n+1} = V_{min} \\ \text{or} \quad p_3^{n+1} = 0 \; . \end{cases}$$

The algorithm for solving this control problem by a projection gradient method consists of the following internal iterations:

$$I_o^n = I^n \; .$$

We compute $(V_3)_{j+1}^n$ from $(V_3)_j^n$ and I_j^n as follows:

(VII.123)
$$\begin{cases} M^n \dfrac{dI_j^n}{dt} + RI_j^n = (V_1^n \, , \, V_p^n \, , \, (V_3)_j^n) \\ \\ I_j^n(0) = I_o \end{cases}$$

(VII.124)
$$\begin{cases} \dfrac{d}{dt} (M^n \, p_j^n) - R \, p_j^n = [(b_e^3)^n \, (I_e)_j^n \\ \qquad + (b_i^3)^n \, (I_i)_j^n + (b_p^3)^n \, (I_p)_j^n \\ \qquad + \dfrac{\mu_o}{4\pi \, R} \, (\text{Log} \, \dfrac{8R}{a} + \xi^n - \dfrac{1}{2}) \, (I_p)_j^n] \, S^n \\ p_j^n(T) = 0 \end{cases}$$

(VII.125)
$$(V_3)_{j+1}^n = \text{Proj}_{U_{ad}} \, [(V_3)_j^n + \rho(p_3)_j^n] \; .$$

The algorithm for determining the optimum of V_3 is shown on the following page.

In practice, three external iterations suffice to guarantee the convergence of the algorithm, and an example of a function $V_3(t)$ so obtained is given in Fig.VII.12 for the start of the discharge with the constraints

$$V_{max} = - V_{min} = 2000V \; .$$

Experimentally it appears that when this voltage is applied the plasma is not perfectly in equilibrium; this is explained by the fact that the model being used is a simplified model and so imperfect. This is the reason for attempting to improve this pre– programming from discharge to discharge.

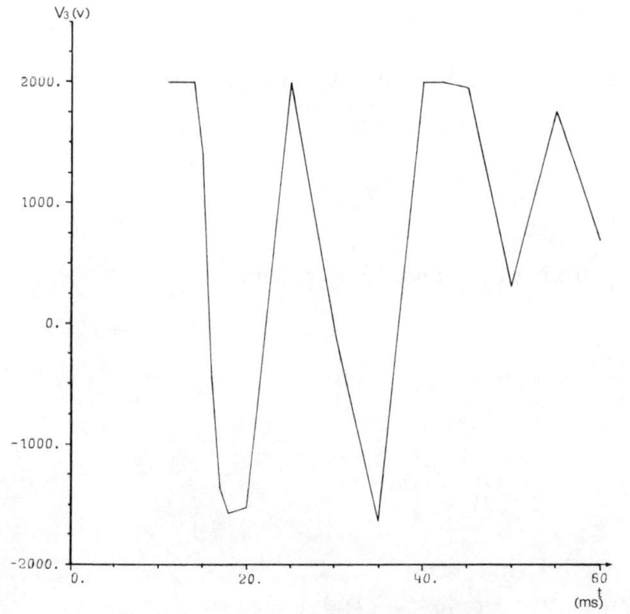

Fig. VII.12 : Preprogramming voltage obtained by optimizing J *at the start of a discharge.*

VII.3.1.3 *The Learning Method*

The feedback system that we shall study in Section VII.3.3 plays a double role: first that of creating the additional vertical field needed for equilibrium that would not have been generated by the pre-programming circuit, and then that of ensuring the stability of horizontal displacements of the plasma. It is important to transfer as much as possible of the first role of the feedback to the pre-programming circuit so that all the power of the feedback can be devoted to its stabilization role. This is why at the $(k+1)^{st}$ discharge we shall construct a reference I_{ref} for the current I_3 which uses the reference for the previous discharge as well as the measurements of the current I_4 of the feedback circuit and the horizontal displacement Δ_H after the kth discharge. By simple linearization at each instant of the kth discharge for $\Delta_H = \Delta_H^k(t)$, we show that in order to have $\Delta_H(t) = \Delta_{ref}(t)$ the current I_3 must after the $(k+1)^{th}$ discharge be equal to

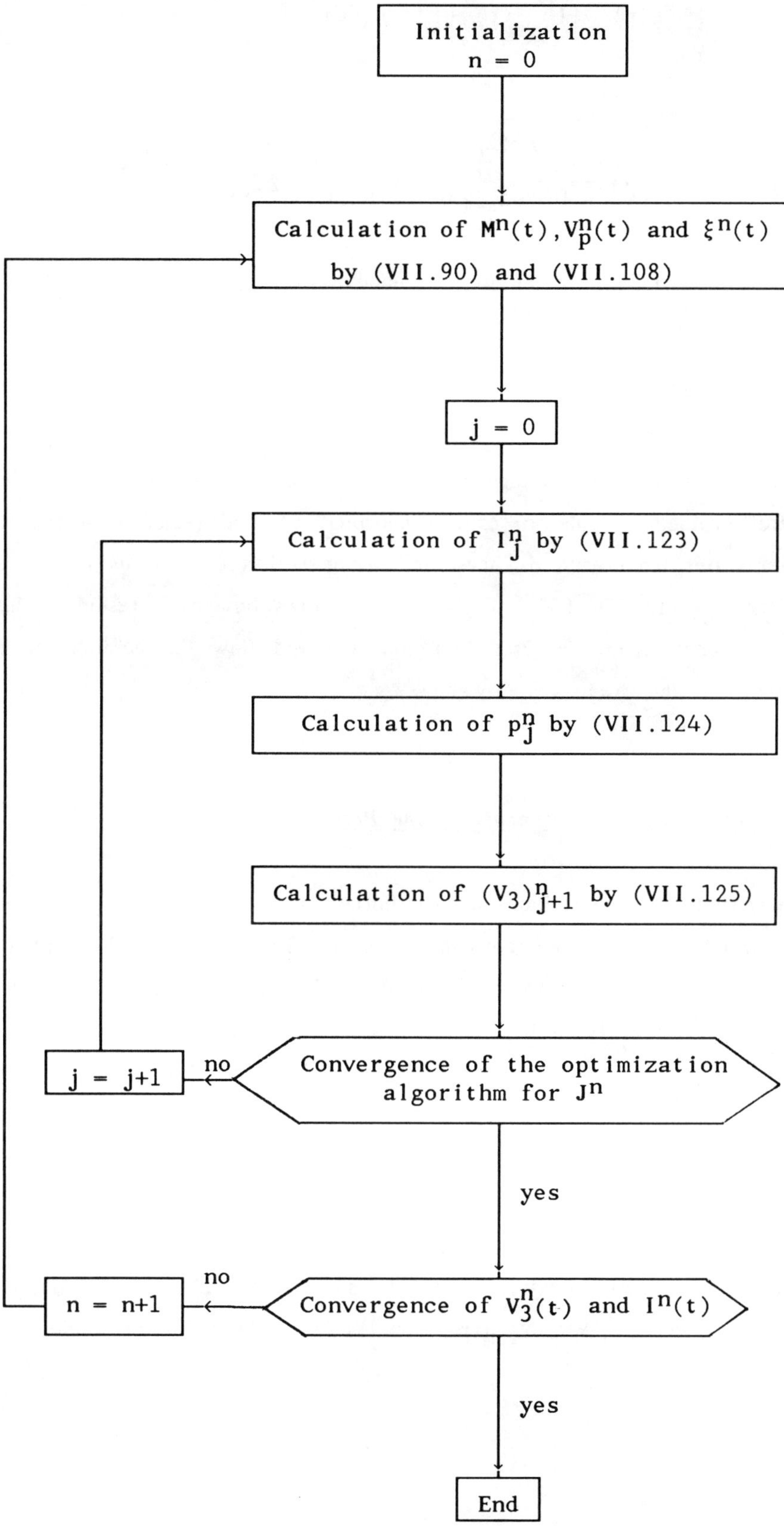
Initialization
n = 0
Calculation of $M^n(t)$, $V_p^n(t)$ and $\xi^n(t)$
by (VII.90) and (VII.108)
j = 0
Calculation of I_j^n by (VII.123)
Calculation of p_j^n by (VII.124)
Calculation of $(V_3)_{j+1}^n$ by (VII.125)
Convergence of the optimization algorithm for J^n
no
j = j+1
yes
Convergence of $V_3^n(t)$ and $I^n(t)$
no
n = n+1
yes
End

$$I_{ref}^{k+1}(t) = I_{ref}^{k}(t) + \frac{n_4}{n_3} I_H^k(t)$$

(VII.126)

$$+ \frac{\left(\frac{\partial B_V^3}{\partial \Delta_H} - \frac{\partial B_{eq}}{\partial \Delta_H}\right)(\Delta_{ref})}{n_3(b_e^3 - b_i^3)(\Delta_{ref})} [\Delta_H^k(t) - \Delta_{ref}(t)] .$$

The current I_3 will be maintained at its reference value I_{ref} by a derivative – proportional voltage V_3 of type

(VII.127) $$V_3^{k+1}(t) = G_3[I_3^{k+1}(t) - I_{ref}^{k+1}(t)] + G_3' \frac{d}{dt} (I_3^{k+1} - I_{ref}^{k+1}) .$$

The process is initialized with the voltage V_3 determined in Section VII.3.1.2 by the solution of the control problem. This so–called "learning" method (since it learns to realize the equilibrium better from each discharge to the next) has enabled us to obtain longer and longer discharges on TFR 600. Figure VII.13 shows how as a result of this learning method the feedback system is used less and less and how the equilibrium is realized better and better as the discharges proceed.

VII.3.2 Stability of Horizontal Displacements of the Plasma

The equation governing the behaviour of the horizontal displacement Δ_H of the plasma is Equation (VII.61). As the time constant τ_V of the liner in TFR 600 is about 300 μs, a perturbation of 3% in the equilibrium field results in a displacement of the plasma with speed of the order of 1 cm/ms.

VII.3.2.1 Stability Criteria

To study the stability of these displacements in the absence of feedback, we linearize (VII.61) about a reference trajectory $\Delta_{ref}(t)$ such that $B_V^3(I_{ref}) = B_{eq}(I_{ref})$, where $I_{ref}(t) = (I_1^{ref}(t), I_p^{ref}(t), I_3^{ref}(t), \Delta_{ref}(t))$. First define indices for the fields B_V^3 and B_{eq} by

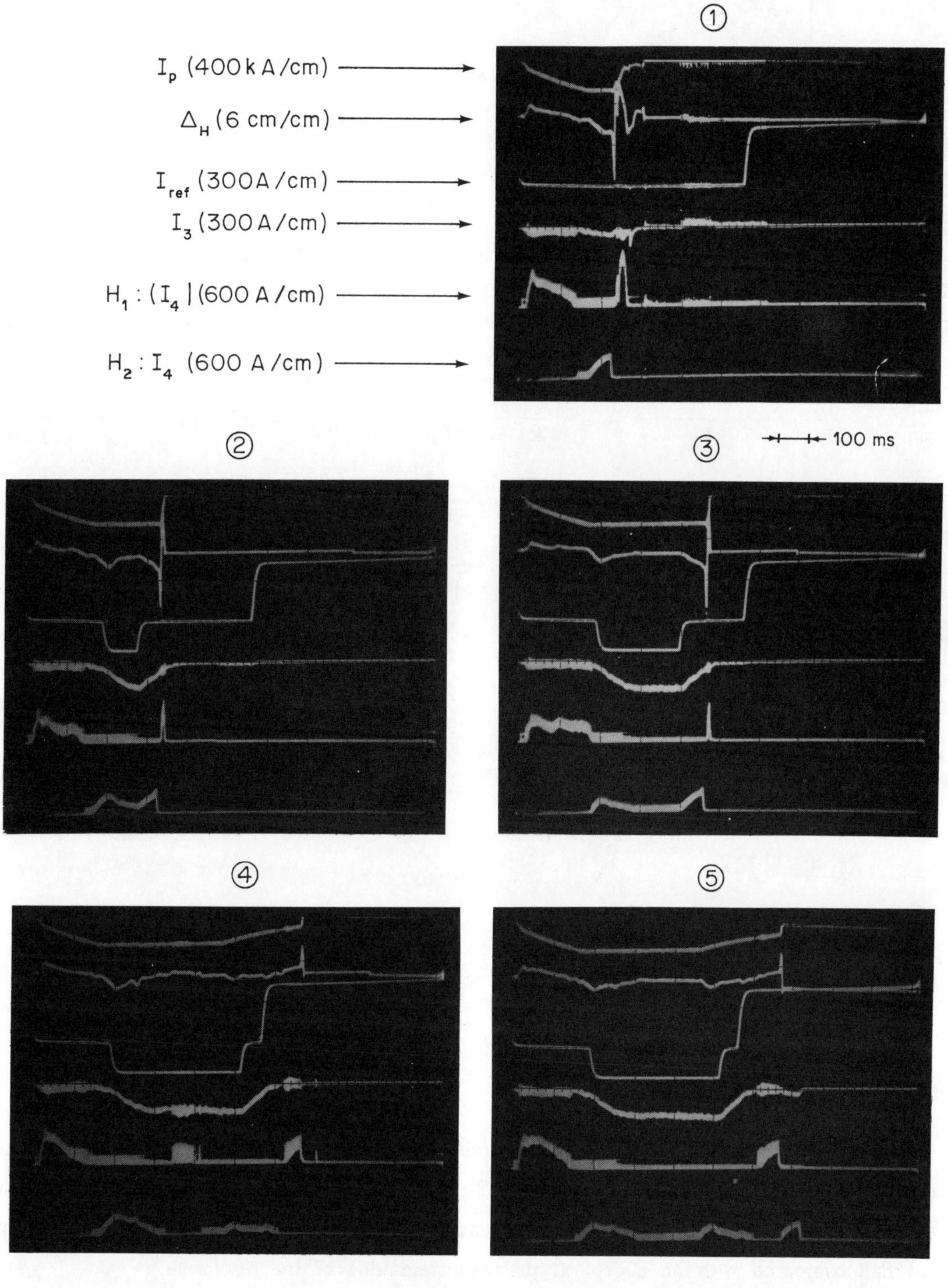

Fig.VII.13 : Successive discharges obtained by using the learning method.

$$\text{(VII.128)} \quad \left\{ \begin{aligned} n_{B_v^3} &= -\frac{R}{B_v^3} \frac{\partial B_v^3}{\partial \Delta_H} \\ n_{B_{eq}} &= -\frac{R}{B_{eq}} \frac{\partial B_{eq}}{\partial \Delta_H} \end{aligned} \right.$$

where the partial derivative of B_v^3 and B_{eq} with respect to Δ_H is taken with t fixed. Let $\Delta B = B_v^3 - B_{eq}$ be a perturbation of the magnetic field. The linearized equation of (VII.61) for $\tilde{\Delta} = \Delta_H - \Delta_{ref}$ is then

$$\text{(VII.129)} \quad \frac{d\tilde{\Delta}}{dt} + \frac{\Delta n + K_2 \frac{dK_1}{dt}}{K_1 K_2} \tilde{\Delta} = -\frac{1}{K_1} \frac{d}{dt} [I_p^{ref} \times (\Delta_{ref} - \Delta_o^{ref})] + \frac{R_{ref}}{K_2 I_p^{ref}} \frac{\Delta B}{B_{eq}^{ref}}$$

with

$$K_1 = I_p^{ref} [1 - (\frac{\partial \Delta_o}{\partial \Delta_H})_{ref}] + (\Delta_{ref} - \Delta_o)(\frac{\partial I_p}{\partial \Delta_H})_{ref} \simeq I_p^{ref}$$

$$K_2 = \frac{\mu_o \tau_v}{2\pi b_o^2} \frac{R_{ref}}{B_{eq}^{ref}}, \quad \Delta n = n_{B_v^3} - n_{B_{eq}} .$$

The behaviour of $\tilde{\Delta}$, governed by the first order differential equation (VII.129), is stable in the sense of Lyapunov if

$$\text{(VII.130)} \quad \frac{\Delta n + K_2 \frac{dK_1}{dt}}{K_1 K_2} > 0 .$$

Since the quantity K_1 is positive with I_p, and K_2 is negative like B_{eq}, the numerator of (VII.130) must be negative. Now, in TFR 600 we have R = 98 cm, a = 20 cm, b_o = 26.5 cm and K_2 dK_1/dt therefore behaves like $(10\tau_v/I_p)\ dI_p/dt$ and is small compared with Δn except at the very beginning of the discharge.

The stability condition (VII.130) is thus simply

$$\Delta n < 0$$

(VII.131) or

$$n_{B_V^3} < n_{B_{eq}} \ .$$

The time constant τ_Δ for $\tilde{\Delta}$ is then

(VII.132)
$$\tau_\Delta = \frac{2}{W} \frac{R^2}{b_o^2} \frac{\tau_V}{|\Delta n|}$$

with

$$W = \text{Log} \frac{8R}{a} + \xi - \frac{1}{2} \ .$$

VII.3.2.2 Computation of the Indices of the Fields B_V^3 and B_{eq} and the Stability Diagram

The main contribution to the field B_V^3 comes from the primary current I_1 and the attraction between the iron and the plasma. From Fig.VII.4 it is clear that, since in TFR the flux ψ_F in the iron core varies between +1 Wb and −1 Wb, the sum Σ I of the Ampère−turns is small compared with I_p (except at the beginning of the discharge). We therefore have approximately

(VII.133)
$$\begin{cases} I_i = -\Lambda_i K I_p \\ I_e = (\Lambda_i - 1) K I_p \end{cases}$$

where Λ_i is the fraction of internal windings in the primary, and K the coefficient in the formula (VII.75).

Hence from (VII.128), (VII.82), (VII.83) and (VII.56) we have

(VII.134)
$$n_{B_V^3} = \frac{4\pi}{\mu_o} \frac{R^2}{W} \left(\frac{b_V}{I_p} \frac{\partial I_p}{\partial \Delta_H} + b'_V\right)$$

where

$$\begin{cases} b_V(\psi_F,\Delta_H) = b_p^3(\psi_F,\Delta_H) - K\Lambda_i b_i^3(\psi_F,\Delta_H) - K(1-\Lambda_i) b_e^3(\psi_F,\Delta_H) \\ b'_V(\psi_F) = b_p'^3(\psi_F) - K\Lambda_i\, b_i'^3(\psi_F) - K(1-\Lambda_i) b_e'^3(\psi_F) \end{cases}$$

where these functions are given by the graphs in Figs.VII.6 and VII.7.

Similarly, using (VII.128), (VII.56) and (VII.73) we obtain

(VII.135) $$n_{B_{eq}} = 1 - \frac{R}{I_p}\frac{\partial I_p}{\partial \Delta_H} - \frac{1}{W}\left[1 + R\,\mathrm{sgn}(\Delta_H)\left(\frac{1}{a} - \frac{\partial \xi}{\partial a}\right)\right] .$$

The variations in β_p, I_p and V_p as the minor radius a of the plasma varies can be simulated using the transport code, as can the variation in I_p via equation (VII.86).

The indices $n_{B_V^3}$ and $n_{B_{eq}}$ so obtained are shown in the diagram of Fig.VII.14, where we observe that in the useful experimental region (-2cm $< \Delta_{ref} < +2$cm) the horizontal displacements of the plasma are unstable. There seems to exist a stable zone for $\Delta_{ref} < -6$cm, but this zone has not been clearly demonstrated experimentally. For $\Delta_{ref} = 0$ we note from Fig. VII.14 that there is no discontinuity in $n_{B_{eq}}$, while the sign of Δ_H comes into (VII.135). This is explained by the fact that if the plasma is eroded so that the minor radius a varies by Δa, with the total current I_p remaining constant, then the current density eroded will be fixed on the plasma boundary so that

$$\Delta \ell_i = \frac{2\Delta a}{a} .$$

The current density does not penetrate into the plasma until after a time related to the resistive time constant of the plasma current. As the variation $\Delta\beta_p$ is very small, we observe that the term $(1/a - \partial\xi/\partial a)$ is also nearly zero, which explains the continuity of $n_{B_{eq}}$ at $\Delta_{ref} = 0$. We mention finally that the contribution to $n_{B_V 3}$ of the attraction between the iron and the plasma is about 2.7 and therefore it is this that has the main destabilizing role. The primary coils themselves make a small stabilizing contribution.

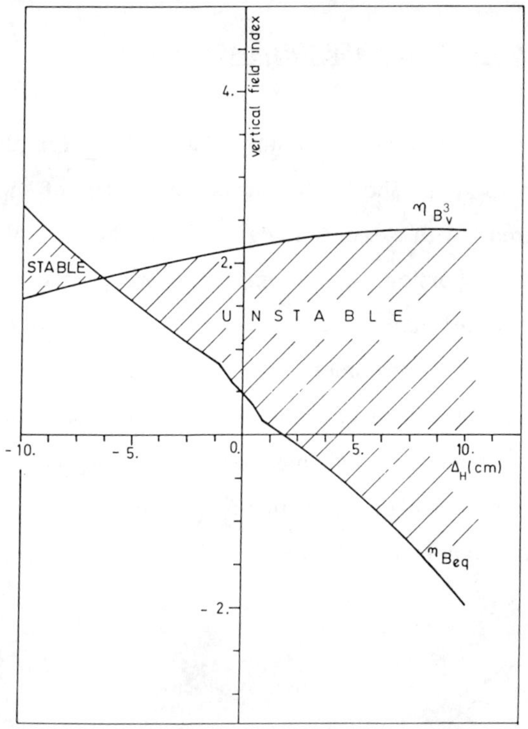

Fig.VII.14 : Stability diagram for horizontal displacements of the plasma.

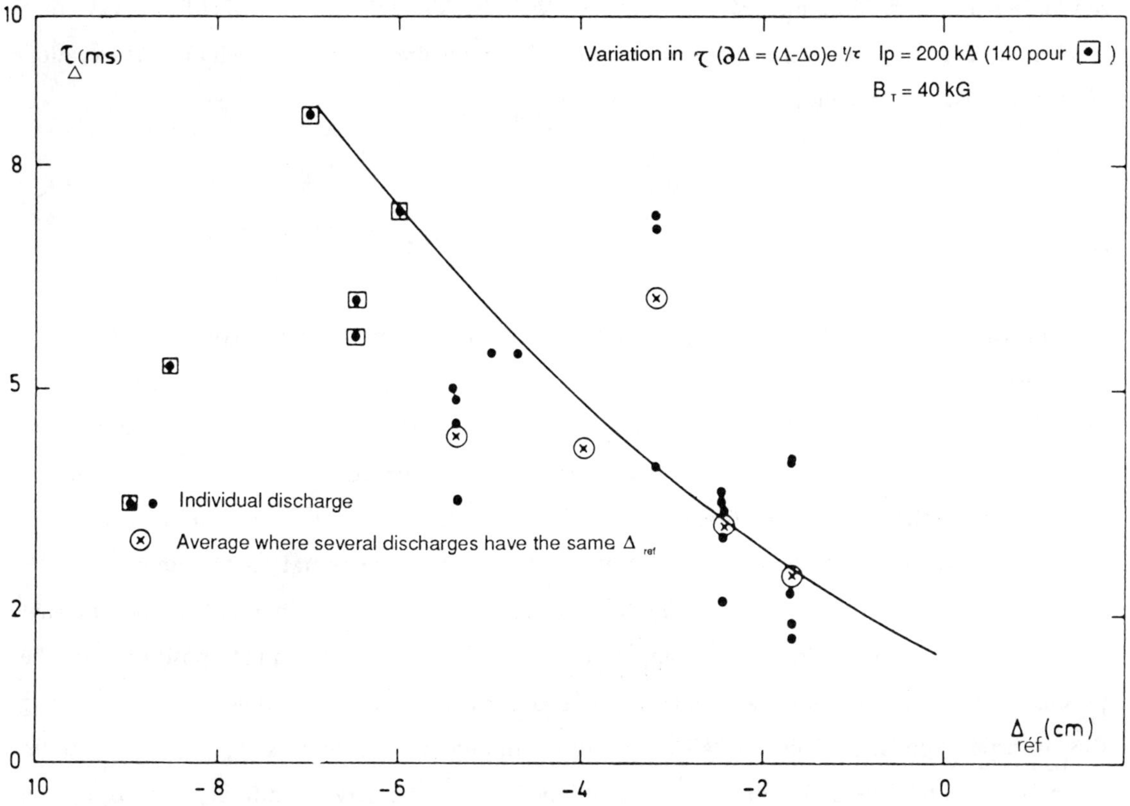

Fig.VII.15 : Diagram relating the time constant for horizontal displacements of the plasma to its reference position : theoretical curve and experimental points.

VII.3.2.3 Time Constant for Horizontal Displacements

In Fig. VII.15 the curve relating the time constant τ_Δ for the phenomenon to the plasma reference position Δ_{ref} is shown for negative values of Δ_{ref} in conformity with (VII.132); it is compared with experimental measurements of this time constant. Although there is a large dispersion of the experimental points, we note that τ_Δ grows as Δ_{ref} decreases; on the other hand, we have not been able to observe the stable zone predicted by Fig.VII.14, since τ_Δ becomes large for these values of Δ_{ref} but at the same time corresponding to unstable displacements. It should be remarked that for highly eroded plasmas impurity phenomena become important; they have not been taken into account in this model, and this could explain the divergence between the theoretical curve and the experimental points for $\Delta_{ref} < -$ 5cm.

Remark VII.5

The stability condition (VII.131) for horizontal displacements of the plasma generalizes the condition $n_{B_v} < 3/2$ of V.S. MUKHOVATOV – V.D.SHAFRANOV that applies to Tokamaks of very high aspect ratio without limiter and in which conservation of poloidal flux is assumed. □

Remark VII.6

In Chapter IV, while studying equilibrium solution branches we have already shown the existence of a stable zone for $\Delta_{ref} < -$ 4cm (cf.Fig. IV.1). The existence of a turning point on this branch was interpreted as loss of stability of horizontal displacements of the plasma. This interpretation is confirmed here from calculations of the indices of the external and equilibrium fields. The key role of the iron was shown in Fig.IV.2 which shows that in the absence of iron the horizontal displacements of the plasma are stable for $\Delta_{ref} < 0$. The only difference is that in Chapter IV the internal distribution of current in the plasma was fixed whatever the radial position of the plasma, while here we take account of variations in β_p and l_i as functions of a, using the transport code. This explains the discontinuity of derivatives for $\Delta_{ref} = 0$ in Chapter IV, while here for the reason mentioned above we have continuity for $\Delta_{ref} = 0$ when the point of contact of the plasma with the limiter changes from the interior point to the exterior point. □

<u>*Remark VII.7*</u>

In all this study we have assumed that the plasma is in contact with the limiter at every instant, so that a and R are given by (VII.73). In R. GRAN – M. ROSSI – F. SOBIERASSKI and in J. HUGILL – A. GIBSON it is the toroidal flux of the plasma that is assumed constant, so that that ratio a^2/R is conserved during displacements of the plasma. A comparative study of these various hypotheses has been made in R. AYMAR – C. LELOUP – M. PARIENTE for the investigation of displacements in Tore 2 (now TORE Supra). □

VII.3.3 <u>*Closed–Loop Control of Horizontal Displacements of the Plasma*</u>

The feedback circuit used in TFR 600 (cf. J. BLUM – R. DEI CAS – J.P. MORERA – P. PLINATE) is supplied by a double thyristor voltage chopper whose voltage V_4 is of "all or nothing" type and which controls the current I_4 linearly so that

$$I_4 = G_4 \; I_p \; \tilde{\Delta} \; . \qquad \text{(VII.136)}$$

The introduction of this feedback modifies the index $n_{B_v^3}$ by an amount

$$\Delta n_4 = - \frac{4\pi \; R^2 \; n_4 \; G_4 \; [b_e^3(\psi_F,\Delta_{ref}) - b_i^3(\psi_F,\Delta_{ref})]}{\mu_o \; W} \; . \qquad \text{(VII.137)}$$

The value of G_4 that ensures the stability of the system, using (VII.131), is then given by

$$G_4 = \frac{\mu_o}{4\pi} \; \frac{W \; \Delta n}{n_4 \; R^2 (b_e^3 - b_i^3)} \qquad \text{(VII.138)}$$

where Δn is the difference between the indices of B_v^3 and B_{eq} in the absence of feedback for $\Delta_H = \Delta_{ref}$. This gain G_4 is shown in Fig.VII.16 as a function of Δ_{ref} for an unsaturated configuration ($\psi_F \simeq 0$). The region where G_4 is negative corresponds to the stability zone for Δ_H in the absence of feedback. In practice, G_4 is chosen to be equal to 10^{-3} cm^{-1} in order to guarantee the stability of horizontal displacements of the plasma around $\Delta_{ref} = 0$.

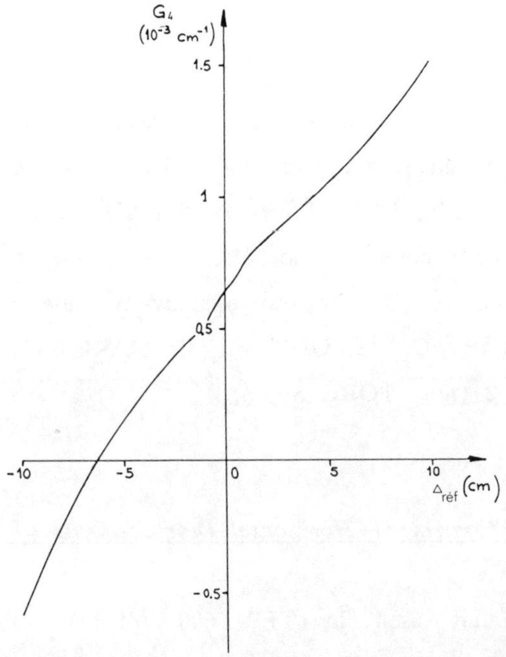

Fig.VII.16 *Gain of the feedback ensuring the stability of the horizontal displacement of the plasma as a function of its reference position.*

The experimental device in reality consists of two voltage choppers H1 and H2 acting as follows:

$$\text{. H2 if } \tilde{\Delta} > 0 \quad \begin{cases} V_4 = V_4^{max} & \text{if} \quad I_4 < G_4 \, I_p \, \tilde{\Delta} \\ V_4 = 0 & \text{if} \quad I_4 > G_4 \, I_p \, \tilde{\Delta} \end{cases}$$

$$\text{. H1 if } \tilde{\Delta} < 0 \quad \begin{cases} V_4 = - V_4^{max} & \text{if} \quad |I_4| < G_4 \, I_p \, |\tilde{\Delta}| \\ V_4 = 0 & \text{if} \quad |I_4| > G_4 \, I_p \, |\tilde{\Delta}| \, . \end{cases}$$

Very complicated commutation problems between these two choppers operating at 1 kHz, as well as the need to decouple completely the pre–programming and feedback circuits, became apparent during the functioning of this control system. For fuller technical information on this system see J. BLUM – R. DEI CAS – J.P. MORERA – P. PLINATE [1] and [2].

Figure (VII.17) shows how control of Δ_H is ensured in practice with the currents I_{H_1} and I_{H_2} of the two choppers H_1 and H_2, the pre–programming current denoted here by I_{B_2}, the plasma current I_p and the radial displacement Δ_H of the plasma.

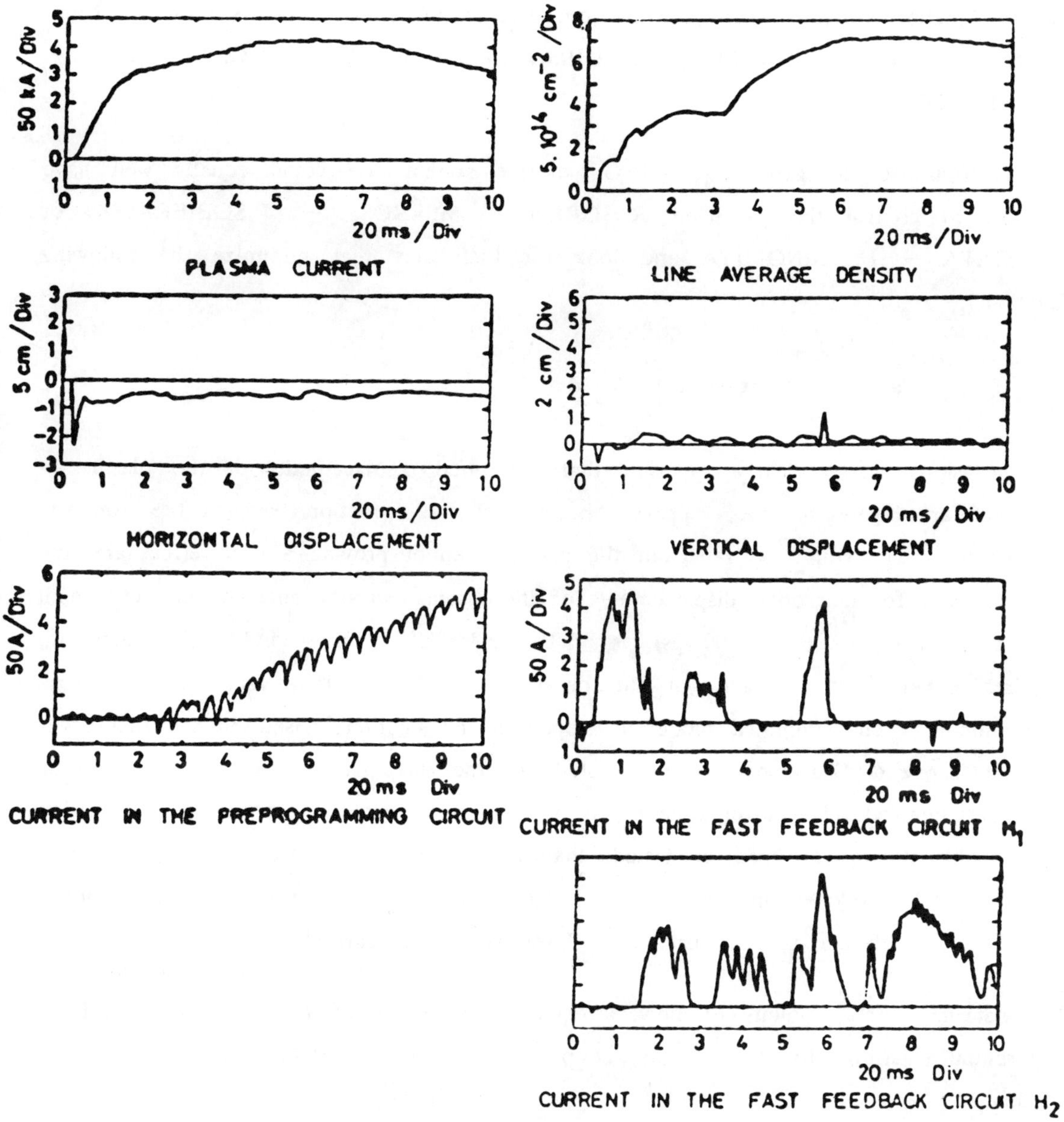

Fig.VII.17 *Example of control of the radial position* Δ_H *of the plasma.*

Remark VII.8

In many Tokamaks it is the voltage V_4 (and not the current I_4) which is a proportional– derivative function of the displacement:

$$V_4 = G_4'(I_p \tilde{\Delta}) + G_4'' \frac{d}{dt} (I_p \tilde{\Delta}) .$$

In this case, the study of stability of displacements requires computation of the transfer function for the system and the use of stability criteria such as those of Routh or Nyquist (cf. J. HUGILL – A. GIBSON, R. AYMAR – C. LELOUP – M. PARIENTE).

To optimize the gains G_4' and G_4'' of the system, we could equally well have recourse to control theory as in R. GRAN – M. ROSSI – F.SOBIERASSKI or A. OGATA – H. NINOMIYA and look for the gains that minimize the following functional:

$$J = \int_0^T \tilde{\Delta}^2 \, dt + N \int_0^T V_4^2 \, dt . \qquad \square$$

In Conclusion It is clearly apparent that horizontal displacements of the plasma in TFR 600 are highly unstable, especially because of the attraction between the iron and the plasma. The coils, too far from the plasma, cannot provide a shell effect and the time constant for horizontal displacements of the plasma, proportional to the very small time constant of the liner (300 μs), is of the order of a millisecond. This therefore requires a very fast feedback (of the order of 1 kHz) and a very accurate pre–programming. The simplified model developed in this chapter, based on the first order analytic theory of Shafranov and on a study of the flux pattern using a magnetostatic code, has enabled this control system to be optimized at a very moderate computation cost. This system can then be tested, thanks to the self–consistent (but much more costly) model developed in Chapter VI. To summarize, optimization of the control system for equilibrium of the plasma in a Tokamak can be carried out in two stages:

1) determine the amounts of power required by means of the system of ordinary differential equations for the circuits, coupled to a simplified transport model for the plasma

2) optimize the pre– programming, the feedback gains and the control of the plasma shape using the self– consistent model in Chapter VI.

Bibliography

ADAMS, R.A.
Sobolev Spaces. Academic Press (1975).

ALBANESE, R. – BLUM, J. – DE BARBIERI, O.
On the solution of the magnetic flux equation in an infinite domain. Proc. 8th Europhysics Conf. on Computational Physics. Eibsee (1986).

ANDREOLETTI, J.
Private communication.

ARTSIMOVITCH, L.A.
Tokamak devices. Nuclear Fusion, 12 (1972), 215–252.

AUERBACH, D.P.
c.f. D.E. SHUMAKER – J.K. BOYD – D.P. AUERBACH – B. McNAMARA.

AYMAR, R. et al
[1] **Le Tokomak TORE Supra.** Report EUR–CEA–FC 1228 (1984).

AYMAR, R. – BLUM, J. – FAIVRE, D. – LE FOLL, J. – LELOUP, C.
[2] **Single null divertor with PF coils outside the toroidal magnet.** European Contributions to the 5th Meeting of the Intor Workshop (January 1981).

AYMAR, R. – BLUM, J. – FAIVRE, D. – GAROT, C. – LE FOLL, J. – LELOUP, C.
[3] **Poloidal magnetic field configuration of Intor in the case of a single null divertor.** NET report n° 6. EUR XII–324/6 (1983).

AYMAR, R. – LELOUP, C. – PARIENTE, M.
Estimation des puissances nécessaires à l'asservissement de la position du plasma dans un Tokomak. *(Estimation of the forces necessary to maintain the position of the plasma in a Tokamak).* Report EUR–CEA–FC–902 (1977).

BERESTYCKI, H. – BREZIS, H.
On a free boundary problem arising in plasma physics. Nonlinear Analysis 4 (1980), 415–436.

BISHOP, C.M. – TAYLOR, J.B.
Degenerate toroidal magnetohydrodynamic equilibria and minimum B. Phys. Fluids, 29/4 (1986), 1144–1148.

BLUM, J.
[1] **Equilibre et diffusion: méthodes analytiques et numériques** *(Equilibrium and diffusion: analytical and numerical methods)* in "La Fusion par Confinement Magnétique". Cargèse (1984). Edited by D. Grésillon and T. Lehner (Editions de Physique, Orsay, 1984).

BLUM, J.
[2] Numerical simulation of the plasma equilibrium in a Tokamak. Computer Physics Reports 6 (1987), 275–298.

BLUM, J. – DEI CAS, R.
Static and Dynamic Control of Plasma Equilibrium in a Tokamak. 8th Symp. on Eng. Problems of Fusion Research. San Francisco (1979), 1873–1878.

BLUM, J. – DEI CAS, R. – MORERA, J.P.
Computation of magnetic flux and currents in a Tokamak with an iron circuit. Proceedings Compumag Conference. Grenoble (1978).

BLUM, J. – DEI CAS, R. – MORERA, J.P. – PLINATE, P.
[1] **Slow and fast feedback circuits for the plasma equilibrium in the TFR600 Tokamak. Numerical simulation of the multitransformer equations.** Proceedings of the 7th Symposium of Engineering Problems in Fusion. Knoxville (1977), 478–484.

[2] **Influence of the iron core on plasma equilibrium and stability in TFR600, a Tokamak without copper shell. Description of the fast feedback system.** 10th Symp. on Fusion Technology. Padova (1978). Fusion Technology (1979) 999–1006.

BLUM, J. – GALLOUËT, T. – SIMON, J,
[1] **Existence and control of plasma equilibrium in a Tokamak.** SIAM Journal of Mathematical Analysis. Vol.17 N°5 (1986), 1158–1177.

[2] **Equilibrium of a plasma in a Tokamak** – in "Free boundary problems : applications and theory" Vol.IV. Research Notes in Mathematics n° 121 – Bossavit, Damlamian, Fremond (Editors), 488–496.

BLUM, J. – GILBERT, J. Ch. – THOORIS, B.
Parametric identification of the plasma current density from the magnetic measurements and pressure profile. Report of JET contract n° JT3/9008 (1984).

BLUM, J. – GILBERT, J. Ch. – LE FOLL, J. – THOORIS, B.
Numerical identification of the Plasma Current Density from Experimental Measurements – Europhysics Conference Abstracts – 8th Europhysics Conference on Computational Physics – Eibsee (1986) – Vol. 10D 49–52.

BLUM, J. – LE FOLL, J.

[1] Plasma equilibrium evolution at the resistive diffusion timescale – Computer Physics Reports 1 (1984), 465–494.

[2] Self consistent equilibrium and diffusion code for JET – Report EUR–CEA–FC–1209 (1983).

BLUM, J. – LE FOLL, J. – LELOUP, C.

Self consistent description of plasma equilibrium evolution in Tore Supra. 12th Symposium on Fusion Technology. Varèse (1984).

BLUM, J. – LE FOLL, J. – THOORIS, B.

[1] **The self–consistent equilibrium and diffusion code SCED.** Computer Physics Communications 24 (1981), 235–254.

[2] **Le contrôle de la frontière libre du plasma dans un Tokomak** *(Control of the free boundary in a Tokamak).* 5ème Conf. Int. sur l'analyse et l'optimisation des systèmes. (INRIA–Versailles 1982).

[3] **Numerical identification of the plasma shape from the magnetic measurements.** JET Workshop on magnetic field measurements. Culham (1980).

BLUM, J.

c.f. R. AYMAR et al [2]. [3].

BONNANS, F. – SAGUEZ, C.

Méthodes numériques d'optimisation *(Numerical optimization methods).* Cours de l'Ecole Nationale Supérieure de Techniques Avancées (1984).

BOUJOT, J.P. – MORERA, J.P. – TEMAM, R.

Applied Math. and Optimization Vol.2, n° 2 (1975), 97–129.

BOUJOT, J.P.

cf. C. MERCIER – J.P. BOUJOT – F. WERKOFF.

BOYD, J.K.

cf. D.E. SHUMAKER – J.K. BOYD – D.P. AUERBACH – B. McNAMARA

BRAAMS, B.J.

The Interpretation of Tokamak Magnetic Diagnostics – Report IPP 5/2, Max–Planck – Institut für Plasmaphysik (1985) – Garching (bei Munchen).

BRAGINSKII, S.I.

In Reviews of Plasma Physics, Vol.1, Consultants Bureau, New York (1965), 205–311.

BREZIS, H.

cf. H. BERESTYCKI – H. BREZIS.

BREZZI, F. – RAPPAZ, J. – RAVIART, P.A.

Finite dimensional approximation of nonlinear problems: Part I: Branches of nonsingular solutions. Num. Math. 36. (1980) 1–25.

Part II: Limit points. Num. Math. 37 (1981), 1–28.

BROWN, B.B.
cf. J.L. LUXON – B.B. BROWN.

BRUSATI, M. et al
Analysis of magnetic measurements in Tokamaks. Computer Physics Reports. Vol.1. N° 7 & 8, p.345.

BYRNE, R.N. – KLEIN, H.H.
Journal of Computational Physics 26 (1978), 352–378.

CAFFARELLI, L.A. – FRIEDMAN, A.
Asymptotic estimates for the plasma problem. Duke Mathematical Journal. Vol.47, N°.3 (1980), 705–742.

CALOZ, G. – RAPPAZ, J. (to be published).

CEA, J.
Optimisation. Théorie et algorithmes *(Optimization Theory and algorithms).* Dunod (1971).

CENACCHI, G. – GALVAO, R. – TARONI, A.
Numerical computation of axisymmetric MHD equilibria without conductivity shell. Nuclear Fusion 16.3 (1976), 457–464.

CENACCHI, G. – SALPIETRO, E. – TARONI, A.
Self–consistent MHD equilibria in Tokamaks with an iron core transformer. 10th Symp. on Fusion Technology. Padova (1978).

CENACCHI, G. – TARONI, A.
The JET Equilibrium–Transport code for Free Boundary Plasmas (JETTO). Europhysics Conference Abstracts. Vol. 10D (1986), 57–60.

CHU, M.S. – DOBROTT, D. – JENSEN, T.H. – TAMANO, T.
Axially symmetric magnetohydrodynamic equilibria with free–boundaries and arbitrary cross section. Physics of Fluids, Vol.17, n° 6 (1974), 1183–1187.

CHRISTIANSEN, J.P. – TAYLOR, J.B.
Determination of current distribution in a Tokamak – Nuclear Fusion, Vol.22, n°1 (1982), 111–115.

CISCATO, D. – DE KOCK, L. – NOLL, P.
11th Symp. on Fusion Technology. Oxford (1980).

CIARLET, P.G.
The finite element method for elliptic problems. North Holland (1978).

CIARLET, P.G. – RAVIART, P.A.
General Lagrange and Hermite interpolation in R^n with applications to finite element methods. Arch. Rat. Mech. Anal. Vol.46, (1972), 177–199.

COLLI FRANZONE, P. – GUERRI, L. – TACCARDI, B. – VIGANOTTI, C.
The direct and inverse potential problems in electrocardiology. Laboratoire d'analyse numérique de Pavie. Report n° 222 (1979).

COURANT, R. – HILBERT, D.
Methods of Mathematical Physics, vol.1–2, Interscience (1962).

CRANDALL, M.G. – RABINOWITZ, P.H.
Bifurcation, perturbation of simple eigenvalues and linearized stability. Arch. Rat. Mech. Anal. Vol.52 (1973), 161–180.

DEGTYAREV, L.M. – DROZDOV, V.V.
An inverse variable technique in the MHD equilibrium problem. Computer Physics Reports. Vol.2 (1985) n°7, 341–387.

DEI CAS, R.
cf. J. BLUM – R. DEI CAS.

DEI CAS, R.
cf. J. BLUM – R. DEI CAS – J.P. MORERA.

DEI CAS, R.
cf. J. BLUM – R. DEI CAS – J.P. MORERA – P. PLINATE [1].

DEI CAS, R.
cf. J. BLUM – R. DEI CAS – J.P. MORERA – P. PLINATE [2].

DELUCIA, J. – JARDIN, S.C. – TODD, A.M.M.
An iterative metric method for solving the inverse Tokamak equilibrium problem. Journal of Computational Physics 37 (1980), 183–204.

DERVIEUX, A.
Perturbation des équations d'équilibre d'un plasma confiné: comportement de la frontière libre, étude des branches de solutions *(Perturbation of the equilibrium equations of confined plasma: behaviour of the free boundary, study of the solution branches.)* INRIA Research Report n° 18 (1980).

DOBROTT, D.
cf. M.S. CHU – D. DOBROTT – T.H. JENSEN – T. TAMANO.

DORY, R.A.
cf. H.R. HICKS – R.A. DORY – J.A. HOLMES.

DURAND, E.
Magnétostatique *(Magnetostatics).* Masson (1968).

FAIVRE, D.
cf. R. AYMAR et al [1], [2], [3].

FENEBERG, W. – LACKNER, K.
Multipole Tokamak equilibria. Nuclear Fusion 13 (1973), 549–556.

FENEBERG, W. – LACKNER, K. – MARTIN, P.
Fast control of plasma surface. Computer Physics Communications 31 (1984), 143–148.

FISCH, N.J.
Confining a Tokamak Plasma with rf–driven currents. Physical Review Letters. Vol.41, n° 13 (1978), 873–876.

FRIEDMAN, A.
cf. L.A. CAFFARELLI – A. FRIEDMAN.

GALLOUËT, T.
Contribution à l'étude d'une équation apparaissant en physique des plasmas *(Contribution to the study of an equation appearing in plasma physics).* Thèse 3ème cycle. Paris VI (1978).

GALLOUËT, T.
cf. J. BLUM – T. GALLOUËT – J. SIMON [1], [2].

GALVAO, R.
cf. G. CENACCHI – R. GALVAO – A. TARONI.

GAROT, C.
cf. R. AYMAR et al

GIBSON, A.
cf. J. HUGILL – A. GIBSON.

GILBERT, J. Ch.
Sur quelques problèmes d'identification et d'optimisation rencontrés en physique des plasmas. *(Some identification and optimization problems arising in plasma physics).* These de l'Universite Paris VI (1986).

GILBERT J. Ch.
cf. J. BLUM – J. Ch. GILBERT – B. THOORIS.
cf. J. BLUM – J. Ch. GILBERT – J. LE FOLL – B. THOORIS.

GLOWINSKI, R. – MARROCO, A.
Analyse numérique du champ magnétique d'un alternateur par éléments finis et sur–relaxation ponctuelle non linéaire *(Numerical analysis of the magnetic field of an alternator by finite elements and nonlinear pointwise over–relaxation).* Computer methods in applied mechanics and engineering 3 (1974), 55–85, North–Holland.

GRAD, H.
Report COO–3077–154–MF93. New York University.

GRAD, H. – HOGAN, J.
Physical Review Letters. Vol. 24, n° 24 (1970), 1337–1340.

GRAD, H. – HU, P.N. – STEVENS, D.C.
Proc. Nat. Acad. Sci. USA, Vol.72, n° 10 (1975), 3789–3793.

GRAD, H. – HU, P.N. – STEVENS, D.C. – TURKEL, E.
Plasma Physics and Controlled Nuclear Fusion Research (1976) Berchtesgaden, 355–365.

GRAD, H.
cf. D.B. NELSON – H. GRAD.

GRAD, H. – RUBIN, H.
Hydromagnetic Equilibria and Force–free Fields. 2nd U.N. Conference on the Peaceful uses of Atomic Energy. Geneva (1958), Vol.31, 190–197.

GRAN, R. – ROSSI, M. – SOBIERASSKI, F.
Plasma position control for TFTR using modern control theory. Proceedings of the 7th Symp. on Engineering Problems of Fusion Research. Knoxville (1977), 104–111.

GRUBER, R.
cf. S. SEMENZATO – R. GRUBER – H.P. ZEHRFELD.

GRUBER, R. – RAPPAZ, J.
Finite Element Methods in Linear Ideal Magnetohydrodynamics – Springer (1985).

GUERRI, L.
cf. P. COLLI FRANZONE – L. GUERRI – B. TACCARDI – C. VIGANOTTI.

GUILLOPPE, C.
Sur un problème à frontière libre intervenant en physique des plasmas *(On a free boundary poblem arising in plasma physics).* Thèse 3ème cycle. Orsay (1977).

HAN, S.P.
Superlinearly convergent variable metric algorithms for general nonlinear programming problems. Math. Prog. 11 (1976). 263–282.

HAZELTINE, R.D.
cf. F.L. HINTON – R.D. HAZELTINE.

HELTON, F.J. – WANG, T.S.
Nuclear Fusion. 18 11 (1978), 1523–1533.

HELTON, F.J. – MILLER, R.L. – RAWLS, J.M.
Journal of Computational Physics. 24 (1977), 117–132.

HICKS, H.R. – DORY, R.A. – HOLMES, J.A.
Inverse plasma equilibria. Computer Physics Reports. Vol.1, nos 7 & 8, (1984), 373–387.

HILBERT, D.
cf. R. COURANT – D. HILBERT.

HINTON, F.L. – HAZELTINE, R.D.
Reviews of Modern Physics, Vol.48, no 2, Part I (1976), 239–308.

HIRSHMAN, S.P. – JARDIN, S.C.
Physics of Fluids, 22 (4), (1979), 731–742.

HOGAN, J.
cf. H. GRAD – J. HOGAN.

HOGAN, J.T.
Nuclear Fusion, Vol.19, no 6 (1979), 753–776 and ORNL–TM 6049 Report (1978).

HOLMES, J.A., PENG – Y.K. – LYNCH, S.J.
Journal of Computational Physics, 36 (1980), 35–54.

HOLMES, J.A.
cf. HICKS – R.A. DORY – J.A. HOLMES.

HOUSEHOLDER, A.S.
The theory of matrix in numerical analysis. Dover ed. (1975).

HU, P.N.
cf. H. GRAD – P.N. HU – D.C. STEVENS.

HU, P.N.
cf. H. GRAD – P.N. HU – D.C. STEVENS – E. TURKEL.

HUGILL, J. – GIBSON, A.
Servo–Control of plasma position in CLEO–Tokamak – Nuclear Fusion 14 (1974), 611–619.

INTOR
International Tokamak Reactor. Phase I. Report IAEA (Vienna) (1982).

JANDL, O.
cf. W. KERNER – O. JANDL.

JARDIN, S.C.
[1] Journal of Computational Physics, 42 (1981), 31–60.

[2] Multiple Time–Scale Methods in Tokamak Magnetohydrodynamics in "Multiple Time Scales". Brackbill–Cohen (Editors), 1985, Academic Press.

JARDIN, S.C.
cf. J. DELUCIA – S.C. JARDIN – A.M.M. TODD.

JARDIN, S.C.
cf. S.P. HIRSHMAN – S.C. JARDIN.

JENSEN, T.H.
cf. M.S. CHU – D. DOBROTT – T.H. JENSEN – T. TAMANO.

JET
The JET project. EUR–JET Report R5.

JOHNSON, J.L. et al
Numerical Determination of axisymmetric toroidal magnetohydrodynamic equilibria. Journal of Computational Physics 32, (1979), 212–234.

KACEMI, T.
Thèse 3ème cycle. Université Paris VI (1986).

KELLER, H.B.
[1] **Numerical Solution of Bifurcation and Nonlinear Eigenvalue Problems in Applications of Bifurcation Theory** (P. Rabinowitz ed.), Academic Press, New York (1977), 359–384.

[2] **Constructive methods for bifurcation and nonlinear eigenvalue problems.** 3ème Coll. Int. sur meth. de Calcul Scientifique et technique. Versailles (1977).

KELLMAN, A.G.
cf. L.L. LAO – H. St.JOHN – R.D. STAMBAUGH – A.G. KELLMAN.

KERNER, W. – JANDL, O.
Axisymmetric MHD equilibria with flow. Computer Physics Communications. Vol.31 N°s 2 & 3, (1984), 269–285.

KIKUCHI, F. – AIZAWA, T.
Finite element analysis of MHD equilibria – Theoretical and applied mechanics. Vol.30 (1981), 513–527.

KLEIN, H.H.
cf. R.N. BYRNE – H.H. KLEIN.

KLEIN, H.H. – BYRNE, R.N.
Journal of Computational Physics 33 (1979), 294–299,

de KOCK, L. – TONETTI, G.
Magnetic measurements on the JET discharge: 11th Eur. Conf. on Controlled Fusion and Plasma Physics. Aachen (1983).

de KOCK, L.
cf. D. CISCATO – L. de KOCK – P. NOLL.

KRUSKAL, M.D. – KULSRUD, R.M.
Equilibrium of a Magnetically Confined Plasma in a Toroid. Physics of Fluids. Vol.1, n° 4 (1958), 265–274.

KUHN, H.W. – TUCKER, A.W.
Nonlinear programming. Proc. 2nd Berkeley Symposium on Math. Stat. and Prob. University of California Press. Berkeley (1961), 481–492.

LACKNER, K.
Computation of ideal MHD equilibria. Computer Physics Communications, 12 (1976), 33–44.

LACKNER, K.
cf. W. FENEBERG – K. LACKNER.

LACKNER, K.
cf. W. FENEBERG – K. LACKNER – P. MARTIN.

LAMINIE, J.
Détermination numérique de la configuration d'équilibre d'un plasma dans un Tokomak *(Numerical determination of the equilibrium configuration of a plasma in a Tokamak).* Thèse de 3ème cycle. Orsay (1977).

LAO, L.L.
Variational moment method for computing magnetohydrodynamic equilibria. Computer Physics Communications 31 n°s 2 & 3, (1984), 201–212.

LAO, L.L. et. al.,

[1] Reconstruction of current profile parameters and plasma shapes in Tokamaks – Nuclear Fusion, Vol.25 n°11 (1985), 1611.

[2] **Separation of β_p and ℓ_i in Tokamaks of non–circular cross–section.** Nuclear Fusion, Vol.25, n°10 (1985), 1421.

LATTES, R. – LIONS, J.L.
Méthode de quasi–réversibilité et applications *(The method of quasi–reversibility and its applications).* Dunod (1967).

LEE, D.K. – PENG, Y.K.
An approach to rapid plasma shape diagnostics in Tokamaks. J. Plasma Physics Vol.25, 1 (1981), 161–173.

LE FOLL, J.
cf. J. BLUM – J. LE FOLL.

LE FOLL, J.
cf. J. BLUM – J. LE FOLL – B. THOORIS [1]. [2]. [3]. and cf. R. AYMAR et al [2], [3].

LE FOLL, J. – THOORIS, B.
Magnetix. Internal Report CISILOG N° 81004 and 81005 (1981).

LE FOLL, J.
cf. J. BLUM – J. LE FOLL – C. LELOUP.

LELOUP, C.
cf. R. AYMAR et al [2], [3].
cf. R. AYMAR – C. LELOUP – M. PARIENTE.
cf. J. BLUM – J. LE FOLL – C. LELOUP.

LERAY, J. – SCHAUDER, J.
Topologie et équations fontionnelles *(Topology and functional equations)*, Ann. Sci. Ecole Norm. Sup. 51 (1934), 45–78.

LIONS, J.L.
cf. R. LATTES – J.L. LIONS.

LIONS, J.L. – MAGENES, E.
Problèmes aux limites non homogènes et applications *(Problems with nonhomogeneous boundary values and applications).* Dunod (1968).

LIONS, J.L.

[1] **Problèmes aux limites dans les équations aux dérivées partielles** *(Boundary value problems in partial differential equations).* Presses de l'Université de Montréal (1965).

[2] **Contrôle optimal de systèmes gouvernés par des équations aux dérivées partielles** *(Optimal control of systems governed by partial differential equations).* Dunod (1968).

[3] **Remarks on the theory of optimal control of distributed systems, in Control Theory of Systems Governed by partial Differential Equations.** Ed. Aziz– Wingate– Balas, Academic Press (1977).

[4] **Contrôle des systèmes distribués singuliers** *(Control of distributed singular systems).* Gauthier– Villars (1983).

LÜST, R – SCHLÜTER, A.
Z. Naturforschung 129 (1957), 850.

LUXON, J.L. – BROWN, B.B.
Magnetic analysis of non– circular cross– section Tokamaks. Nuclear Fusion, vol.22, n° 6 (1982).

LYNCH, S.J.
cf. J.A. HOLMES – Y.K. PENG – S.J. LYNCH.

McNAMARA
cf. D.E. SHUMAKER – J.K. BOYD – D.P. AUERBACH – B. McNAMARA.

MAGENES, E.
cf. J.L. LIONS – E. MAGENES.

MARDER, B. – WEITZNER, H.
A bifurcation problem in E– layer equilibria. Plasma Physics Vol.12 (1970), 435– 445.

MARTIN, P.
cf. W. FENEBERG – K. LACKNER – P. MARTIN.

MARROCCO, A.
cf. R. GLOWINSKI – A. MARROCO.

MASCHKE, E.K.
Plasma Physics 13 (1971), 905.
Plasma Physics 14 (1972), 141.

MASCHKE, E.K. – PANTUSO SUDANO, J.
Etude analytique de l'évolution d'un plasma toroidal de type Tokomak à section non circulaire *(Analytic study of the evolution of a toroidal plasma of Tokamak type with non–circular section),* Report EUR– CEA– FC– 668 (1972).

MERCIER, C.
Lectures in Plasma Physics. **The MHD approach to the problem of plasma confinement in closed magnetic configurations** (1974). Commission of the European Communities. Luxembourg.

MERCIER, C. – BOUJOT, J.P. – WERKOFF, F.
Computation of Tokamak transport. Computer Physics Communications 12 (1976), 109–119.

MERCIER, C. – SOUBBARAMAYER.
Plasma Physics and Controlled Nuclear Fusion Research. Vol. IAEA. Vienna 1975, p.403.

MIGNOT, F. – PUEL, J.P.
Sur une classe de problèmes non linéaires avec nonlinéarité positive, croissante, convexe *(On a class of nonlinear problems with positive, increasing, convex nonlinearity).* Comm. in Partial Differential Equations, 5 (8), (1980), 791–836.

MILLER, R.L.
Nuclear Fusion, vol.20, n° 2 (1980), 133–147.

MILLER, R.L.
cf. F.J. HELTON – R.L. MILLER – J.M. RAWLS.

MORERA, J.P.
cf. J. BLUM – R. DEI CAS – J.P. MORERA.
cf. J. BLUM – R. DEI CAS – J.P. MORERA – P. PLINATE [1], [2]/
cf. J.P. BOUJOT – J.P. MORERA – R. TEMAM.

MOSSINO, J. – TEMAM, R.
Directional derivative of the increasing rearrangement mapping and application to a queer differential equation in plasma physics. Duke Mathematical Journal 48 (1981), 475–495.

MUKHOVATOV, V.S. – SHAFRANOV, V.D.
Plasma equilibrium in a Tokamak. Nuclear Fusion. 11 (1971), 605–633.

MURAT, F. – SIMON, J.
Contrôle par un domaine géométrique *(Control in a geometry domain).* Publication du Laboratoire d'analyse numérique. Univ. Paris VI (1976).

NECAS, J.
Les méthodes directes en théorie des équations elliptiques *(Direct methods in the theory of elliptic equations).* Masson (1967).

NEILSON, G.H.
cf. D.W. SWAIN – G.H. NEILSON.

NELSON, D.B. – GRAD, H.
Heating and transport in Tokamaks of arbitrary shape and beta. Oak Ridge Report ORNL/TM–6094 (1978).

NINOMIYA, H.
cf. A. OGATA – H. NINOMIYA.

NOLL, P.
cf. D. CISCATO – L. de KOCK – P. NOLL.

OGATA, A. – NINOMIYA, H.
Use of modern control theory in plasma control at neutral beam injection. Proceedings of the 8th Symp. on Engineering Problems of Fusion Research. San Francisco (1979), vol.4, 1879–1883.

PANTUSO, J.
cf. E.K. MASCHKE – J. PANTUSO SUDANO.

PARIENTE, M.
cf. R. AYMAR – C. LELOUP – M. PARIENTE.

PENG, Y.K.
cf J.A. HOLMES – Y.K. PENG – S.J. LYNCH.
cf D.K. LEE – Y.K. PENG.

PIRONNEAU, O.
Variational methods for numerical solutions of free boundary problems and optimum design problems: in Control Theory of Systems Governed by Partial Differential Equations. Editors A.Z. AZIZ et al. Academic Press (1977), 209–229.

PFEIFFER, W.
cf. L.L. LAO et al [1]. [2].

PLINATE, P.
cf. J. BLUM – R. DEI CAS – J.P. MORERA – P. PLINATE [1], [2].

POLAK, E.
Computational Methods in Discrete Optimal Control and Nonlinear Programming: a unified approach. Academic Press (1970).

POST, D.E. et. al.,
Steady–state radiative cooling rates for low–density high–temperature plasmas. <u>Atomic Data and Nuclear Data Tables</u>. 20, (1977), 397–439.

PUEL, J.P.
Un problème de valeurs propres non linéaire et de frontière libre *(A nonlinear eigenvalue problem with free boundary).* C.R.A.S. Paris, 284, série A (1977), 861–863.

PUEL, J.P.
cf. F. MIGNOT – J.P. PUEL.

RABINOWITZ, P.H.
cf. M.G. CRANDALL – P.H. RABINOWITZ.

RAPPAZ, J.
cf. F. BREZZI – J. RAPPAZ – P.A. RAVIART.

RAVIART, P.A.
cf. F. BREZZI – J. RAPPAZ – P.A. RAVIART.
cf. P.G. CIARLET – P.A. RAVIART.

ROBILLARD, G.
Détermination de la frontière du plasma à partir des mesures magnétiques *(Determination of the plasma boundary from magnetic measurements).* Département de Recherche sur la Fusion Contrôlée. C.E.A. Fontenay– aux– Roses (1984) unpublished.

ROSSI, M.
cf. R. GRAN – M. ROSSI – F. SOBIERASSKI.

ROUBIN, J.P. : private communication.

ROUGEVIN– BAVILLE, P.
Résolution numérique d'un problème mal posé: identification de la frontîre du plasma dans un Tokomak circulaire à partir des mesures magnétiques *(Numerical solution of an ill–posed problem: identification of the plasma boundary in a circular Tokamak from magnetic measurements).* Student Report, Ecole Centrale (1979).

RUBIN, H.
cf. H. GRAD – H. RUBIN.

St.JOHN, H.
cf. L.L. LAO et al [1], [2].

SAGUEZ, C.
cf. F. BONNANS – C. SAGUEZ.

SALPIETRO, E.
cf. G. CENACCHI – E. SALPIETRO – A. TARONI.

SAMAIN, A.
Les Tokomaks permettront– ils la fusion contrôlée *(Will Tokamaks allow controlled fusion?).* Annales de Physique, Vol.4 (1979), 395– 446.

SATTINGER, D.H.
Stability of Solutions of Nonlinear Equations. Journal of Mathematical Analysis and Applications, 39 (1972), 1–12.

SEMENZATAO, S. – GRUBER, R. – ZEHRFELD, H.P.
Computation of symmetric ideal MHD flow equilibria. Computer Physics Reports. Vol.1, n°s 7 & 8 (1984), 389– 426.

SERMANGE, M.
Etude numérique des bifurcations et de la stabilité des solutions des équations de Grad– Shafranov *(Numerical study of the bifurcations and the stability of the solutions of the Grad–Shafranov equations).* Computing methods in applied Sciences and engineering. E. R. Glowinski – J.L. Lions. North– Holland (1980).

SCHAEFFER, D.G.
Non uniqueness in the equilibrium shape of a confined plasma. Comm. in Partial Differential Equations, 2 (6), (1977), 587– 600.

SCHAUDER, J.
cf. J. LERAY – J. SCHAUDER.

SCHLÜTER, A.
cf. R. LÜST – A. SCHLÜTER.

SCHNEIDER, F.
A novel method for the determination of the plasma position. Proceedings of the 10th Symp. on Fusion Technology. Padova (1978), p.1013.

SHAFRANOV, V.D.
[1] **On magnetohydrodynamical Equilibrium configurations.** Soviet Physics JETP. Vol.6 (33), Number 3 (1958), 545–554.

[2] **Determination of the parameters β_p and ℓ_i in a Tokamak for arbitrary shape of plasma pinch cross–section:** Plasma Physics. Vol.13 (1971), p.757.

[3] **Equilibrium of a toroidal plasma in a magnetic field.** Plasma Physics (Journal of Nuclear Energy Part C) 1963 – Vol.5, p.251–258.

[4] **Plasma Equilibrium in a Magnetic field.** In Reviews of Plasma Physics Ed. Leontovitch. Vol.2, p.103–151. Consultants Bureau – New York (1965).

SHAFRANOV, V.D.
cf. V.S. MUKHOVATOV – V.D. SHAFRANOV.
cf. L.E. ZAKHAROV – V.D. SHAFRANOV.

SHAFRANOV, V.D. – ZAKHAROV, L.E.
Use of the virtual–casing principle in calculating the containing magnetic field in toroidal plasma systems. Nuclear Fusion, 12 (1972), 599–601.

SHUMAKER, D.E. – BOYD, J.K. – AUERBACH, D.P. – McNAMARA, B.
Journal of Computational Physics, 45 (1982), 266–290.

SIMON, J.
[1] **Differentiation with respect to the domain in boundary value problems.** Numer. Funct. Anal. and Optimiz. 2 (5 et 8), (1980), 649–687.

[2] On a result due to L.A. Caffarelli – A. Friedman concerning the asymptotic behaviour of a plasma. Collège de France Seminar n°4. Research Notes in Mathematics. Pitman (1983), 214–239.

[3] Asymptotic behaviour of a plasma induced by an electric current. Nonlinear Analysis, theory, methods and applications, Vol.9, n°2 (1985), 149–169.

SIMON, J.
cf. J. BLUM – T. GALLOÜET – J. SIMON [1], [2].
cf. F. MURAT – J. SIMON.

SOBIERASSKI, F.
cf. R. GRAN – M. ROSSI – F. SOBIERASSKI.

SOLOVEV, L.S.
Sov. Phys. JETP 26 (1968), 400.

SPITZER, L.
Physics of fully ionized gases. Interscience Publishers (1962).

STAMBAUGH, R.D.
cf. L.L. LAO et al [1], [2].

STAMPACCHIA, G.
Equations elliptiques du second ordre à coefficients discontinus *(Elliptic second order equations with discontinuous coefficients).* Presses de l'Université de Montréal (1965).

STEVENS, D.C.
cf. H. GRAD – P.N. HU – D.C. STEVENS.
cf. H. GRAD – P.N. HU – D.C. STEVENS – E. TURKEL.

STORER, R.G.
cf. A.D. TURNBULL – R.G. STORER.

SWAIN, D.W. – NEILSON, G.H.
An efficient technique for magnetic analysis for non–circular, high–beta Tokamak equilibria. Nuclear Fusion, Vol.22, n° 8 (1982), 1015–1030.

TACCARDI, B.
cf. P. COLLI FRANZONE – L. GUERRI – B. TACCARDI – C. VIGANOTTI.

TACHON, J.
Confinement magnétique par Tokomaks: aspects physiques. Revue Générale Nucléaire, n°6 (1980).

TAMANO, T.
cf. M.S. CHU – D. DOBROTT – T.H. JENSEN – T. TAMANO.

TARONI, A.
cf. G. CENACCHI – R. GALVAO – A. TARONI.
cf. G. CENACCHI – E. SALPIETRO – A.TARONI.

TAYLOR, J.B.
cf. C.M. BISHOP – J.B. TAYLOR.
cf. J.P. CHRISTIANSEN – J.B. TAYLOR.

TEMAM, R.
[1] **A non linear eigenvalue problem: equilibrium shape of a confined plasma.** Arch. Rat. Mech. Anal., Vol.60 (1975), 51–73.

[2] **Remarks on a free boundary problem arising in plasma physics.** Comm. in Partial Diff. Equations 2 (6) (1977), 563–585.

[3] Int. Conf. on "Recent methods in nonlinear analysis and applications", Roma (1978).

TEMAM, R.
cf. J.P. BOUJOT – J.P. MORERA – R. TEMAM.
cf. J. MOSSINO – R. TEMAM.

EQUIPE TFR
Le Tokomak TFR 600. Report EUR–CEA–FC–916 (1977).

THOORIS, B.
cf. J. BLUM – J. Ch. GILBERT – B. THOORIS.
cf. J. BLUM – J. LE FOLL – B. THOORIS.

TODD, A.M.M.
cf. J. DELUCIA – S.C. JARDIN – A.M.M. TODD.

TONETTI, G.
cf. L. de KOCK – G. TONETTI.

TONON, G. – MOULIN, D.
Lower hybrid electron heating and current drive in large Tokamaks: application to Tore Supra. Proc. of the Fourth Int. Symp. on Heating in Toroidal Plasmas. Roma (1984), Vol.1, 721.

TUCKER, A.W.
cf. H.W. KUHN – A.W. TUCKER.

TURKEL, E.
cf. H. GRAD – P.N. HU – D.C. STEVENS – E. TURKEL.

TURNBULL, A.D. – STORER, R.G.
Journal of Computational Physics, 50 (1983), 409–435.

VIGANOTTI, C.
cf. P. COLLI FRANZONE et al.

VIGFUSSON,
Ph.D. Thesis (1977). New York Univ.

WEITZNER, H.
cf. B. MARDER – H. WEITZNER.

YANENKO,
Méthodes à pas fractionnaires *(Fractional step methods)*. Armand Colin. Paris. Intersciences (1968).

YVON, J.P.
Contrôle optimal d'un four industriel *(Optimal control of an industrial furnace)*. Research report IRIA n° 22 (1973).

ZAKHAROV, L.E.
Numerical methods for solving some problems of the theory of plasma equilibrium in toroidal configurations. Nuclear Fusion, 13 (1973), 595–602.

ZAKHAROV, L.E. – SHAFRANOV, V.D.
Equilibrium of a toroidal plasma with noncircular cross–section. Sov. Phys. Tech. Phys. Vol.18, n°2 (1973), 151–156.

ZEHRFELD, H.P.
cf. S. SEMENZATO – R. GRUBER – H.P. ZEHRFELD.

ZIENKIEWICZ, O.C.
The finite element method in engineering science. McGraw–Hill (1971).